Public Space and the Ideology of Place in American Culture

Architecture Technology Culture 3

Editors

Editorial Address:
Prof. Dr. Klaus Benesch
Department of English/American Studies
University of Munich
Schellingstrasse 3/VG
80799 Munich

Public Space and the Ideology of Place in American Culture

Edited by
Miles Orvell & Jeffrey L. Meikle

Rodopi

Amsterdam - New York, NY 2009

Photo cover: Atta Kim, "On Air #110-8, from the New York Series," 2005.

Cover design: Pier Post

The paper on which this book is printed meets the requirements of "ISO 9706: 1994, Information and documentation - Paper for documents - Requirements for permanence."

ISBN: 978-90-420-2574-5

Printed in The Netherlands

ACKNOWLEDGMENTS

This volume originated in a conference—"Public Spaces and the Ideology of Place in American Culture"—sponsored by the Bavarian American Academy in Munich on November 10-11, 2006, and we are most grateful to the Academy for its support. The editors also wish to acknowledge the indispensable contributions of Klaus Benesch, who organized the conference and has supported this volume. Likewise, we thank our colleague, David Nye, who helped at key points in planning the conference and the volume. Eleven of these essays were presented in draft form at the conference, and the remainder were solicited for this volume. Our thanks as well to Bridie Chapman, of Temple University, who copyedited and helped prepare the volume for publication. Special thanks also to Sascha Pöhlmann, whose computer skills untied the Gordian knot in the end and released the manuscript.

ACKNOWLEDGMENTS

[illegible]

Contents

Public Space and the Ideology of Place in American Culture

Introduction

Miles Orvell and Jeffrey L. Meikle

"Public Space" seems like an obvious and straightforward term, denoting areas where anyone—the public—might go. At the same time, the functional meaning of "public space" is dependent on its opposite, private space, for without the latter, all space would be "public," as once it might have been in some putative golden age before property lines were drawn and when all things were held in common. As lines of ownership are drawn, establishing private property, what's left is, perforce, "public space."

Yet we use the term not so much to signify everything that is *not* private space; we use it to imply space that has been deliberately created as a public amenity, space that has some deliberate public use, be it ceremony, recreation, celebration, or commerce. Public space, in this sense, is functional. The distinction between some sort of designated functional public space and, for want of a better term, unclaimed space, is even more evident when we think of who "owns" public space. Ownership of such space seems oxymoronic: who can own what is already free? Yet public space, in the sense of functional space, is always a construction, and as such an expression of someone's will—a ruler, a representative government, an owner—some person or persons who name the space, give it purpose, and monitor its existence.

The analysis of public space, as represented in the essays here collected, can take a variety of approaches, themes, and subjects; but we can reduce this variety, for the sake of gaining a purchase on what is inevitably an unruly interdisciplinary field, to several key questions: Who has the right to define public space? How do such places generate and sustain symbolic meaning? Is public space unchanging, or is it subject to our subjective perception? Do we, given the public nature of public space, have the right and power to subvert public space?

These broad questions motivate the essays that follow, which develop particular case studies in considerable detail. They all are concerned with public space in an American context and as such they are examples of the spatial turn that American studies generally has taken in the last twenty-five years. Together, they offer a compelling argument for the necessity of re-visioning American culture as the history of place-making and of the instantiation of meaning in the structures, boundaries, and configurations of space. Place, as it emerges in these essays, is the record of assertion and displacement, of authority and the subversion of authority, in short, of the symbolic investment of cultural meaning in space and of the contestation of public space.

The range of topics here is broad, as is the time-span under consideration, from the initial majestic conceptualization of Washington, D.C. by Pierre L'Enfant to the equally majestic postmodern realization that public space in the United States is more and more a matter of waste. More particularly, the topics range from parks to cities to small towns, from open-air museums to airports, and they encompass the marketing of place by corporate authority and power as well as the repossession of public space by otherwise disenfranchised subgroups. In the end, public space is variously imagined here as the site of powerful forces of social, political, and aesthetic change.

We open the volume with a section devoted to "Public Space as Symbol." Addressing the imagining of public space for the new American republic, Anna Minta describes how L'Enfant's scheme for the new capitol city would make use of public monuments as a way of physically embodying a national identity and patriotic history. Yet L'Enfant's plans, as Minta demonstrates through the history of the Washington Mall, have been modified by a public that would define its own history, through demonstrations, against the official history laid down on the map and monuments of the city. Turning to a more specific instance of Washington's public history, John F. Sears traces the creation of the "Castle" on the Mall to house the original Smithsonian Institution, which was intended as an expression of the new nation's faith in the diffusion of scientific knowledge as an essential component of a democratic republic. Drawing on the utopian tradition of New Harmony, the Owen brothers, scientists and educators, saw in the Norman architectural style a more appropriate expression of the scientific spirit of pragmatism and flexibility than the more austere classical edifices and quasi-temples otherwise being built in Washington to represent the ideals of America.

The imaging of space in America in ideal terms is the subject of the next three essays as well, though they shift the argument to the twentieth century. Juggling multiple ironies, Torben Larsen situates the Museum of Appalachia, located near Norris, Tennessee, within both the idealized conceptual space of Leo Marx's pastoral "middle landscape" and a physical pastoral landscape that was not actually realized until mid-century with the modernizing reforms of the Tennessee Valley Authority. Privately founded and composed of buildings moved from elsewhere into an imaginary setting, the museum, in Larsen's view, satisfies widespread desire for a public history embodied in spaces whose pastoral calm seems to have eluded the ravages of modernity. Turning from rural settings to the small town as a focus of popular historical nostalgia, Miles Orvell examines a series of models for the mythical American Main Street, ranging from reconstituted New England villages, through John D. Rockefeller's Colonial Williamsburg and Henry Ford's Greenfield Village, to Disneyland and the instant towns of the New Urbanism. While these popular reconstructions simulate the messy vitality and makeshift arrangements of less predictable public spaces, Orvell suggests that their emphasis on orderly perfection threatens the survival of commonly accessible public space. Finally, turning to printed representations of public space, Jeffrey L. Meikle examines the American scene of the 1930s and 1940s as portrayed in colorized view postcards based on heavily retouched black-and-white photographs. These idealized images, churned out by the millions by Curt Teich & Co. of Chicago, applied a generic stylizing and streamlining to thousands of unique places and thereby submerged regional diversity in a homogeneous sense of place. All three authors ultimately lament the diminishment of public space and suggest a loss of the real in the search for the ideal.

Moving on from the realm of symbolic representation to the realm of politics, the second section centers on the idea of "Contesting Public Space." The lines of the argument are clearly drawn in two essays examining conflicts over public and private space as they have played out in the histories of two major cities, Montreal and Chicago. Nadine Klopfer surveys Montreal's Mount Royal as it evolved from a distant geological feature into a public park totally engulfed by the urban fabric. Mount Royal Park offered the citizens of Montreal an opportunity to experience "nature" in the midst of the city, thereby defining an ideal conception of urban public space, but the park proved so popular that competing social classes contested whether it should serve as a healthful retreat for the elite or be made widely accessible by public transportation. As much as we might like to imagine a unitary meaning for the word "public," Klopfer demonstrates that it inevitably conceals social divisions and conflicts. Similar issues are at stake in Peter B. Hales's wide-ranging historical survey of the evolution of public space in Chicago from the late nineteenth century to the present. Grounding the conflict between public and private in the original platting of Chicago's street grid in 1830, Hales then analyzes such episodes as the use by striking workers of the quasi-public Haymarket in 1886, the creation of a faux-public park-like ambience for the World's Columbian Exposition in 1893, the construction of private stores and restaurants so vast as to assume the role of public space, and the contemporary struggle of Critical Mass bicycle riders to seize access to the avowedly public Millennium Park from which they have been excluded. Both Klopfer and Hales remind us of the complex interpenetration of public and private space throughout any major city.

Three contributors offer interpretations of particular contemporary types of space that cannot quite be assigned to the category of public or private because their functions are contested, shifting, obscured, or conflicted. Laura Lawson, for example, examines the community garden movement in an essay illustrated with her own documentary photographs. Drawing upon recent concepts of "everyday urbanism," she shifts the grounds of discussion to the way urban neighborhoods, as they change demographically, have evolved new ways of appropriating public space, although not without conflict from other potential users of public space, who see the "common" withdrawn from common use in such functions as a community garden. Lawson maintains that while community gardeners give expression to a widely shared desire to return to nature, their possession of this moral high ground falters because they benefit economically by obtaining "free produce" from public space. While Lawson examines public ground that is otherwise temporarily undeveloped, Kay F. Edge focuses on one of the most complex of developed public spaces, the contemporary airport. She finds, however, that architects and planners typically design this major public transportation utility, so much like a city in so many ways, as if it were in fact a private shopping mall. The behavior of an airport's temporarily captive population is channeled through demographics, market research, and architecture into patterns of consumption familiar to a leisure-driven society. Rather than examining publicly owned space that has been colonized as private space, Bryant Simon focuses on the Starbucks coffee chain for its success in making wholly private space feel public to its users. Selling itself as a "third space" between home and office, a place where local community can develop, the firm works hard to impose a certain order upon the consumer, who is persuaded that he or she has the freedom to innovate within a structure that is actually crafted to avoid spontaneity and community. After considerable participant observation, Simon concludes that little significant interaction occurs among customers, though they might think they are communing as at a neighborhood bar.

The section on debate and contestation over public space concludes with a meditation by Rickie Sanders on contemporary urban communities experiencing tensions of race, ethnicity, or class. Rejecting traditional static notions of public space as parks and public buildings, she focuses instead on fluid definitions emanating from actual control and use. Although streets, sidewalks, and festival markets are presumably intended for all citizens, the label "public" does not prevent exclusion based on non-membership in an ethnically or economically dominant demographic. Aware of the ironies of this situation, she opens with an anecdote regarding young black males interfering with the access of a white jogger to a public sidewalk in South Philadelphia. Sanders observes that rather than working to restore the concept of public access, contemporary society tends toward abandoning public space entirely in favor of virtual electronic communities (most in fact actually privately mediated).

The third and final section of the volume highlights "The Mutability of Public Space." It is clear by now that our experience of public space is as much a matter of perception as of any purely physical or material configurations. The meanings of any particular space, and the relative predominance of its public or private qualities, shift over time not only in response to obvious social, economic, and political pressures but also in response to changing subjective constructions of that space. Contemporary public space often tends to dissolve, partially in the face of increasing social investment in non-physical online virtual worlds such as MySpace, Facebook, and Second Life. More insidious, however, is a widespread, historically unprecedented, and mostly unconscious recognition of the provisional, even irrelevant status of the various kinds of physical public space this volume addresses, in the face of privatization, commercialization, and what might be referred to as virtualization.

The next five essays, taking phenomenological approaches, focus our attention on the act of perception as the key to the experiencing of place. Eric J. Sandeen examines the transformation of the High Line Canal in the Denver metropolitan region, including the suburb of Aurora. Constructed in the late nineteenth century to facilitate and regulate irrigation according to legal strictures which sorted out public and private water rights much as Progressive-era cities sought to sort out public and private access to land, the High Line has now been reconfigured as a recreational greenbelt whose meandering linear extent affords definition to the otherwise featureless suburban face of Aurora. Equally engaged by the act of perceptual definition as an organizer of space, Kerstin Schmidt revisits the classic Chicago architecture of Louis Sullivan and Frank Lloyd Wright, playing their spatial innovations off against Stephen Kern's historical interpretation of the culture of the early twentieth century as defined by space/time considerations of the new physics. In Schmidt's revisioning of Sullivan's dynamic skyscrapers and Wright's flowing interior spaces, their innovative genius proceeds from their manipulation of voids and immateriality itself as positively charged environmental elements.

In contrast to Sandeen and Schmidt, who describe perceptual shifts occurring in response to cultural and intellectual developments of major historical proportion, Timothy Davis and David E. Nye analyze situations in which perceptions of public space shift almost moment-by-moment, depending on who is using that space or the immediate conditions of use. Following the insights of Michel de Certeau on appropriation by marginalized populations, Davis exposes what he calls the "invisible geographies" of Rock Creek and Potomac Parkway in Washington, D.C. Ostensibly a meandering linear arrangement of meadows, glades, pedestrian paths, and recreational facilities surrounding a genteel roadway intended to move motorists through a series of picturesque natural vistas, the parkway offers the city's office workers and apartment dwellers the respite of a traditional urban park. However, Davis offers evidence, through ethnographic description and documentary photography, of three types of users—the homeless; gay cruisers; and suicides jumping from bridges—whose appropriation of public space subverts traditional norms and often abruptly intersects with more "normal" users. Focusing on infrequent, unexpected, even catastrophic disruptions of public order, Nye offers a historical survey of the shifting responses of New Yorkers to electrical blackouts over the years from 1936 to 2003. He finds that loss of light and power not only suspends normal use of such public and quasi-public spaces as streets, shops, restaurants, hallways, and elevators. Extraordinary limitations on sight, movement, and transport also create liminal spaces in which communities develop and momentarily thrive, exhibiting a euphoric heterotopian mood that temporarily dissolves normal boundaries of class and status. Nye suggests that over the years, however, with each new blackout, this communal indulgence has tended to dissolve in public fear of such criminal behavior as looting, thereby highlighting the ongoing erosion of faith in the safety of public space.

The ambiguities of public space, evident in so many ways throughout these essays, come to the foreground in three concluding essays emphasizing the fluidity and dissolution of space in the face of a new sensibility of impermanence that has haunted artists and writers from the modernists to the present. Sarah Luria, for example, describes the implosion of the modernist bird's-eye view of the city, a magisterial perspective exemplified not only by Le Corbusier's urban plans and models but by the opening scene of the film *West Side Story* as the camera zooms down from an airplane-oriented urban overview to the limited, enclosed terrain of a tenement basketball court. More recently, she maintains, such technologies as cell phones, global positioning devices, and even the zoom function of video games have taught us to experience space simultaneously from both air and ground, thus liberating us but rendering space also nebulous and uncertain. Luria finds this new multiperspectival consciousness in the architecture of Zaha Hadid and the photography of Atta Kim. Although in Luria's interpretation a de-spatializing simultaneity seems potentially euphoric, Andrew S. Gross sympathetically examines Henry Miller's excoriating American travelogue, *The Air-Conditioned Nightmare*, with its vision of an "interstate" landscape reduced to bland, commercially hyped homogeneity. Similar observations characterized post-World War II literature and social commentary, according to Gross, who points out the irony that the automobile both allowed individuals to encompass vast stretches of space and eventually rendered that space as completely abstract. Ironically, given Jean Baudrillard's postmodern reworking of Miller's perspective, Klaus Benesch argues in the volume's concluding essay that many contemporary American writers are attempting to restore an awareness of the physical specificity of the human ecosphere. Maintaining that such classic American writers as Emerson used or exploited nature by creating a romanticized *space* for the working out of a national ideology, he finds a refreshing turn toward representing specific urban *places*, the used and abused as well as the idealized—junkyards, waste ground, and even garbage itself—in recent works by Jonathan Franzen, Paul Auster, A. R. Ammons, and Don DeLillo. While not directly addressing public space in precisely the sense in which most of our essays define it, Benesch, writing as a self-defined ecocritic, seeks to remind us in a larger sense of the public nature of all the earth's space even if any generalized meaning must necessarily be dissolved when one engages the physicality of particular places.

Beginning with the ideology of nation-building and ending with the ideology of consumption (with its ineluctable consequence of waste), the essays in this volume comprise a narrative of decline that may seem all too familiar. Yet "public space" has suffered more by neglect than by attention, for the mainstream of cultural discourse has been largely unaware of the spatial context that structures our ideology. Lawrence Buell makes this point when he argues for our recognition of an "environmental unconscious," a concept he compares to Fredric Jameson's "political unconscious." For Buell, an "embeddedness in spatio-physical context is even more intractably constitutive of personal and social identity."[1] It is our hope that the essays here bring to our conscious awareness the role of public space in shaping ideology and the way that ideology can, conversely, shape public space. As part of the larger discourse of the politics of culture, this volume aims to open up vital areas of inquiry that will yield even richer results in the years ahead.

---

[1] Lawrence Buell, *Writing for an Endangered World: Literature, Culture, and Environment in the U.S. and Beyond*, Cambridge, MA: Harvard, 2001, 24.

Part One: Public Space as Symbol

Planning a National Pantheon: Monuments in Washington, D.C. and the Creation of Symbolic Space

Anna Minta

Since the founding of Washington, D.C., as the capital of the United States in 1791, various initiatives have been taken to crown the city's public space with a church for national purposes—a national pantheon: a monument as the culmination of commemorating and celebrating patriotic history and its most prominent protagonists. Although state and religion are separated by the United States constitution, there have been governmental and private as well as denominational ambitions motivating the establishment of such a monument, each group thereby claiming for itself a hegemonic status in interpreting and representing national history and identity. This paper discusses competing strategies of institutionalizing a patriotic and sacred space in projects like the Washington Monument, the Washington National Cathedral, and the Basilica of the National Shrine of the Immaculate Conception as well as proposals for the Beacon of Progress, a National Hall of Fame, and a National Church and Shrine.

The City of Washington was founded in 1790 and chosen to be the capital of the United States for its geographic position as a mediator between the southern and northern states. The French born engineer Charles Pierre L'Enfant, who had worked closely with George Washington and Thomas Jefferson before, presented a first master plan for the city in the following year, 1791. Being aware of the importance and the representational function of the city, he decided on a large scheme. In a letter to Washington he explained that his plans were "laid out on a dimension proportioned to the greatness which [. . .] the Capital of a powerful Empire ought to manifest."[1] Grown up in Paris, he used Versailles and other European cities of grand scale as a model for the axial layout of streets as well as for locating important buildings as focal *points-de-vue* in a complex spatial design of streets, public

---

[1] Quote from L'Enfant's explanations for the master plan, n.d. [ca. 1791] in Kite, 47. See also: Passonneau; Reps; Longstreth; and Meyer.

spaces, and parks.[2] His design for the federal city was meant to represent the fundamental and constitutional principles of the American republic in the general layout and the location of the democratic institutions. The balance of legislative and executive power is represented by the placement of the Congress and the White House on two distinctive hills within the rigid grid of city streets with major diagonal roads radiating from them. It is interesting to note that L'Enfant hardly gave any details for public, governmental, or educational institutions and he even omitted a separate site for the third constitutional power. The Supreme Court did not receive a prominent location by itself and was housed in the Capitol building until 1935, when it finally moved to the Supreme Court Building on East Capitol Hill.

L'Enfant concentrated on installing the center as a symbolic space, commemorating history and celebrating the nation—a public space which, as will be shown in this paper, was vividly contested for its form and contents by his contemporaries as well as following generations. L'Enfant himself developed a scheme for visualizing patriotic history and national identity by providing numerous public sites and monuments for celebrating national memory. An equestrian statue of George Washington, which Congress voted for in 1783 when the treaty of peace at Paris granted America's independence, was supposed to mark the meeting point of the two major axes that shape the Mall with the White House and Capitol as reference points (fig. 1, letter A). Additionally, in his master plan he developed a series of squares—"most advantageously and reciprocally seen from each other and [. . .] equally distributed over the whole City district, and connected by spacious avenues."[3] L'Enfant gave a detailed description of possible designs and functions of these squares:

> The center of each Square will admit of Statues, Columns, Obelisks, or any other ornament such as the different States may choose to erect: to perpetuate not only the memory of such individuals whose counsels or Military achievements were

[2] Thomas Jefferson, who kept a large library, sent him plans from Amsterdam, Bordeaux, Frankfurt, Karlsruhe, Lyon, Marseilles, Milan, Montpellier, Orleans, Paris, Strasbourg, and Turin (Reps, 15).

[3] Quote from L'Enfant's 1791 captions for the map in Kite, 65-66.

> conspicuous in giving liberty and independence to this Country; but also those whose usefulness hath rendered them worthy of general imitation, to invite the youth of succeeding generations to tread in the paths of those sages, or heroes whom their country has thought proper to celebrate. (Kite, 65)

Monuments in L'Enfant's understanding of memorial culture do not only represent history and past events but serve as models for patriotic engagement for future generations. In addition to the idea of integrating history through sites of memory throughout the city by placing monuments on every square, L'Enfant prepared for a major ceremonial space in the form of a patriotic church:

> This Church is intended for national purposes, such as public prayer, thanksgiving, funeral orations etc., and assigned to the special use of no particular Sect or denomination, but equally open to all. It will be likewise a proper shelter for such monuments as were voted by the late Continental Congress for those heroes who fell in the cause of liberty, and for such others as may hereafter be decreed by the voice of a grateful Nation. (Kite, 64)

Placed in approximately the same distance to the White House and the Capitol (fig. 1, see letter D; E is a grand fountain), the national church was projected to become the focus in celebrating American history and its protagonists. Although L'Enfant called it a church, it was not supposed to have any denominational affiliations. He presumably was inspired by buildings such as Westminster Abbey in London or the Panthéon in Paris (fig. 2). The latter was begun in 1757 as the church Sainte-Geneviève, but in consequence of the French Revolution was secularized and turned into a national pantheon, a burial place for the great of the nation and a sanctuary for collective celebrations. At about the same time, the French architect Etienne-Louis Boullée worked on various designs for vast pantheon-like churches and monuments that became prototypes for sublime, awe-inspiring structures celebrating human history. Boullée presented models for funerary monuments, cenotaphs, basilicas, and a monument of public gratitude. A nation, Boullée wrote in his only posthumously published treatise *Architecture, Essai sur l'art*, will be eager to demonstrate its love and gratitude toward its benefactors in the form of a monument.[4] These monuments, he thought, should be of noble, majestic proportions in order to induce a state of profound reverence. For location, Boullée recommended choosing a high place, so that the monument would dominate the city and be a memorial lasting for posterity. References to the Divine Being and to the beauty of nature, not in Christian, but in enlightened understanding, would add an aura of sacredness and enhance the character of magnificence and grandeur. Following Edmund Burke's widespread treatise on aesthetics, *A Philosophical Enquiry into the Origin of Our Ideas of the Sublime and Beautiful* (1757), these characteristics would turn architecture into sublime monuments, providing for long-lasting gratitude and reverence. This poetry of architecture, wrote Boullée, would fill the hearts of a nation with joy and mystic delight and be a true earthy Paradise: "Here before you are all the riches with which the Nation would like to prolong and brighten the days left to its benefactors" (Rosenau, 88).

These ideas of aesthetics and enlightened understanding of world history were widespread and well received in architectural debates in America, particularly in the context of national monuments. Still, L'Enfant's church for national purposes was never built in Washington. But through the years, various proposals have been made for establishing just such a symbolic space for commemorating American history, which was conceived as a culmination of collective memory by comprising and exceeding all other individual monuments. Although the constitution clearly states the separation of state and religion, the idea of a national church—a sacred and patriotic public space—inspires republican and private as much as religious initiatives to campaign for such a national monument. The cultural phenomena of "civil religion"[5] in the United States, the institutionalized collection of sacred beliefs about the American nation, seems to be fundamental for the respective hegemonic claims on representing the nation by designing its public monuments and defining its symbolic contents. In the following, I will not discuss in detail the many monuments which were erected on and along the Mall in Washington. Instead, I am introducing some of the efforts for appropriating the idea of a national

[4] Rosenau, 88. Boullée's treatise translated by Rosenau.
[5] Bellah, "Civil Religion in America," 1-21. See also: Bellah and Hammond.

pantheon, as envisioned by L'Enfant, and the debates which were raised by those efforts. Who was competing over filling this space with meaning, installing a specific version of national history, and defining common memory, values, and identity? What tensions developed among the individual, the national, the religious—as initiators of public monuments and memorials or as objects of commemoration? Who claimed to be legitimized as the official interpreter of history and representative of national identity? How do political and religious spheres interpenetrate each other? What were the ambitions for promoting projects such as the National Cathedral, the National Shrine, the National Hall of Fame, and the Beacon of Progress as denominational and non-denominational sites for national representation and collective memory? And which ideas and strategies were used for translating history into monuments and enhancing them with national significance?

The first major conflict about designing monuments and defining their historical, political, and symbolic contents occurred during a discussion about a statue for George Washington, which implied the question of how to define the nature of the republic and to find an appropriate form for adequately representing its founding father. In 1783, under the influence of the military impression of the War of Independence, the Continental Congress voted for an equestrian statue of a triumphant conqueror presenting Washington as a general in Roman dress, holding a truncheon in his right hand and with martial groups at each corner of the pedestal.[6] The monument was to be executed by the Roman neoclassical sculptor Giuseppe Ceracchi (Caemmerer, 199). Because of its closeness to imperial prototypes such as the antique equestrian statue of Marc Aurelius on the Capitol in Rome, this design was accused of using monarchical, imperialist imagery instead of representing republican ideals. Washington himself, pointing at the tense financial situation of the States, objected to the idea of a monumental equestrian statue (Federal Writers' Project, 319). During the following years the image of Washington as a victorious general was superseded by the reputation of a wise and successful politician. Federalist John Jay therefore proposed to Congress in 1785

[6] H.J. Brock, "Now to Finish Washington's Monument. Congress Is Asked to Provide a Proper Base for the Noble Shaft," *New York Times*, 1 January 1928. See also: Savage.

to substitute the pedestal of the equestrian statue with its bas-relief depicting Washington's battle victories by a simple, unadorned book including the inscription "Life of General Washington. Stranger read it. Citizens imitate his example" (Savage, 228). Jay wanted a "laconic" monument, establishing Washington as a noble person incarnating moral authority and republican virtues instead of solely glorifying his military deeds. Again, no consent was found in Congress and the project was rejected, this time for being too modest. It was only after Washington's death in 1799 that new initiatives for a national monument were taken (Morales-Vázquez, 152). Despite Washington's express wish for burial in his residence on Mount Vernon, the General Assembly of Virginia in Richmond and Congress in Washington competed about transferring the tomb to the respective town and erecting an honorable, monumental mausoleum. Although Judge Bushrod Washington, nephew of George Washington and proprietor of Mount Vernon, refused to consent to the removal of Washington's remains from the family's vault, plans for a crypt in the center of the Capitol's dome and rotunda were prepared. Benjamin Latrobe even developed the idea of a crypt further to the dimensions of a monumental mausoleum in form of a huge, 100-feet tall pyramid with four porticos on a podium outside the Capitol building (Peatross, 13). (See fig. 3.) The design follows a long tradition of colossal funeral monuments from the ancient pyramids of Egypt to Etienne-Louis Boullée's renderings for pyramidal cenotaphs and burial sites that he suggested for great men and national heroes. The architectural reference to the ancient pharaohs' mausoleums and the monumentality of the pure, geometric form are employed in Latrobe's design to cause awe-inspiring, sublime feelings hereby communicating assertions about Washington's everlasting fame. Again, complaints against monarchical imagery were raised as being inappropriate to commemorate the founder of the American republic. Additionally, Latrobe's design in its giant dimensions was considered too massive: it promoted idolizing and transfiguring Washington into a divine figure unapproachable for common man. The monument accordingly would lose its model character that was intended to inspire imitation.[7] During years of

[7] Peter Force, mayor of Washington, proposed another design of a colossal pyramid for a Washington monument in 1872 (Peatross, 13-14).

debates in Congress no agreement was found for defining an appropriate honorable and also Republican form for a national monument to George Washington and to the state in general. In 1832 the most noted American sculptor, Horatio Greenough, was commissioned by Congress to execute a sculpture of George Washington.[8] Trained in Italy and inspired by the classical revival in the United States, he took ancient statuary of Greece and Rome as a model and created a 12-feet (nearly 4 meters) high seated statue out of Carrara marble (fig. 4). The naked figure, only dressed by a Roman toga, recalls statues of Greek and Roman gods such as Zeus. When it was shipped to Washington in mid-1841 it did not find broad approval either, although about twenty years earlier another neoclassical statue of Washington (1820) by the Italian sculptor Antonio Canova in the statehouse of Raleigh, North Carolina was highly praised. It was not only the colossal size, the antique costume, and the semi-nakedness of Greenough's statue that roused protest, but its indistinctive military or civil character. Instead of a standing military posture or a figure dressed in Roman armor like Canova's statue, Greenough presented Washington sitting on an antique throne, his right hand pointing to a moral sublime, his left hand holding a reversed sheathed sword—symbolically handing it over to the people. The statue emphasizes Washington's civic virtues at the end of his military career: a person of moral integrity aware of transcendent supremacy as well as democratic principles. The immense weight of the statue in the Capitol's rotunda that endangered the structural statics of the entire building served as a convenient argument to have it removed, already, in 1844. It was then placed in front of the Capitol where it stayed until 1908 when the statue was transferred into the Smithsonian museum collection (Caemmerer, 321). Only in 1976, for the nation's bicentennial celebrations, was the statue given a place of honor in the National Museum of History and Technology (today the National Museum of American History).

With this background of unsuccessful and unsatisfying design proposals for a Washington monument a group of private citizens founded the Washington National Monument Society in 1833, first headed by Chief Justice John Marshall, succeeded by former President

---

[8] Wills, 420-441; Encyclopaedia Britannica Online, s.v. "Greenough's Statue of Washington," http://www.britannica.com/eb/article-9385901 (accessed 9 June 2007).

James Madison. They intended to collect one million dollars by private contributions for building "the highest edifice in the world, and the most stupendous and magnificent monument ever erected to man" (Savage, 233). Among various proposals in 1836, the design of the renowned architect Robert Mills was chosen for execution because of its simple, unaffected, and impressive but honest and therefore truly republican character (fig. 5). Mills presented a vast obelisk of 500-feet (152 meters) in height, rising from a 100-feet (30.5 meters) high circular colonnaded building. He used the Doric order which emblematically represented republican character and tradition. The temple-like building was to be accessed through a colonnaded portico crowned by an equestrian group: a chariot led by six horses, driven by a figure to be interpreted as George Washington.[9] Mills envisioned the rotunda as a national pantheon where revolutionary heroes and historic figures would be represented in statues and murals with a colossal statue of George Washington in the center. Mills imagined the temple as a repository of patriotic icons in which Washington would literally be surrounded within his national historic context, thereby turning the monument in honor of Washington into a national monument. Due to the slow collection of donations, Mill's design was reduced to the monumental obelisk only. The corner stone was laid on Independence Day: July 4, 1848, close to the site L'Enfant had reserved for Washington's equestrian statue in his master plan. Speaker of the House of Representatives Robert Winthrop delivered the oration:

> Lay the corner stone of a monument which shall adequately bespeak the gratitude of the whole American people to the illustrious Father of his country. Build it to the skies; you can not outreach the loftiness of his principles! Found it upon the massive and eternal rock; you can not make it more enduring than his fame. [. . .] Exhaust upon it the rules and principles of ancient and modern art; you can not make it more proportionate than his character. (Caemmerer, 221)

---

[9] Proceedings of the Board of Managers, Records of the Washington National Monument Society, RG 42, National Archives and Records Administration (hereafter: NARA), Washington D.C. See also: Harvey; Bryan, 291; and Savage, 233. Mills's design is undated and he reworked it with minor changes.

With the interruption of the Civil War, the monument was dedicated on February 21, 1885. During the process of construction further conflicts over defining national history and its influential participants aroused. Protest escalated when the Vatican sent a block of marble originally from the Temple of Concord in Rome to Washington to be integrated among the other 187 memorial stones which were sent by the States, by cities of the Union, by individual persons, and by other nations worldwide to demonstrate solidarity and support for the Republic. The anti-Catholic party of the Know-Nothings collected signatures against this "gift of a despot [. . .] with his animosity to republican institutions," on which "true Americans" will look upon "with feelings of mortification and disgust."[10] Claiming that the Pope's stone would violate the commemoration of American liberty, the block was stolen and destroyed. Moreover, in 1855, the monument site was stormed and occupied by the Know-Nothings in the vain hope to complete the obelisk under its auspices and ideological interpretations. The Know-Nothings failed to collect funds too and collapsed as a political body, hence the monument was given back to the official Monument Society. These episodes show the symbolic impact of monuments to define national identity. The history of the Washington Monument demonstrates how little consensus existed about the interpretation of national history as well as the form and the content of its visualization in public space. Intrinsically tied together and contested were questions about political liability, public responsibility, and religious influences as well as the relationships between commemorations of George Washington and other national heroes. Nevertheless, as conceptualized by L'Enfant, the Mall in Washington became the most important public space for national representation, even without realizing the plans for a national pantheon.

The Washington Monument, when finally finished nearly forty years later, was celebrated not only as a worthy monument in commemoration of George Washington, but also as a tribute to the American people. It was considered an icon of modern technology and American progress for being the tallest structure built on earth at that time and equipped with advanced technology like the interior elevator.

---

[10] Collection in NARA, RG 42, Records of the Office of Public Buildings and Public Parks of the National Capital, Box 1.

An article in the *New York Times* in December 1884 praised it as "something characteristically American" by towering above even "the highest cathedral spires designed by the devout and daring architects of the Middle Ages" (Savage, 241). By referring to the medieval cathedrals, America claimed to surpass the Old World of Europe.

Only some years later, around 1900, the architect Désiré Despradelle, professor at the MIT, resumed this idea of glorifying and representing America by the image of progress and grandeur and carried it to the extreme. He designed a giant pyramidal tower, a "Beacon of Progress [. . .] typifying the apotheosis of American civilization"[11] to be erected in Chicago, at the site where in 1893 America celebrated its World Exposition. If realized, it would have been the tallest man-built structure on earth, taller than the Eiffel Tower, St. Peter's Basilica in Rome, or the Great Pyramids in Egypt; taller than medieval gothic cathedrals or modern skyscrapers—all structures to which Despradelle liked to compare his Beacon. As the history of Rome was inscribed upon Trajan's column, the 1500-feet (nearly 500 meters) tall monument was planned as a "glorious pantheon." In architecture and iconography, as well as through the presentation of the museum's collections in the interior, it was intended to depict "the greatness and achievements of the American people from its birth to the present day," hereby installing America as the culmination of civilization.[12] Despradelle described his never realized Beacon of Progress as "the most tremendous structure of its kind ever planned by man since the days of the Tower of Babel."

Here again, reference is made to religious, particularly Christian, history and its architectural representations of progress in human civilization. The Washington Monument was praised for surpassing medieval cathedral spires and the Beacon of Progress was planned to finally fulfill human desires for building the highest structure on earth—a modern tower of Babel. Despite these religious references criticism on the profane understanding of creating a secular pantheon celebrating modern, progressive America was raised. An article in the *Washington Post* from June 20, 1903 critically stated: "Washington is rapidly becoming the center of educational, religious,

---

[11] "The Beacon of Progress," 306. See also: Oechslin, 17-18.
[12] "Greatest Monument in World's History," *Washington Post*, 21 June 1903.

and social life. The material side of our greatness is given adequate recognition, but the spiritual [. . .] is not fittingly acknowledged."[13] More than twenty years later, the same complaint about the absence of religious monuments of national dimension was published in the *New York Times* in July 1926:

> It is true that in Washington one sees everywhere 'national affirmations'; the power of a national Government is affirmed by the majestic Capitol; patriotism is affirmed by impressive monuments to the great leaders of the nation; art, literature, science, are visibly honored by appropriate buildings; but one finds no national affirmation of the religion which is the only foundation for life, individual or national.[14]

As a consequence, various denominations took the initiative to develop their version of a pantheon in the form of a national church. It was Bishop Henry Yates Satterlee who eagerly pursued the project of not only building an Episcopalian church in Washington, but institutionalizing it as the "national American cathedral."[15] Quoting L'Enfant's original idea of a national pantheon and referring to Westminster Abbey in London which is a burial site for the greats of the nation and the church for monarchical coronations, he wanted his cathedral to be an "American Westminster Abbey." Going beyond the idea of a civil religion in the United States, a fundamentally religious sentiment but denominationally undefined sacred belief, Satterlee emphasized the church's responsibility for a national pantheon. Speaking for the Episcopalian church, he claimed a hegemonic position among Christian denominations and recommended Washington cathedral as a burial site and a ceremonial space for statues of important public persons. "The cathedral," as Satterlee reminded the architects Henry Vaughan und George Frederick Bodley, "will inspire combined religious and patriotic feelings." He asked for sufficient space for:

---

[13] "On Site of Cathedral," *Washington Post*, 20 June 1903.
[14] Walden Myer, "Washington Cathedral," *New York Times*, 2 July 1926.
[15] Feller and Fishwick, 7. For more on the history of Washington Cathedral see: Feller; Minta, "'Nearer, my God, to thee.'" I am grateful to Dr. Richard Hewlett and Diane Ney, archivists of Washington National Cathedral, for much information and many inspiring discussions.

> statues, bas-reliefs and other works, commemorating great American heroes and statesmen of the United States, and historical incidents of Colonial times and after the Revolution, which are dear to the hearts of American people.[16]

By integrating, "some intensely interesting and religious scenes of American History [. . .] Washington Cathedral would not only be religious but also national."[17]

Although former president Woodrow Wilson was the only politician or person of public interest to be buried in Washington Cathedral, the Episcopalian church succeeded in creating a close relation to the government as well as to the political sphere in general by inviting politicians and other representatives for important ceremonies. President William McKinley came for the peace-cross ceremony, the official announcement of the cathedral project in 1898. President Theodore Roosevelt joined the cornerstone ceremony in 1907 and illustrious public personalities like General Pershing (fig. 6) were campaigning for the cathedral project that was completed in 1990 (finishing the west towers). Even today Washington National Cathedral, as it is commonly called, is used for presidential funeral ceremonies, like for Ronald Reagan in 2004 and for Gerald R. Ford in 2007. In 2001, when President George W. Bush asked to have the official interfaith service on the National Day of Remembrance for the victims of the terrorist attacks on September 11 at Washington Cathedral, the church was confirmed as a public sacred space for national ceremonies.

In addition to this political-religious mutual involvement, an iconographic program was installed that integrates the cathedral into the course of national history and hereby communicates its claims to embody a national church for all people. Sculpture, needlepoint memorials, and, particularly, stained glass windows depict historic events and personalities like the statesmen window showing Thomas Jefferson and James Madison (fig. 7) and others like the Puritan

[16] Explanation of Preliminary Drawings. Submitted by Henry Vaughan and George Frederick Bodley, 1907, reprinted in Row, 251.

[17] Letter by Henry Y. Satterlee to George Frederick Bodley, 11 January 1907, Archives of Washington National Cathedral (hereafter: WNC), Washington, D.C. 162-7-2.

Reverend John Eliot "Apostle to the Indians." Referring to more modern history, a war memorial chapel and a space window including a piece of moon rock are part of the pictorial program. These representations of American history and personalities are integrated into a comprehensive program of Christian iconography. In this context, Washington National Cathedral uses a complex system of symbolic references to Christian and American history and identity, thereby transcending the idea of a profane pantheon and embedding American history and progress into a Christian, specifically Episcopalian, setting. This national narrative ignores the constitutional principle of separating state and religion, but promotes an amalgamation of both spheres under the supremacy of Christian and, precisely, Episcopalian interpretation.

This Episcopalian interpretation of a national church obviously raised protest by other denominations and non-religious groups. Among various cathedral projects—all claiming national relevance—it was particularly the Roman Catholic church which, due to its constantly growing parishes, challenged the status of the Episcopalian community as a national church by claiming this status for itself. In 1920 the cornerstone for the National Shrine of the Immaculate Conception was laid (fig. 8).[18] However, in this time of widespread anti-Catholic feeling, the Catholic church was not able to receive the same support by politicians and therefore had to develop a different strategy for communicating its claims to represent the nation. There are no statues of politicians or so-called national heroes dominating within the Shrine. Pictorial references to American history are installed in subordinated places like war mosaics above minor side entrances to the Shrine. Hidden behind interior galleries are stained glass windows (fig. 9) with patriotic sites and symbols likc thc Capitol, the Statue of Liberty, and the heraldic eagle of the Great Seal of the United States. A crown as a symbol of the Blessed Mother Mary, the Catholic patroness of America, is placed above these American icons to express her protection and care for the nation. Instead of a prominent patriotic

---

[18] See: Kennedy; McKenna; Minta, "National Cathedral und National Shrine"; and Tucker. I am grateful to the Dr. Geraldine Rohling, archivist of the Basilica of the National Shrine of the Immaculate Conception, for much information and inspiring discussions.

iconographic program, numerous Marian chapels and statues all dedicated by different ethnic groups within the United States such as Lithuanian, Polish, Austrian, German, Chinese, Vietnamese, among others, are placed in the Shrine. As a result, the attempt to represent the nation was not followed by establishing a pantheon for national heroes, but by reflecting the diversity of the American people and consequently presenting a widespread—public—support for national ambitions, instead of relying on political support as the Episcopalians primarily did.

Alarmed by these denominational activities and aspirations for a ceremonial space to represent and commemorate the nation, private initiatives developed concepts for a secular, non-denominational pantheon as L'Enfant had envisioned it in his master plan. In 1925, a patriotic society promoted the construction of a:

> National Church and Shrine – A Temple belonging to the People of the Nation. A Church for all Religious Denominations. A national Memorial Shrine bearing witness to the United Patriotic Serve and The Religious Aspirations of the American People. (Remey, pamphlet)

Charles Mason Remey, active in the Baha'i community, proclaimed that his project would be a "great demonstration of patriotism [. . .] [and] not only by far the largest church in the world, but the most beautiful shrine and place of worship ever erected by man" (pamphlet). (See fig. 10.) The Memorial Shrine was not to be reserved to the Baha'i community only, but in analogy to the community's faith of merging Christian, Moslem, Hindu, and Buddhist traditions and beliefs, the shrine was to be a true national church for the entire nation.

A multilayered crypt and subcrypt with a series of catacombs, mortuary chapels, and burial vaults was to extend under the entire building, thus affording ample place for the burial of the nation's illustrious dead for centuries to come. Crypt and church were supposed to be in monumentality and permanency "unequaled in modern times, and unsurpassed even by the ancients" (Remey, 36). First renderings that were presented showed a giant cathedral being "an American interpretation of the Gothic style." This style was chosen, on the one hand, to capture the spirituality ascribed to Christian cathedrals, and, on the other, for functional reasons, because the Gothic structure allowed for a flexible and complex system of chapels, crypts, and vaults. Although the patriotic society promised prominent mortuary chapels for the builders, endowers, and founders of the National Church and Shrine, it was not able to raise sufficient funds for further developing these plans.

At about the same time—and this demonstrates how powerful the desire to glorify the American nation in a built structure was—the artist Frederic Wellington Ruckstull, in communication with the curator of the Capitol Charles E. Fairman, promoted the idea of a national "Hall of Fame" that Ruckstull had developed earlier in 1915.[19] Driving this were rather pragmatic discussions within the Capitol: Statuary Hall[20] (fig. 11) showed static problems because of the over abundance of statues, and various options were discussed for moving them out of the Capitol building. Ideas for a national pantheon which had been suggested in 1902 by the Senate Park Commission (alternatively called the McMillan Commission) were revived. The Commission had called for a grand city development scheme for celebrating the capital's centennial. Architect Daniel H. Burnham, chairman of the Commission that consisted of well renowned members like the architect Charles F. McKim, landscape architect Frederick Law Olmsted, Jr., and sculptor Augustus Saint-Gaudens, proposed an imposing master plan, which in its monumentality and grandeur was supposed to communicate the pride, glory, and dignity of the nation's capital. "Make no little plans. They have no magic to stir men's blood, and probably themselves will not be realized," is a quote attributed to Burnham in reference to his master plan for the World's Columbian Exposition in Chicago in 1893, which could also be used to interpret the grand scheme of the Senate Park Commission for Washington (Federal Writers' Project, 117). The Commission reserved a national memorial site at the south end of the White House axis with the Washington Monument at the crossing point

[19] Art and Reference Files, Proposal for Hall of Fame, Collection of documents in Archives of the Architect of the Capitol (hereafter: AoC), Washington, D.C. I am grateful to Dr. Barbara Wolanin, curator of the Capitol, and her staff at the archives for unreserved support during my research.

[20] Statuary Hall, the two story, semi-circular room south of the Capitol's rotunda, was used as the U.S. House of Representatives from 1807–1857 and turned into a space honoring prominent citizens of each state in 1864.

with the rectangular axis from the Capitol to the future Lincoln Memorial.

Ruckstull eagerly campaigned for his idea of a pantheon, a Hall of Fame, a Walhalla, as he alternately named his project. In contrast to Westminster Abbey in London or the Panthéon in Paris he wanted his Hall of Fame to be purely secular, honoring the greats of the nation ([Ruckstull], 158-162). Addressing members of House and Senate, he called the situation in Statuary Hall because of its congestion with statues a "National Nuisance" and a "Chamber of Horrors":

> Statuary Hall is so crowed, with its three circles of statues, all more or less gnome-like in the pale, pallid light of that Hall, that it is a national disgrace, and a joke even in Tokyo. We are the but of ridicule all over the world. [. . .] what the now thoroughly Jazzised and vulgarized American Ho-poloi badly needs is, a recrudescence of Reverence, for many things, above all for the great men of the past. [*sic*][21]

In the tradition of Boullée's monumental, awe-inspiring church designs, he envisioned his pantheon to become "one of the most majestic monuments in the world," consisting of a square building with a huge dome and two wings. Ruckstull gave no details for the exterior design, but only pointed at the domed Capitol in Washington or monumental designs for cupola buildings in the Beaux-Arts tradition of Paris for inspiration. He focused on conceptualizing an appropriate floor plan (fig. 12). There would be fifty-two galleries to house the statues— one for each of the forty-eight states, and four extra for future states that may be created in years to come. Underneath the dome and the lateral halls would be placed "statues of our National heroes, like Washington, Franklin, Jefferson, Marshall, Paine, Paul-Jones, Lincoln, Grant, Lee [. . .] etc."[22] Such a building, Ruckstull wrote,

> would soon pay for itself – in the patriotism and energy it would let loose among the citizens of our land. [. . .] Nothing will so

[21] Letter by F. W. Ruckstull to Robert Underwood Johnson, 20 July 1929, AoC, Art and Reference Files, Proposal for Hall of Fame.

[22] Letter by F.W. Ruckstull to Charles E. Fairman, curator of the Capitol, 28 April 1915, AoC, Art and Reference Files, Proposal for Hall of Fame.

> stimulate and raise our whole national life up to a higher plan, and to bind our people in a closer union, both north and south.[23]

Ruckstull's perception of a national pantheon basically points back to L'Enfant's original idea of a "church for national purposes." Both envisioned a national monument that would serve not only for commemorating the past, but also as an educational, inspiring model for future generations to follow the presented ideals of patriotism, heroism, and a longing for liberty and unity.

Through all these years of debating a national pantheon, various designs for monumental and sometimes giant buildings have been presented to glorify the American nation and its achievements. None of them were able to acquire unified support either financially or ideologically and none of them, finally, were realized. Too diverse and conflicted was the understanding of representing the nation and finding an adequate form for it. Individual statues and memorials along the Mall and its wider region, which includes Arlington on the other side of the Potomac, were likewise criticized, causing fierce debates about the design and its interpretation. The Lincoln Memorial, for example, was accused of being an imitation of a Greek temple suggesting Lincoln was a "Greek Deity;" therefore, the memorial had no connection historically with Lincoln's character of Americanism, or with America's democracy or independence.[24] Similarly, the Jefferson Memorial with its reference to the Pantheon in Rome was considered a Roman-imperialist statement. Women chained themselves to the cherry trees on the memorial site to protest against the "tree slaughter."[25] The modern designs for the Franklin Delano Roosevelt Memorial by Petersen and Tilney as well as by Marcel Breuer and Herbert Beckhard

---

[23] Ruckstull to Andrew W. Mellon, Secretary of the Treasury, 1 May 1930, AoC, Art and Reference Files, Proposal for Hall of Fame.

[24] Letter by the Illinois Chapter of the American Institute of Architects to the Committee on the Library and members of the House of Representatives, Washington D.C., 14 January 1913, Archives of the American Institute of Architects, Washington D.C., RG 801, Series 3, Box 8, Folder Illinois Chapter, 1913-14.

[25] This has been noted in numerous articles and protest letters in major newspapers. See, for example, Edward T. Folliard "Jefferson Memorial Row Recalls Disputes Over Other Monuments," *Washington Post*, 1 May 1938, B5; *Washington Star*, "Civic Group Protests Tree Slaughter at Memorial Sight," 9 December 1938, A6.

were avidly disapproved and nicknamed "Instant Stonehenge."[26] Obviously, it is not only the design but generally the iconographic and symbolic meaning of the monuments that was heavily contested. Monuments are interpretations of the past: they acknowledge and appreciate historic events and personalities and assign them certain places and commemorative status in the course of history. Furthermore, and probably as important, they extend their normative influence into the present: protagonists in planning and campaigning for a monument instrumentalize history for their own interests. By placing themselves into specific historic traditions, they intend to legitimize their claims for historical, political, and moral authority. Participating in erecting a public monument and defining its content legitimizes the protagonists, too, and establishes them as important interpreters of national history as well as presumably justifiable representatives of national identity. This explains the intensive discussions and struggles about the memorials and the shaping of the Mall as the most important public space. As a consequence of these memorial debates and protests the fundamental understanding and functioning of the Mall has changed. L'Enfant had envisioned the Mall as a pure representational space for an official—undisputed— version of national history and identity with the governmental buildings, various monuments, and the church of national purposes as architectural elements of a national memorial culture. This idea of a capital as a place for undisturbed, unified historic representation and calm reflection on the nation's identity was reconfirmed in 1882, when the legislation on the use of the Capitol Grounds was passed.[27] It prohibited speeches, banners, and all kinds of organized public protest within the range of the Mall. Despite these regulations, in the course of time this perception changed radically: monuments and buildings were understood as references to American history and politics which have to be struggled for and negotiated in their contents and visual form. Instead of being a site for a national pantheon, the Mall has been transformed into an open public and symbolic space—an open, outdoor pantheon—hereby providing adequate background space for popular political debate and protest.

[26] See: Kohler, 86-91; Hyman; and Gatje, 227.
[27] Barber; Mauch.

Bibliography

Archives of the American Institute of Architects, Washington, D.C.
Archives of the Architect of the Capitol, Washington, D.C.
Archives of Washington National Cathedral, Washington, D.C.
Barber, Lucy G. "Marches on Washington and the Creation of National Public Spaces, 1894 to the Present." In *Berlin, Washington, 1800-2000*, edited by Andreas W. Daum and Christof Mauch, 285-303. Cambridge: Cambridge University Press, 2005.
"The Beacon of Progress." *Technology Review* 2, no. 4 (Oct. 1900): 305-308.
Bellah, Robert N. "Civil Religion in America." *Daedalus* 96 (Winter 1967): 1-21.
— and Phillip E. Hammond. *Varieties of Civil Religion*. San Francisco: Harper & Row, 1980.
Bryan, John M. *Robert Mills: America's First Architect*. New York: Princeton Architectural Press, 2001.
Caemmerer, H. Paul. *A Manual on the Origin and Development of Washington*. Washington, D.C.: Government Printing Office, 1939.
Federal Writers' Project. Works Progress Administration. *Washington: City and Capital*. Washington, D.C.: Government Printing Office, 1937.
Feller, Richard T. *Completing Washington Cathedral: For They Great Glory*. Washington, D.C.: Washington Cathedral, 1989.
Feller, Richard T. and Marshall W. Fishwick. *For Thy Great Glory*. n.p., 1965.
Gatje, Robert F. *Marcel Breuer: A Memoir*. New York: The Monacelli Press, 2000.
Harvey, Frederick L. *History of the Washington National Monument and Washington National Monument Society*. Washington, D.C.: Government Printing Office, 1903.
Hyman, Isabelle. "Marcel Breuer and the Franklin Delano Roosevelt Memorial." *JSAH* 54, no. 4 (Dec. 1995): 446-458.
Kennedy, William P. *National Shrine of the Immaculate Conception*. Washington, D.C.: Salve Regina Press, Catholic University, 1945.
Kite, Elizabeth S. *L'Enfant and Washington 1791–1792*. Baltimore: Johns Hopkins University Press, 1929.
Kohler, Sue A. *The Commission of Fine Arts. A Brief History 1910–1990*. Washington, D.C.: Government Printing Office, 1991.
Longstreth, Richard, ed. *The Mall in Washington, 1791–1991*. Hanover: University Press of New England, 1991.
Mauch, Christof. "Capital Gardens: The Mall and the Tiergarten in Comparative Perspective." In *Berlin, Washington, 1800-2000*, edited by Andreas W. Daum and Christof Mauch, 201-216. Cambridge: Cambridge University Press, 2005.
McKenna, Bernard A. *Memoirs of the First Director of the National Shrine of the Immaculate Conception 1915-1933*. Washington, D.C., 1959.
Meyer, Jeffery F. *Myths in Stone: Religious Dimensions of Washington, D.C.* Berkeley: University of California Press, 2001.
Minta, Anna. "National Cathedral und National Shrine: Konkurrenz der Großkirchen um architektonische Präsenz in Washington D.C." In *Building America. Die Erschaffung einer neuen Welt*, edited by Anke Köth, Anna Minta, and Andreas Schwarting, 151-180. Dresden: Thelem Verlag, 2005.

—. "'Nearer, my God, to thee:' Die Nationalkathedrale in Washington D.C. im Machtgefüge von Staat und Kirche." In *Institutionelle Macht. Genese – Verstetigung – Verlust*, edited by André Brodocz, Christoph O. Mayer, Rene Pfeilschifter, and Beatrix Weber, 39-55. Köln: Böhlau Verlag, 2005.

Morales-Vázquez, Rubil. "Redeeming a Sacred Pledge: The Plans to Bury George Washington in the Nation's Capital." In *Establishing Congress: The Removal to Washington, D.C., and the Election of 1800*, edited by Kenneth R. Bowling and Donald R. Kennon, 148-189. Athens: Ohio University Press, 2005.

National Archives and Records Administration, Washington, D.C.

Oechslin, Werner. "Apotheose einer monumentalen Architektur." *Archithese* 18 (1976): 13-21.

Passonneau, Joseph R. *Washington Through Two Centuries: A History in Maps and Images*. New York: Monacelli Press, 1994.

Peatross, C. Ford. *Capital Drawings: Architectural Designs for Washington, D.C., from the Library of Congress*. Baltimore: Johns Hopkins University Press, 2005.

Remey, Charles Mason. *The National Church and Shrine of the United States of America to Be Built in the City of Washington*. Washington D.C., 1929.

—. Pamphlet (1928). In *The National Church and Shrine of the United States of America to Be Built in the City of Washington*. Washington D.C., 1929.

Reps, John William. *Monumental Washington: The Planning and Development of the Capital Center*. Princeton: Princeton University Press, 1967.

Rosenau, Helen. *Boullée & Visionary Architecture, Including Boullée's Architecture. Essay on Art*. New York: Harmony Books, 1976.

Row, Christopher Dean Hamilton. *World Without End. Philip Hubert Frohman and the Washington National Cathedral*. PhD diss., Harvard University, 1999.

[Ruckstull, F.W.], "A Palace of Fame for Washington." *The Art World* (Dec. 1916): 158-162.

Savage, Kirk. "The Self-Made Monument: George Washington and the Fight to Erect a National Memorial." *Winterthur Portfolio* 22, no. 4 (Winter 1987): 225-242.

Tucker, Gregory W. *America's Church: The Basilica of the National Shrine of the Immaculate Conception*. Huntington: Our Sunday Visitor Publishing, 2000.

Wills, Garry. "Washington's Citizen Virtue: Greenough and Houdon." *Critical Inquiry* 10, no. 3 (Mar. 1984): 420-441.

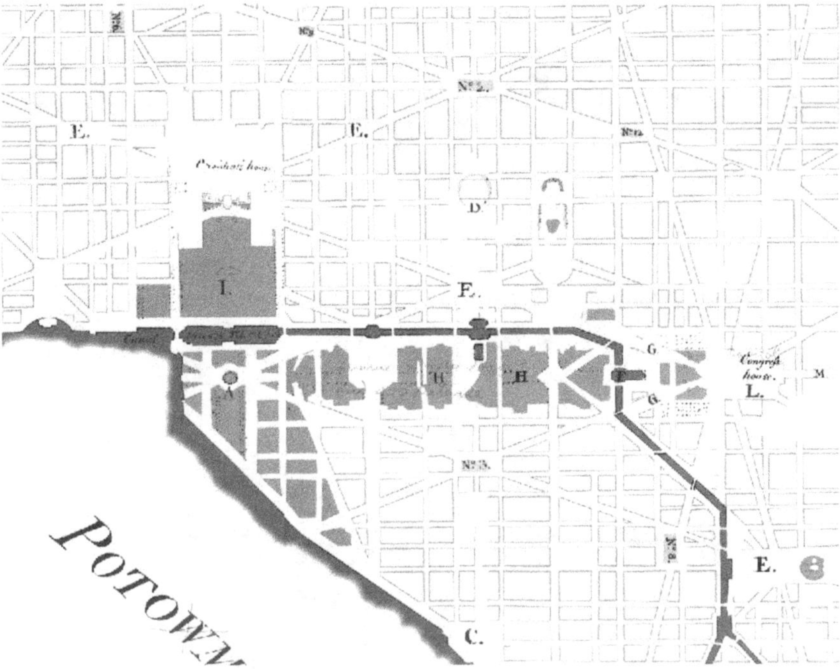

Fig. 1. *Plan of the city intended for the permanent seat of the government of the United States* [. . .] Pierre Charles L'Enfant's manuscript (1791). Computer assisted reproduction, detail. 1991. U.S. Geological Survey, Geography and Map Division, Library of Congress, Washington D.C.

Fig. 2. Panthéon, Paris (1757-1790) by Jacques-Germain Soufflot. 2004. Photograph by author.

Fig. 3. Proposed monument for George Washington, Washington, D.C. Graphite, ink, and watercolor on paper. 1799. Benjamin H. Latrobe. Latrobe Papers, Library of Congress, Washington, D.C.

Fig. 4. Statue of George Washington by Horatio Greenough (1841), following its removal from Capitol rotunda to East Capitol grounds. 1899. Photograph by Frances B. Johnston. Prints and Photographs Division, Library of Congress, Washington D.C.

Fig. 5. *Proposed Washington National Monument by Robert Mills*. Lithograph. 1846. Charles Fenderich. Prints and Photographs Division, Library of Congress, Washington, D.C.

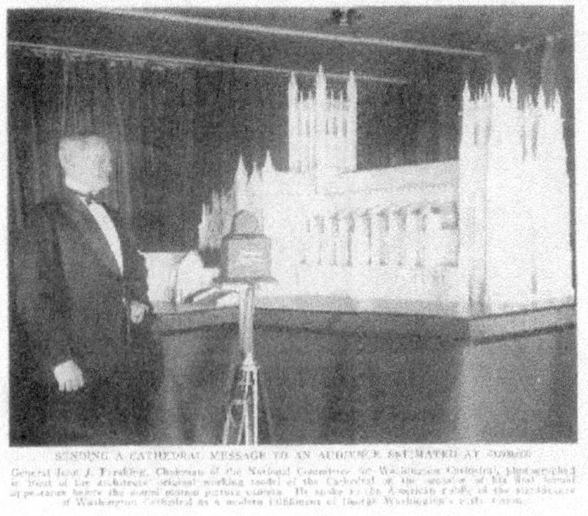

Fig. 6. *General John J. Pershing campaigning for Washington National Cathedral.* Photograph. 1930. *The Cathedral Age*, Easter (1930): 38.

Fig. 7. Statesmen Window showing Thomas Jefferson and James Madison. Stained glass window at Washington National Cathedral. 2002. Photograph by Elody R. Crimi. Elody R. Crimi and Diane Ney, *Jewels of Light: The Stained Glass of Washington National Cathedral*, Washington D.C.: Washington National Cathedral Guidebook, 2004, 103.

Fig. 8. Basilica of the National Shrine of the Immaculate Conception, Washington, D.C., 1920-1959, west façade with campanile "Knights of Columbus Tower." 2005. Photograph by author.

Fig. 9. Stained glass windows, east and west transept galleries, Basilica of the National Shrine of the Immaculate Conception, Washington, D.C. (completed 1965). 2004. Photograph by author.

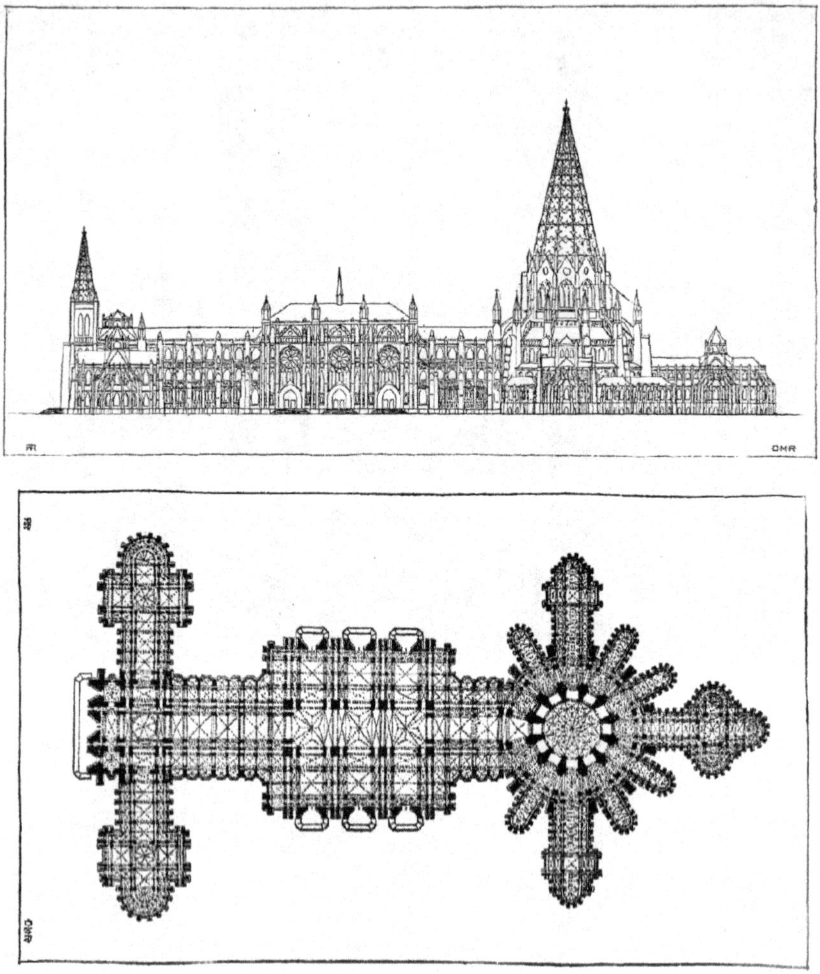

Fig. 10. *Proposed National Church and Shrine of the United States of America by Charles Mason Remey to be built in the City of Washington. Side elevation and main floor plan.* 1929. Charles Mason Remey, *The National Church and Shrine of the United States of America to Be Built in the City of Washington*, Washington D.C., 1929.

Fig. 11. *Statuary Hall, Capitol, Washington, D.C.* Photograph. 1920. Charles Moore, *The Story of Washington: A Glance at the History and Along the Vista of the Future of the Nation's Capital*, Washington, D.C.: The National Geographic Society, 1923, 57.

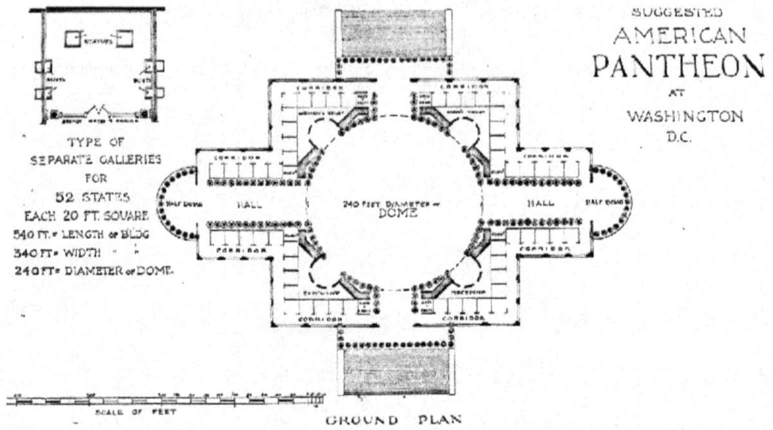

Fig. 12. *Proposed American Pantheon, Washington, D.C. by Frederick Wellington Ruckstull. Ground floor plan*. 1916. F.W. Ruckstull, "A Palace of Fame for Washington," *The Art World* (Dec. 1916): 160.

"How the Devil It Got There": The Politics of Form and Function in the Smithsonian "Castle"

John F. Sears

The irregular profile and dark reddish hue of the Smithsonian Institution building, known today as "the Castle," clash with the symmetrical white classicism of much of the rest of Washington architecture. The choice, in 1846, of the late Norman style for the building, reflected the ideas of Indiana Congressman Robert Dale Owen, who introduced the bill establishing the Smithsonian, and his brother, the geologist David Dale Owen, about the purposes the Smithsonian should serve. The Owen brothers believed the Smithsonian should diffuse practical scientific knowledge as widely as possible through a museum, research, lectures, inexpensive publications, and teacher training. Their vision originated in the educational philosophy of Johann Pestalozzi and the scientific and educational community of New Harmony, Indiana, and, more immediately, in David Dale Owen's work as scientist and educator. As a leader of federal and state geological surveys, David Dale Owen renovated a series of buildings in New Harmony for use as a chemical laboratory, museum for the display of geological specimens, and lecture hall for public education. These multi-use facilities functioned on the regional level as the Owen brothers hoped the Smithsonian would on the national level. The Owen brothers chose the late Norman style of architecture for the Smithsonian building because they believed its solidity, simplicity, and flexibility made it perfectly suited to the practical scientific and educational purposes they envisioned the Smithsonian Institution would serve. In defending this choice, Robert Dale Owen condemned the Greek Revival style of Girard College, another building constructed to serve an educational function. Unlike the Norman style, Owen argued, the Greek Revival could not adapt its external form to serve the internal functions of the building. Although the Smithsonian building as actually constructed did not completely fulfill the Owen brothers' vision of how the Smithsonian should function, it marked a milestone in the transformation of America into a more democratic culture in which scientific research and public education played central roles. The architecture of "the Castle" expresses an effort by the Owens to find a national architectural style more suited to the modern scientific age and "our matter-of-fact country."[29]

[29] I am grateful to Anne D. Owen for access to the Owen Family Papers and to Jane Blaffer Owen and the Robert Lee Blaffer Trust for generous support of two research trips to New Harmony. I am also grateful to Steven Conn, Cynthia Field, and Carla Yanni, who read earlier versions of this essay, for their suggestions.

In 1851, the American sculptor Horatio Greenough recorded his impression of the Smithsonian Institution building, which was then under construction:

> Suddenly, as I walked, the dark form of the Smithsonian palace rose between me and the white Capitol, and I stopped. Tower and battlement, and all that medieval confusion, stamped itself on the halls of Congress, as ink on paper! Dark on that whiteness—complication on that simplicity! It scared me [. . .] the connoisseurs may well 'wonder how the devil it got *there*.'[30]

The Smithsonian "Castle," as it has been affectionately known since the 1960s (Field, Stamm, and Ewing, xiv), continues to be a prominent feature of the Washington Mall, a landmark partly because it is so out of harmony with its fellow buildings. When the Smithsonian building was built between 1847 and 1855, the Mall, which Pierre Charles L'Enfant laid out in 1791 as part of his visionary design for the nation's capital, was still raw and undeveloped. The Capitol, where the long work of adding wings and the large dome we know today got under way in 1851, stood at the eastern end; the Washington Monument, the only other significant structure on what would become the visual and symbolic axis of the nation's capital, was under construction to the west. Planning a building to house the Smithsonian Institution helped stimulate the development of the Mall as the great democratic public space it would later become. When he drew a design for a Smithsonian building in 1841, Robert Mills also proposed a plan for landscaping the whole Mall, and in 1851, Andrew Jackson Downing created a landscape design for the Mall, including the Smithsonian grounds, at the request of President Millard Fillmore and the Smithsonian Building Committee. Although only Downing's plan achieved partial realization and was much altered later on, these designs began the process of transforming the Mall into a national

---

[30] Greenough, 36, 38. Greenough adapted the quote "wonder how the devil it got there" from Alexander Pope's "Epistle to Dr. Arbuthnot" (1735) in which Pope dismisses his minor critics as "hairs" in amber: "Pretty! in amber to observe the forms / Of hairs, or straws, or dirt, or grubs, or worms; / The things, we know, are neither rich nor rare, / But wonder how the devil they got there?" (Representative Poetry Online. Alexander Pope. http://rpo.library.utoronto.ca/poem/1633.html.).

gathering place where strolling in a grand park and museum-going share the stage with ceremonies, marches on Washington, concerts, folk festivals, touch football, and other activities open to all citizens (Scott, 48-49; O'Malley, 61-73).

When James Renwick designed the Smithsonian building, its architecture signaled that medieval revival styles were challenging the Greek and Roman classical styles that many people still regarded as the most appropriate styles for public buildings. The irregular, medieval profile and dark reddish hue of the Smithsonian violated the symmetrical white classicism of the rest of Washington architecture. Indiana Congressman Robert Dale Owen, the person most involved in planning the first Smithsonian building and choosing its architect, believed, however, that the early English round arch style of the Smithsonian building that he called "Norman" far better expressed the democratic ideals of the United States than the Greek and Roman classical styles so prevalent in public buildings in his day. To him, the Norman style possessed a solidity, simplicity, boldness, and, above all, flexibility that made it perfectly suited to the practical scientific and educational purposes he envisioned the Smithsonian Institution would serve fig. 1).[31]

James Smithson, the English chemist whose bequest to the United States government made the establishment of the Smithsonian possible, asked that the funds be used to create an institution for "the increase and diffusion of knowledge among men." Robert Dale Owen and his brother, the geologist David Dale Owen, who worked closely with him on preliminary plans for the building, believed the primary purpose should be the diffusion of practical scientific knowledge as

[31] I have chosen to use the term "Norman," as the Owens defined it, throughout this essay rather than "Romanesque" or another term for the Smithsonian building's medieval revival style because that is the word the Owens themselves used and the term "Romanesque" was not widely used at the time. The Owens' use of the term "Norman" derives from Thomas Rickman and J.A. Picton (see note 13). In retrospect, it is possible to regard the Smithsonian Institution building as an important contribution to the Romanesque revival in America, as Kathleen Curran does. She connects the introduction of the Romanesque style in America to its earlier revival in Germany and documents Renwick's borrowings from German Romanesque models in his design of the Smithsonian. At the time of the Smithsonian's design and construction, however, eclectic borrowings from different periods and nations and debates about how to classify medieval architecture confused the stylistic categories (Curran).

widely as possible. Eventually, their views came into conflict with those of Joseph Henry, the first secretary of the Smithsonian, who believed that the Smithsonian should function primarily as a scientific research institution and objected to the size, design, and expense of the proposed building. As Kenneth Hafertepe and other scholars have noted, Henry was a modern scientist who did much to promote pure science in America (Hafertepe, 40-41). The Owen brothers expressed a significant, alternative vision of the place of scientific knowledge in what Robert Dale Owen called "our matter-of-fact country" (*Hints on Public Architecture*, 65-66), one that was, in fact, more in keeping with the emerging role of natural history museums in nineteenth-century American culture than Henry's. After Henry's tenure as Secretary (1846–78), the Smithsonian evolved into a multifaceted institution, similar, in the way it seeks to diffuse knowledge to a wide audience, to the one the Owens envisioned. The complex of free museums the Smithsonian now operates along the Mall, for which the Castle serves as the widely recognized visual marker, draws people from all over the world to what many regard as the most significant public space in America.

The Smithsonian Institution building can be seen as an artifact of the struggle among the Owen brothers, Joseph Henry, and others over the role of science and education in America, the question of how broad a segment of the population the nation's public institutions would serve, and the kind of architecture that would best express American political ideals. Several historians have written excellent accounts of the architectural history of the Smithsonian Castle (Hafertepe; Field, Stamm, and Ewing; Curran, 240-258). By providing a fuller understanding of the Owen brothers' vision of the form and function of the Smithsonian, I am seeking greater insight into the way the Smithsonian project reflected the culture wars of mid-nineteenth century America over education, science, politics, and architecture.

The Owen brothers lived in New Harmony, Indiana, one of the most remarkable communities in antebellum America. Their father, the British socialist Robert Owen, failed in the two-year effort he began in 1825 to establish a utopian community in that town. But in the process the elder Owen and William Maclure, another philanthropist who collaborated with Owen, brought together a group of talented people who stayed on after the senior Owen returned to England and Maclure moved to Mexico. These included Owen's four sons (Robert Dale, David Dale, Richard, and William) and three teachers trained in the methods of the Swiss educational reformer, Johann Heinrich Pestalozzi (1746–1827). One of these teachers, Joseph Neef, had written *Sketch of a Plan and Method of Education* (1808), the first significant account of the Pestalozzian method published in America.[32] The three educators established schools at New Harmony based on Pestalozzian principles. These experiments in education lasted for several years after the brief utopian phase of the community and continued to influence the activities of the Owen family throughout their careers.[33]

Pestalozzi dismissed the Greek and Latin language curriculum followed in his time by European and American educational institutions as outdated. The growth of scientific knowledge and the needs of an industrial society, he thought, made such learning irrelevant to most people. Going a step further than Pestalozzi, Maclure, Neef, and the Owens believed that popular education could also bring about social and economic reform. Knowledge is power, Maclure argued, and the diffusion of knowledge through industrial and scientific education would free the working class from exploitation (Gutek, 36, 43-44).

Pestalozzi believed that people acquired knowledge directly through their senses from concrete objects. Actual experience, not the second-hand experience furnished by books, was the best teacher (Gutek, 74-75). It was much more effective, Maclure argued in 1824, to convey an idea to the mind with the aid of an object or visual representation of an object than with a description (Maclure, 188-190). Maclure wanted students to derive knowledge not from the teacher, but directly from things or images of things—a revolutionary concept in education at the time. Collections of specimens and scientific demonstrations were, therefore, essential tools in the teaching process. The Pestalozzian educational method harmonized perfectly with the major activities being carried out by the natural scientists of the day:

---

32 Gutek, 36, 43-44. The other two teachers were Marie Fretageot and Phiquepal d'Arusmont.

33 Neef maintained close familial, as well as philosophical, ties to the Owen family. His daughter Caroline married David Dale Owen and his daughter Anna Eliza married Richard Owen. Neef left New Harmony for six years after a disagreement with Maclure, but returned to spend the remainder of his life near his daughters (Hendrickson, 23-24; Gutek, 61).

the collecting, identifying (or naming), depicting, and classifying of natural objects.

All four Owen brothers attended the Hofwyl School of Emanuel von Fellenberg in Switzerland, which offered an object-centered curriculum similar to that taught in the schools founded by Pestalozzi. Robert Dale Owen took a particular interest in the educational methods to which Hofwyl exposed him. After the end of the utopian experiment in New Harmony, he became a radical journalist in New York City and then an Indiana State Legislator before being elected to Congress. Throughout his career he wrote and spoke on behalf of equal, practical education for all (e.g., Robert Dale Owen, Remarks of Mr. Owen of Posey). His brother David Dale Owen became an internationally known geologist, making his reputation as the leader of the state and federal geological surveys he began conducting in 1837. He used the specimens he collected to train other geologists and educate the public (Carmony and Elliott). When Robert Dale Owen won election to the House of Representatives in 1843, he went to board at the same rooming house where Ohio Senator Benjamin Tappan stayed when in Washington. Tappan had taken a strong interest in the educational experiments in New Harmony and become a friend of Joseph Neef.[34] In Washington, Tappan and Robert Dale Owen no doubt discussed the Smithson bequest, which Congress had accepted in 1836 but not yet agreed on how to spend.[35] In 1840 Tappan had written to Neef asking for suggestions on how to use the bequest. Neef sent back two proposals: one written by Richard Owen calling for the establishment of a normal school that would train teachers to disseminate a national system of education; the other by William Owen that proposed a "a great National Repository" comprised of botanical and zoological gardens, mineralogical and geological cabinets, an observatory, a library, and an art gallery, much like the multifaceted institution the Smithsonian would eventually become (Richard Owen, Tappan Papers). When, in 1844, Tappan introduced a Smithsonian bill, it directed that the bequest be used to establish a museum, library, chemical laboratory, and professorships in

[34] *American National Biography*, s.v. "Tappan, Benjamin." Tappan's letter has not been found but both Neef and Richard Owen refer to it in their replies.

[35] Robert Dale Owen served two terms in Congress from 1843 to 1847.

geology, chemistry, astronomy, and natural history. The professors would direct their research and lectures toward providing practical knowledge to the public. The bill passed the Senate, but with amendments that would have directed most of the funds to the establishment of a library, and the bill died in the House (Rhees, 338, 344; Hafertepe, 9-10).

Tappan retired from the Senate after serving one term, but in December 1845, Robert Dale Owen, who had become chairman of the House Select Committee on the Smithson Bequest, introduced his own bill. The bill incorporated all of the features of Tappan's bill, but added two significant provisions: one for a national normal school for the training of teachers (a simplified version of Richard Owen's 1840 proposal), the other for original research in the natural sciences (Rhees, 344).

Both Tappan and Owen wanted the Smithsonian to provide practical education for the common man. The opposition favored a more specialized institution serving a more highly educated clientele. John Quincy Adams proposed the creation of an observatory, but his idea did not receive strong support. Senator Rufus Choate of Massachusetts and Representative George Perkins Marsh of Vermont pushed for the establishment of a national library that would rival the best libraries of Europe. The fight that developed between their plan and the Tappan/Owen proposal was in part a contest between the egalitarian, populist, and utilitarian values of the Midwest and the more elitist, literary values of the Northeast. It was also a struggle between those who saw words and those who saw objects as the basis of knowledge, between those who regarded a library of books and those, like the Owens, who regarded museums of specimens, laboratories, and scientific demonstrations as providing the best tools for learning. In supporting the idea of a national library, Choate asked his fellow senators, is not a substantial library "one of the surest, most constant, most permanent, and most economical instrumentalities to increase and diffuse knowledge?" By contrast, Marsh argued in the House, a laboratory "is a charnel house, chemical decomposition begins with death, and experiments are but the dry bones of science" (Rhees, 287, 385).

Robert Dale Owen, on the other hand, attacked Choate and Marsh as "bibliomaniacs." The Library of Congress already existed, he argued, and already owned forty to fifty thousand volumes. Did America need still more books? Americans possessed far nobler resources in its free and aspiring population than Europeans did in their books, statues, and paintings. Although he recognized that by nature men would always be unequal in intelligence, education could reduce the artificial inequalities introduced by society. Therefore, he argued, it is "a republican obligation" to "elevate," the level "of our common-school instruction. I hold this to be a far higher and holier duty than to give additional depth to learned studies" (Rhees, 340, 343-345). As a Congressman, Robert Dale Owen was continuing to promote his father's agenda and the principles of Pestalozzi, Maclure, and Neef: practical, object-based education would give power to the productive classes and increase equality. He was not opposed to increasing knowledge, but only the diffusion of knowledge would achieve his reformist goals: "To effect permanent good," he said, "we must diffuse knowledge [. . .] even to Tom, Dick and Harry [. . .]" (Rhees, 345, 348-349).

Owen's eloquence, however, failed to achieve all his ends. On August 10, 1846, Congress passed a compromise bill establishing the Smithsonian Institution. It called for a building with rooms to display "objects of natural history, including a geological and mineralogical cabinet," as well as a chemical laboratory, library, gallery of art, and lecture rooms (Rhees, 426). It did not provide for a normal school. The bill left a lot of room for determining what programs the Smithsonian would ultimately conduct, and the battle in Congress shifted to a struggle between Joseph Henry, who gave up a professorship of natural philosophy at Princeton to become the first Secretary of the Institution, and Robert Dale Owen. Henry, like Adams, Choate, and Marsh, believed that the Smithsonian's main function should be the increase of knowledge; diffusion was a lesser goal. But Henry wished to focus on research, resisting both the champions of the library and those like Robert Dale Owen who wanted to place the emphasis on education, public lectures, the museum collections, and the publication of inexpensive pamphlets (Hafertepe, 40).

After the passage of the Smithsonian bill, Robert Dale Owen, who had been appointed to the Board of Regents of the new institution, turned his primary attention to the building's architecture. Like other social reformers of their time, including Fourier and the Shakers, the

Owen family had a long-standing interest in the potential of architecture for achieving social, economic, and political ends. Robert Owen senior participated in the planning of model mill buildings at New Lanark, Scotland. Later, working with the architect Stedman Whitwell, he designed the "quadralanular village" that he intended to construct at New Harmony (fig. 2). It expressed his belief that surrounding children from birth with "superior external arrangements" would reshape society (Robert Owen, *Robert Owen's Journal*, I, 21, 116). Owen abandoned the plan when his communitarian project failed, but as Curran argues persuasively, the arrangement and function of the buildings in Whitwell's design, which embodied Robert Owen and William Maclure's conception of New Harmony as "a center of popular education, scientific research, and the dissemination of knowledge," are a prototype of the multifaceted program Tappan and Robert Dale Owen proposed for the Smithsonian (Curran, 242-249). Robert Dale Owen, however, relied on a more immediate source in generating plans for the Smithsonian building.

In August 1845, optimistic that the Smithsonian bill he planned to introduce would pass the following year, Robert Dale Owen wrote to the architect Robert Mills, who had drawn plans in 1841 for a Smithsonian building in the medieval "Saxon" (round arched) style, requesting his advice. Unable to find his original drawings, Mills sent a "sketch" of his 1841 design.[36] On August 15, 1845, Robert Dale Owen wrote to his brother, David Dale Owen, asking him for "a well digested plan of suitable buildings." He enclosed the "rough ground plan" supplied by Mills, some guidelines on what functions the rooms would play and their approximate dimensions. Following Mills' choice of a medieval revival design, Robert Dale Owen suggested that the building be in what he called the "pure Norman" style "with its Saxon arches and simple forms."[37] The reason he was asking David Owen, rather

[36] Gallagher, 193. Mills wrote to Robert Dale Owen after the passage of the Smithsonian bill, probably in September since he mentions that it is about three months before Christmas. The sketch Mills sent did not survive. See also: Field, Stamm, and Ewing, 4-8; Hafertepe, 6-8, 18.

[37] Mills defined "Saxon" as round-arched, "Anglo-Saxon" as Gothic or pointed-arched, and "Norman" as a mixture of round and pointed arches. The "stylistic muddle," as Curran calls it, caused by "the difficulties of medieval terminology" created confusion at the time, as it does today (Curran, 250-251).

than “a regularly bred architect,” to draw up a plan, he added, “is, that I know you will consult utility first, in the various internal arrangements, and let architectural elegance follow” (Owen and Owen, “Correspondence Explanatory”).

Although Curran has suggested Whitwell’s design for the proposed “quadralanular village” in New Harmony as a prototype of the Smithsonian, no one has recognized the key role played in the Smithsonian planning process by David Dale Owen’s experience in constructing buildings in New Harmony based on Pestalozzian principles. Although David Dale Owen had no training as an architect or builder, he had considerable experience adapting buildings to the purposes of scientific research and public education.[38] Over the years, as his need for space increased, David Owen had converted three successive buildings in New Harmony for use as a chemical laboratory, a museum for the display of specimens, and a lecture hall for public education (fig. 3).[39] The third of these buildings, which he acquired in 1843, was a large stone and brick building known as the Granary that had been built by the followers of Father George Rapp, the original founder of New Harmony (“More Liberality”). During the two years before his brother wrote requesting a plan for the Smithsonian, David Dale Owen invested considerable effort in converting the Granary into an expanded version of the integrated, multi-use educational facility that he used to analyze, organize, and display the geological specimens he collected in the field, train his field workers, and educate the public. A visitor to Owen’s museum in 1859 noted the systematic way in which the minerals, rocks, and other specimens were arranged to convey lessons about the history of the earth (“To the Editors”). As a facility in which scientific research and public education were

[38] On October 30, 1845, in explaining to Tappan why he had chosen to consult his brother, Robert Dale Owen wrote, “The Doctor [. . .] knows, what no professed architect knows or cares about, the interior arrangements required, in the various scientific departments [. . .] No one but a scientific man with some knowledge of architecture can do it properly” (Tappan Papers).

[39] Robert Dale Owen, “Circular Addressed.” Finally, in 1859, David Owen designed and built a new laboratory in New Harmony in the Norman style, a building that once again integrated museum, research, and educational functions. Owen died the following year, however, and the Owen family converted it into a home.

seamlessly combined, David Dale Owen's New Harmony laboratory was designed to function on the regional level very much as Robert Dale Owen hoped the Smithsonian would on a national level.

In the way David Owen's laboratory displayed specimens and used them both for research and public education, it was among the prototypes of the natural history museums that became major institutions in American life after the Civil War. These institutions included the museum of the Academy of Natural Sciences in Philadelphia, first opened to the public in 1826, and Louis Agassiz's Museum of Comparative Zoology in Cambridge, which opened in 1860, as well as the museum organized in the 1850s by Spencer Baird as part of the Smithsonian. As Steven Conn argues in his book, *Museums and American Intellectual Life, 1876-1926*, these museums reflected an "object-based epistemology." Like Pestalozzi, their leaders believed that knowledge came from objects. Scientists studying the specimens in the collections of these museums generated new knowledge and students and the general public acquired scientific understanding by observing the museums' displays, carefully organized to impart scientific facts visually. These museums functioned as the most important centers of scientific knowledge in the nation until colleges and universities began taking over this role toward the end of the century.[40]

David Owen's architectural plan for the Smithsonian, which he submitted to his brother on October 10, 1845, reflected his determination to integrate scientific research and public education. The drawings that accompanied his description of the plan have not been found, but his letter to his brother provides a detailed description of the interior arrangement of the building. The structure, as David Owen conceived it, would be a machine for teaching through objects and demonstration, and, therefore, a reflection of the educational principles of Pestalozzi, Maclure, and Neef and his own practices as a science educator in New Harmony. He proposed five lecture halls, each of them designed to meet the needs of instruction in a particular subject. He planned a room in the west wing of the building, for example, to be "fitted up with a low stage, so arranged as to permit the suspension behind the lecturer of large illustrations on canvass, after the manner of

[40] Conn, 4-5, 15-18, 38-39, 43. See also: Winsor, 119-120 and Yanni, 31-32.

drop scenes in a theatre." He intended this room for lectures on geology or any other course "demanding the aid of very large diagrams, maps, paintings or similar illustrations."[41]

As to architectural style, David Dale Owen noted that he had "adopted, as suggested, the Norman style of architecture," but in its modified form as it "appears towards the close of the twelfth century, associated with occasional pointed arches" (fig. 4). David Owen said he had access in New Harmony to only a limited number of works on architecture, but was convinced, nonetheless, that the Norman "best admits the introduction of all sorts of conveniences: it recommends itself, also, for public buildings, on account of its simplicity & massiveness of structure." In language and in thought, David Dale Owen's argument in favor of the Norman style draws heavily from an article by J.A. Picton on "the Anglo-Norman Style of Architecture" in J.C. Loudon's innovative British *Architectural Magazine*. Picton's article suggests why the Owens preferred the Norman to the Gothic. Picton emphasizes the "massiveness and strength" of the Norman style, its economy, and its lack of ornamentation compared to the Gothic: "It requires no intricate tracery, no crocketed pinnacles or canopied niches; and no rich paneling or heraldic devices."[42] These qualities fit the practical, scientific program of the Smithsonian as the Owens envisioned it (Owen and Owen, "Correspondence Explanatory").

In the fall of 1846, after the Smithsonian bill had passed, the Board of Regents Building Committee, which Robert Dale Owen

---

[41] Owen and Owen, "Correspondence Explanatory." Both Robert Dale Owen's letter of August 15, 1845 to David Owen and David Owen's lengthy reply, which includes his plan for the Smithsonian building, are handwritten with sentences crossed out and additions made between sentences and in the margins. David Arnot, one of the architects who participated in the competition, published a slightly revised version of Robert Dale Owen's letter to David Dale Owen and substantial excerpts from David Dale Owen's response.

[42] Picton, "A Few Observations," 289-290. In his *Attempt to Discriminate the Styles of English Architecture* (1817), Thomas Rickman divided the Gothic into four periods, the Norman being the earliest. The Owens would have found a summary of Rickman's classification in Picton's essay on Gothic architecture, including the following description of the Norman phase of the Gothic: "1st, The Semicircular, or Norman; extending in its pure state from the time of the Conquest to the reign of Stephen, A.D. 1136; and, with the mixed or transition style which succeeded, to about the year 1190" (Picton, "An Attempt to Explain," 330).

chaired, conducted an architectural competition to choose an architect and design. The competitors were asked to follow David Dale Owen's proposals as a guide. They may also have been shown a new design for a Smithsonian building in a medieval revival style prepared by Robert Mills after studying David Dale Owen's plans, which included the staircase towers and mixture of round and pointed arches Owen had proposed (Field, Stamm, and Ewing, 7-8). As a result of these directions, all but one of the thirteen known submissions was in a medieval revival style. Most of the designs that survive are Gothic or include both round and pointed arches, perhaps, as Kathleen Curran suggests, because the architects were more familiar with the Gothic. Few examples of buildings built in the Norman style existed at that time in the United States. James Renwick, however, had designed an important building in the Norman style (the Church of the Puritans in New York City) and was therefore in a strong position to render a building in the style Robert Dale Owen had in mind (Curran, 251). Nevertheless, to cover all the bases, Renwick prepared two designs, one in the pointed, Gothic style, one in the round arch, Norman style. Of the surviving plans only his Norman design features round arches exclusively, except for the lancet windows within the round-arched window openings on the second floor of the main building (fig. 5). Determined to realize the vision of the Smithsonian that he and his brother shared, Robert Dale Owen was largely responsible for the committee's choice of Renwick's Norman design, the only plan, the Building Committee concluded, that took into account "all the accommodations demanded by the charter."[43]

Probably one of the reasons the Owen brothers preferred the medieval revival style to the classical styles more often adopted for public buildings up to that point is that they would have associated classical architecture with the Greek and Latin educational curriculum they regarded as elitist, impractical, and unsuited to a modern egalitarian and utilitarian society. It is difficult, at first, to see why a medieval architectural style would not have evoked equally

[43] Field, Stamm, and Ewing, 11. For detailed accounts of the architectural competition, the various designs submitted, and the controversy that arose over the manner in which the Building Committee conducted the competition, see Hafertepe, 21-61; Field, Stamm, and Ewing, 4-17; and Curran, 250-253.

objectionable associations. Although Renwick drew on ecclesiastical and collegiate models, not fortresses, for his design, both contemporary and modern observers of the building frequently associated it with castles because of its battlement-like parapets, keep-like tower on the south side, and cruciform, slit windows in one of the towers in the west wing (fig. 6). When Robert Dale Owen later wrote *Hints on Public Architecture* (1849) in defense of the Smithsonian design, Robert Carey Long, Jr. wrote a mocking review in *The Literary World* in which he noted that while Owen reminded his readers that the romantic castles of the Rhine were once the strongholds of cruel tyrants, it is from "these hindrances to the happiness of the million" that "the accessories of bastion and battlement in the Smithsonian building have been most largely drawn" (Long, 511).

Associations, however, are by their very nature in the mind of the beholder. In the article by J.A. Picton, on which David Dale Owen relied when arguing the virtues of the Norman, Picton wrote, "The ancient architecture of England is closely identified with the brightest periods of our history; with the triumph of liberty at Runnymede, and with the glories of Cressy, Poictiers (sic), and Agincourt" and with "the advancement of civilization and the arts" (Picton, "An Attempt to Explain," 329). The Norman and Gothic styles were also closely associated with the collegiate architecture of Smithson's native land. Although Robert Dale Owen does not refer in *Hints on Public Architecture* to the use of the medieval style in British collegiate architecture, the architects involved would certainly have been aware of this connection. Mills, for example, drew on this tradition when he employed Tom Tower at Oxford University's Christ Church (College) as a model for the dome intended to contain the observatory in his 1841 Smithsonian design (Field, Stamm, and Ewing, 6).

Robert Dale Owen wrote *Hints on Public Architecture* at the request of the Smithsonian Building Committee, but no doubt at his own initiative. In it he argues that a national, American style of architecture must be flexible: "Its external forms must not control and dictate its internal adaptations, but must yield and lend itself to these. That freedom, the vital principle of our political system, must breathe its spirit also over our architectural code" (8). When Horatio Greenough looked at the Smithsonian, however, he did not see a relationship between its external forms and interior purposes. After partially recovering from his alarm at first seeing the building, he wrote, "There is still a certain mystery about those towers and steep belfries that makes me uneasy. This is a practical land. They must be for something? Is no *coup d'etat* lurking there?" (Greenough, 37-38). Although Greenough could not detect any evidence of form following function in the Smithsonian, Robert Dale Owen believed the Norman style of the building exemplified that very ideal. It was not the picturesque effects of Norman architecture that mattered, but its adaptability: "The flexibility which the Norman and Gothic manners possess," he wrote in *Hints on Public Architecture*, "the facility with which they assume whatever external forms may be suggested by interior purpose" are "essential characteristics in an Architecture that is to attain, in our utilitarian age and in our matter-of-fact country, to the character of national" (12, 65-66).

David Dale Owen had expressed the same architectural philosophy in the plan for the Smithsonian building he sent to his brother in 1845. Although to the modern eye the Norman style hardly appears to be an example of form following function, David Dale Owen asserted that in his plan "the exterior has been made to conform, & to be subservient to, convenience and utility." He reported that before drawing any elevations he had carefully considered the interior organization of the building and prepared floor plans. Some of the towers, for example, housed the staircases, allowing for uninterrupted gallery space within the main structure (Owen and Owen, "Correspondence Explanatory").

In *Hints on Public Architecture*, Robert Dale Owen describes how the process of external form interpreting internal purpose operated in practice in the design of the chemical lecture room in the east wing of the Smithsonian (fig. 7). In order to protect delicate laboratory apparatus set up on the floor, the room required an outer staircase that would enable the audience to enter the slanted lecture gallery from above. This stairway:

> was obtained within a porch projecting from the eastern front; and as, in a porch of suitable proportions, the requisite height could not be gained without making the stairs too steep, a small outer porch was added, with a few steps therein. Thus the peculiarities of

> internal adaptation in this wing stamped upon its eastern elevation the exterior it now presents[.][44]

As the antithesis to the flexible Norman design of the Smithsonian, Robert Dale Owen points to Girard College in Philadelphia, another building intended to serve an educational function but designed by Thomas Ustick Walter in the Greek temple style. This style of building, Robert Dale Owen argues, could not adapt its external structure to its internal needs. The "rigid and uncomplying forms" of the classical style are the furthest removed from modern needs, such as heating, ventilation, and windows for lighting. As a result, the building could barely count "one really comfortable or convenient recitation room," according to David Owen (Robert Dale Owen, *Hints on Public Architecture*, 43; Owen and Owen, "Correspondence Explanatory").

Girard College was also very much on Robert Dale Owen's mind because the history and politics of the project had much in common with the Smithsonian. Like the Smithsonian, Girard College came into being as the result of a bequest from a philanthropic individual to a government for the purpose of establishing an educational institution. When he died in 1831, Stephen Girard, the most prominent banker in America, left the bulk of his fortune to the City of Philadelphia and specified that it be used to establish a boarding school for at least three hundred "poor white male orphan boys" (Laverty, Lewis, and Taylor, 11). As with the Smithson bequest, conflict about how to carry out the benefactor's wishes had long delayed the Girard College project. In fact, though an expensive Parthenon-like building had been constructed to house it, the college did not open until 1848 (fig. 8).[45]

Ironically, Girard, like the Owens, believed in a practical, scientific education rather than the traditional classical one. The

---

[44] Robert Dale Owen, 42. Renovations to the east wing beginning in 1855 altered this arrangement (Field, Stamm, and Ewing, 24-28).

[45] Rhees, 333. Part of the problem stemmed from Girard's will, which left little freedom to the architect designing the school's building. Girard did not specify a particular architectural style, but his rigid specifications for building materials and for the dimensions and arrangement of rooms created what Michael Lewis calls "a technological and functional catastrophe" (Laverty, Lewis, and Taylor, 23).

students at Girard College should be taught "Reading, Writing, Grammar, Arithmetic, Geography, Navigation, Surveying, Practical Mathematics, Astronomy, Natural, Chemical and Experimental Philosophy," French and Spanish, he directed in his will. He even shared with the Owen brothers a commitment to "object education": "I would have them taught facts and things, rather than words or signs" (Girard, 150). From the Pestalozzian perspective of the Owens, the classical style of the Girard College building clashed with the educational methods its founder wished to see implemented within its walls.

In writing *Hints on Public Architecture* and making the Smithsonian an architectural response to Girard College, Robert Dale Owen intended to do more than justify the architectural choices he and his colleagues had made; he wished to make a political point. Nicholas Biddle, who was elected president of the board of trustees of Girard College in 1833 and played as active a role in determining the architectural design of the college as Owen would later play in shaping the architecture of the Smithsonian, was also president of the Bank of the United States before President Andrew Jackson's veto of the bill re-chartering the bank caused its demise when the charter expired in 1836. Biddle was a leading proponent of the "American System," which featured a national bank as a key component and which Jacksonian Democrats like Owen adamantly opposed.[46] Carried on in the midst of the battle between Biddle and President Jackson, the debate about the architectural style of Girard College became highly politicized. Greek architecture, which had become a popular style for public buildings, particularly for banks, by the 1830s, became associated in the north with the Whig party, which formed in opposition to Jackson's policies. As the architectural historian Michele Taillon Taylor puts it, Greek architecture "constituted a red flag to Jacksonian Democrats who identified it as a symbol of the elite." The Greek style was "directly connected in the public mind with Biddle's pro-bank policies and the 'American System.'" A cartoon depicting Jackson's destruction of the economic power of the Bank of the United States shows Jackson

[46] Robert Dale Owen argued the case against a national bank in his "Address to the People of Indiana," for example, which he delivered at a meeting of the Democratic Party of Indiana on February 10, 1838.

holding a scroll that emits bolts of lightning. The lightning bolts are toppling and breaking in two the Greek Doric colonnade of Biddle's Bank (fig. 9). Although Biddle praised the Greek style of Girard College for its simplicity and utilitarianism, the Jacksonian Democrats attacked it as ornate, extravagant, and unfit for a building housing a school for the practical education of orphan boys.[47] By implication, the "American System" was also inflexible and totally unsuited to the needs of the country.[48]

By asserting the superiority of the Norman over the Greek Revival style, Robert Dale Owen confirmed the triumph of Jacksonian Democrats over the pro-bank forces. In writing *Hints on Public Architecture* he took part in the culture war of the time in which the Jacksonians argued that what was at stake was whether national policy and institutions should serve primarily a moneyed elite or ordinary working people. As Taylor has pointed out, as the Jacksonians gained political power, they displaced the wealthy, classically educated leaders like Biddle who had led the nation up to that point. Fashions in rhetoric and architecture shifted along with political power. Ornate speeches full of Greek and Latin quotations became targets of ridicule and architectural taste moved away from the classical toward the eclectic styles of the mid-nineteenth century (Laverty, Lewis, and Taylor, 85). The design of the Smithsonian marked a milestone in this cultural transformation.

---

[47] Laverty, Lewis, and Taylor, 71-73, 79-80. In another irony, which indicates how inconsistent the connections between politics and architectural taste were at the time, architects hired by Andrew Jackson in 1831 and after a fire in 1834 transformed his Federal-style home, the Hermitage, into a Greek Revival mansion ("History of the Hermitage Farm").

[48] In a provocative article entitled "The Greek Revival: Americanness, Politics and Economics," W. Barksdale Maynard questions the commonly repeated idea that the Greek Revival appealed to nineteenth-century Americans because of its associations with Greek democracy: "the historical record offers scant evidence that nineteenth-century Americans viewed the Greek Revival primarily in political terms [. . .] [T]he Greek Revival as described by contemporaries was overwhelmingly about aesthetics and cultural sophistication." Maynard makes a persuasive argument, but in the case of Girard College and Robert Dale Owen's defense of the Smithsonian in *Hints on Public Architecture*, politics did enter into discussions of the appropriate architectural style. The issues were more complex, however, than an identification of the Greek Revival style with Greek democracy (Maynard, 137).

In the end, the Smithsonian building only partially fulfilled the vision the Owen brothers shared of its form and function. Its outward style remained as they intended, but the interior contained only one lecture hall, the chemical lecture room in the east wing (fig. 10). That room featured a laboratory that opened directly into the lecture hall from behind the podium, as David Owen had proposed, thus facilitating scientific demonstrations and integrating the building's research and teaching functions. Uninterested in Owen's pedagogical objectives and dissatisfied with the room's lighting, ventilation, and size, Joseph Henry ordered this room gutted one year after the completion of the east wing in 1849, replacing it with a hall that seated nearly a thousand people and eliminating the laboratory behind the podium. In 1855, Henry replaced that lecture hall with an even larger one on the second floor, but after a fire destroyed the second floor in 1865, Henry eliminated the lecture hall altogether, thus reducing the role of the Smithsonian building as public space.[49] Despite Henry's efforts to thwart their intentions, however, the Owens had succeeded in creating a distinctive work of architecture that contributed after the end of Henry's tenure as Secretary to the development of the Mall as a forum for public education.

Ironically, Henry thought that the Smithsonian Institution building was just as egregious an example of architecture ill suited to its purpose as Girard College. This was partly because of his view of the Smithsonian primarily as a research institution and his lack of interest in the public education functions the Owens envisioned for it, but also because Henry felt that imitations of earlier architectural styles that made sense for cathedrals or castles did not serve contemporary needs. "The edifice is copied from the structures along the Rhine," he asserted. "These structures were admirably devised for the purpose of their erection. Every part is in keeping with the ideas and strategy of feudal warfare, for the use of the matchlock and the crossbow" (Henry, "Thoughts on Architecture," 30, 34). But these architectural features did not suit the purposes of a modern scientific institution like the Smithsonian.

---

[49] Field, Stamm, and Ewing, 25-28, 72-75, 78-80. Joseph Henry described the acoustical and visual design of the second floor lecture hall in an article entitled "On Acoustics Applied to Public Buildings."

Girard College and the Smithsonian stood as prominent object lessons in an ongoing search in mid-nineteenth-century America for an architectural style that would fit the practical, matter-of-fact American context as the nation moved into an urban, industrial, and scientific age. The lesson drawn varied according to who looked at them. In "Form and Function," Horatio Greenough wrote that the architecture of the Greek temple "when so modified as to serve for a customhouse or bank, departs from its original beauty and propriety as widely as the crippled gelding of a hackney coach differs from the bounding and neighing wild horse of the desert." He found the adaptation of the Norman style to the Smithsonian still more ungainly. Greenough found true examples of form following function only in modern structures such as ships and steam engines where the "stern organic requirements of the works" freed the designers from the authority of tradition (56, 116-117). America had not yet found an architectural style that expressed the same principles. Although David Dale Owen had, in fact, carefully planned the interior of the building to meet the needs of the institution as he and his brother envisioned them, and tried to adapt the Norman style to satisfy these needs, neither Greenough nor Joseph Henry could discern the functions of the Smithsonian in its form.

Although Robert Dale Owen did not succeed in replacing the Greek Revival style with the Norman as the dominant American form of American public architecture, as he hoped to do through the example of the Smithsonian building and his defense of its design in *Hints on Public Architecture*, the building expresses his effort to find a style more suited to a democratic, down-to-earth America. An architectural style that truly fit the requirements of the modern industrial and scientific age did not yet exist. Robert Dale Owen and David Dale Owen chose the available style that seemed to them most adaptable to the scientific and educational program they intended the building to serve. Moreover, Robert Dale Owen's assertion that the Norman provided a model for an American national style was not entirely off the mark. American architects designed many public buildings and some private homes in the Norman or emerging Romanesque revival style before the Civil War, and after the war the master architect Henry

Hobson Richardson created a distinctive American version of the Romanesque that left a lasting imprint on the national landscape.[50]

Bibliography

American National Biography. New York: Oxford University Press, 1999.

Arnot, David. *Animadversions on the Proceedings of the Regents of the Smithsonian Institution in their Choice of an Architect*. New York: Published for Circulation, 1847.

Carmony, Donald F. and Josephine M. Elliott. "New Harmony, Indiana: Robert Owen's Seedbed for Utopia." *Indiana Magazine of History* 76 (Sept. 1980): 181.

Conn, Steven. *Museums and American Intellectual Life: 1876–1926*. Chicago: University of Chicago Press, 1998.

Curran, Kathleen. *The Romanesque Revival: Religion, Politics, and Transnational Exchange*. University Park: Pennsylvania State University Press, 2003

Field, Cynthia R., Richard E. Stamm, and Heather P. Ewing. *The Castle: An Illustrated History of the Smithsonian Building*. Washington, D.C.: Smithsonian Institution Press, 1993.

Gallagher, H.M. Pierce. *Robert Mills*. New York: Columbia University Press, 1935

Girard, Stephen. "Will of Stephen Girard: Excerpts Relating to the Founding of Girard College." 1830. In *Monument to Philanthropy: The Design and Building of Girard College, 1832–1848*, edited by Bruce Laverty, Michael J. Lewis, and Michele Taillon Taylor, 147-152. Philadelphia: Girard College, c1998.

Greenough, Horatio. *Form and Function: Remarks on Art by Horatio Greenough*. Edited by Harold A. Small. Berkeley: University of California Press, 1947.

Gutek, Gerald Lee. *Joseph Neef: The Americanization of Pestalozzianism*. University: University of Alabama Press, 1978.

Hafertepe, Kenneth. *America's Castle: The Evolution of the Smithsonian Building and Its Institution, 1840–1878*. Washington, D.C.: Smithsonian Institution Press, 1984.

Hendrickson, Walter Brookfield. *David Dale Owen: Pioneer Geologist of the Middle West*. Indianapolis: Indiana Historical Bureau, 1943.

Henry, Joseph. "On Acoustics Applied to Public Buildings." In *Writings of Joseph Henry*, II, *Smithsonian Miscellaneous Collections*, 30 (Washington, D.C., 1886): 419-421.

—. "Thoughts on Architecture." n.d. *A Scientist in American Life: Essays and Lectures of Joseph Henry*. Edited by Arthur P. Molella, et al. Washington, D.C.: Smithsonian Institution Press, 1980.

"History of the Hermitage Farm." http://thehermitage.com/index.php?option=com_content&task=view&id=81&Itemid=103

---

[50] The Richardsonian Romanesque, Curran writes, "has claims to being a truly national style of American architecture" (257-258).

Laverty, Bruce, Michael J. Lewis, and Michele Taillon Taylor. *Monument to Philanthropy: The Design and Building of Girard College, 1832–1848*. Philadelphia: Girard College, c1998.
Long, Robert Carey, Jr. Review of *Hints on Public Architecture* by Robert Dale Owen. *The Literary World* 4 (16 June 1849): 511.
Maclure, William. "Memoranda, extracted from a Letter to the Editor, dated Alicante, (Spain,) March 6, 1824, from William Maclure, Esq. President of the American Geological Society, &c. &c." *American Journal of Science and Arts* 8 (1824): 188-190.
Maynard, W. Barksdale. "The Greek Revival: Americanness, Politics and Economics." In *American Architectural History: A Contemporary Reader,* edited by Keith L. Eggener. London: Routledge, 2004.
"More Liberality." *Indiana Statesman* II (2 Sept. 1843): 2.
O'Malley, Therese. "'A Public Museum of Trees': Mid-Nineteenth Century Plans for the Mall." In *The Mall in Washington, 1791–1991*, edited by Richard Longstreth. Washington D.C.: National Gallery of Art, 2002.
Owen Family Papers, New Harmony, IN.
Owen, Richard. An act to dispose of the Smithsonian Legacy according to the wishes of the Testator. Handwritten MS. Tappan Papers. This document, though unsigned, follows Richard Owen's letter to Tappan of May 6, 1840 in the Tappan Papers and is in his handwriting.
Owen, Robert. *Robert Owen's Journal, 1850–52*. London: James Watson, 1851.
Owen, Robert Dale. "Address to the People of Indiana." February 10, 1838. Clippings and letters concerning work and speeches in the Indiana legislature. Indiana State Library, Indianapolis, IN.
—."Circular Addressed to the friends of Liberal Education in general, and to the former readers of the Free Enquirer, in particular." 1835. Indiana State Library, Indianapolis, IN.
—. Facts in Regard to Dr. Owen's Laboratory. Caroline D.B. Allen Papers. Owen Family Papers. Unsigned, but the handwriting appears to be Robert Dale Owen's.
—. *Hints on Public Architecture*. New York: Putnam, 1849.
—. Remarks of Mr. Owen of Posey. At the Education Convention, held at the Capitol, on 27 December 1837. Clippings and letters concerning work and speeches in the Indiana legislature. Indiana State Library, Indianapolis, IN.
Owen, Robert Dale and David Dale Owen. Correspondence Explanatory of the details of a plan of building for a Smithsonian Institution prepared by David Dale Owen, M.D. MS. 1845. Workingmen's Institute, New Harmony, IN.
Picton, J.A. "An Attempt to Explain the Elements and Principles of Gothic Architecture to the General Reader." *Loudon's Architectural Magazine* I (Nov. 1834): 328-333.
—. "A Few Observations on the Anglo-Norman Style of Architecture, and Its Applicability to Modern Ecclesiastical Edifices." *Loudon's Architectural Magazine* I (Oct. 1834): 289-290.
Rhees, William Jones, ed. *The Smithsonian Institution: Documents Relative to Its Origin and History*. Vol. 1. Washington, D.C.: Government Printing Office, 1901.

Scott, Pamela. "'The Vast Empire': The Iconography of the Mall, 1791-1848." In *The Mall in Washington, 1791–1991*, edited by Richard Longstreth. Washington, D.C.: National Gallery of Art, 2002.
Tappan, Benjamin. Papers, microfilm. Manuscript Division, Library of Congress, Washington, D.C.
"To the Editors of the Louisville Journal," 11 August 1859 news clipping in David Dale Owen. Correspondence, 1846–1858. Owen Family Papers.
Winsor, Mary P. *Reading the Shape of Nature: Comparative Zoology at the Agassiz Museum*. Chicago: University of Chicago Press, 1991.
Yanni, Carla. *Nature's Museums: Victorian Science and the Architecture of Display*. Baltimore: Johns Hopkins University Press, 1999.

Fig. 1. Smithsonian Institution today. 2005. Photograph by author.

Fig. 2. Stedman Whitwell's plan for Robert Owen's "quadralanular village" in New Harmony, IN (never built). Woodcut in *Co-operative Magazine*. 1828. Frank Podmore, *Robert Owen: A Biography*, London: George Allen & Unwin, 1906, opposite.

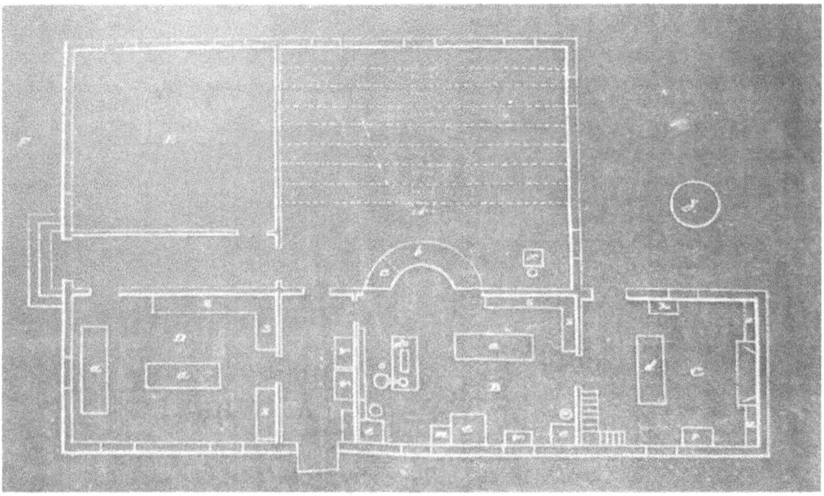

Fig. 3. David Dale Owen's Laboratory and Lecture Room, New Harmony, IN (a) lecture room, (b) laboratory, (c) mineralogical cabinet, (d) apparatus room, (e) intended as a museum. Robert Dale Owen. "Circular Addressed to the friends of Liberal Education in general, and to the former readers of the Free Enquirer, in particular." 1835. Courtesy of the Pamphlet Collection, Indiana Division, Indiana State Library.

Fig. 4. Towers on the north façade of Smithsonian Institution. Wood engraving by W. Wade, after a sketch by James Renwick, Jr. Robert Dale Owen, *Hints on Public Architecture*, New York: Putnam, 1849.

Fig. 5. Lancet windows on second floor of Smithsonian Institution. 2005. Photograph by author.

Fig. 6. Keep-like tower and battlements on the southern façade of Smithsonian Institution. 2005. Photograph by author.

Fig. 7. Detail of the Smithsonian Institution from the southeast showing two projecting porches constructed to house the stairway to the lecture hall. Lithograph by Sarony & Major after drawing by Weingärtner and Sarony. Robert Dale Owen, *Hints on Public Architecture*, New York: Putnam, 1849.

Fig. 8. *Girard College for Orphans*. Engraving by A.W. Graham after drawing by Thomas U. Walter, Architect. 1835. Courtesy of the Athenaeum of Philadelphia.

Fig. 9. *The Downfall of Mother Bank. Draw'd off from Natur by Zek. Downing, Neffu to Major Jack Downing. Printed & Published by H.R. Robinson; 52 Courtland Stt. N. York.* Lithograph cartoon. 1833. Courtesy of the Library Company of Philadelphia.

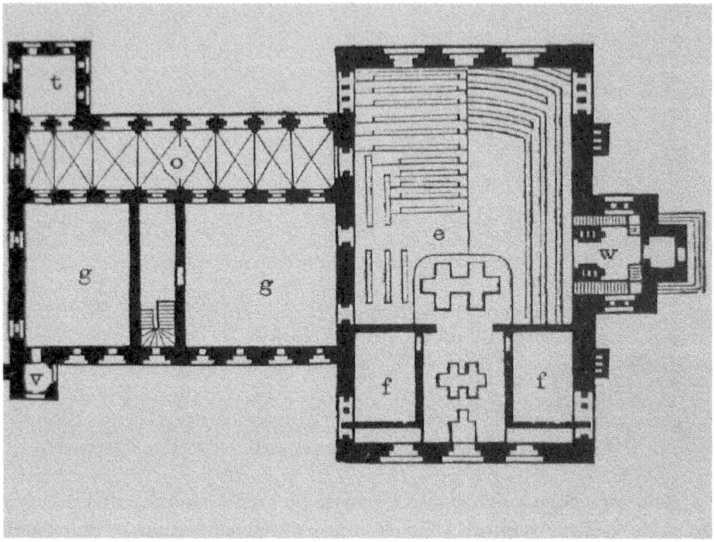

Fig. 10. Ground-Plans, Smithsonian Institution by James Renwick, Jr. Detail of east wing showing (e) lecture hall, laboratory directly behind podium, (f) adjoining laboratories, and (g) apparatus rooms. Wood engraving. J.H. Hall. Robert Dale Owen, *Hints on Public Architecture*, New York: Putnam, 1849.

The Museum of Appalachia and the Invention of an Idyllic Past

Torben Huus Larsen

John Rice Irwin's Museum of Appalachia narrates the history of the region in pastoral terms, evoking the sense of simplicity and authenticity that is so appealing to the tourist seeking a momentary escape from the complexities of modernity. However, such claims of authenticity are problematic both in terms of the individual exhibits and also in relation to the narrative structure of the place. The museum fits well into the category of "staged authenticity," Dean MacCannell's term for an impression of authenticity created specifically for a tourist audience. Although somewhat less than genuine, it is interesting that the staged experience seems to be enough for many tourists, suggesting that what they may in fact be searching for is a sense of anti-modernity.

Located on the outskirts of the small town of Norris, Tennessee, John Rice Irwin's Museum of Appalachia is a mere half-hour's drive from the hustle and bustle of Knoxville. The non-profit museum that began as a small private collection of Americana with the purpose of preserving regional heritage and culture has at present become a major local tourist attraction drawing thousands of yearly visitors from all over the world. Since it opened its doors in 1969, the museum has even been portrayed in media as diverse as *Reader's Digest*, a centerfold article in *Parade* magazine, and before a national audience on NBC's *Today Show* (Irwin, 5).

Undoubtedly the museum's popularity is, at least partly, due to the narrative it tells. By portraying the regional past as a simpler and more harmonious time the museum appeals to a modern cultural nostalgia while offering visitors a chance to escape, albeit briefly, from everyday life. The sense of nostalgia permeates the museum grounds and the lamentation for the disappearance of a culture in the face of modernity is emphasized on the museum's website, where the description of a "true breed of diminishing mountain folk [. . .] [who]

are among the most admirable people in the world [. . .] [and who] are kind, gentle and compassionate, yet tough and resourceful" conjures up a popular image of the pioneer farmer, an often mythologized character, who is now reduced to a memory.[1]

After having visited the museum on the internet, one naturally suspects a sentimentalist approach to history, and walking about the museum grounds such suspicions are confirmed. The tourist is surrounded by pastures, log cabins, and split-rail fences, and the idyllic atmosphere is enhanced by wandering peacocks, grazing sheep, and live country music. Hinting at the process by which such a garden would have been created in the heavily wooded mountains of southern Appalachia, a handwritten sign next to the display of the American Axe reads: "Our most important tool, without which the howling wilderness would have so remained" (fig. 1). Irwin further elaborates on this explanation in his book, *The Museum of Appalachia Story*, where he points out that the axe was the tool "responsible for converting the wilderness into the tranquil, domestic and agricultural countryside capable of feeding and clothing a growing and vibrant nation" (30). Using the Puritan rhetoric of William Bradford's "howling wilderness," Irwin's American Axe exhibit, combined with the open and green museum grounds, evokes an epic story of nation-building and further suggests an age-old narrative of subduing the wilderness in order to create a civilization. This is a version of American history that is highly relevant to understanding the museum itself.

As mentioned, an important reason for the Museum of Appalachia's success has to do with its portrayal of the past in such pastoral terms, evoking the open and controlled garden environment that traditionally has signified both mankind's desire to dominate nature, as well as his quest to reintegrate the human and the natural in a balanced Edenic coexistence (figs. 2 and 3). As Leo Marx pointed out in his classic study, *The Machine in the Garden*, as far back as Virgil's *Eclogues* the pastoral poet has sought "a resolution of the opposed worlds of nature and art" (Marx, 22). The pastoral landscape served as an expression of that resolution by merging the natural qualities of the wilderness with civilization's need for control. Just as Virgil's fictional

---

[1] Museum of Appalachia.
http://www.museumofappalachia.com/About%20the%20Museum.htm.

Arcadia served as a refuge from the political world of Rome, the idea of escaping the corruptions and injustices of early modern Europe into a world of balance became a powerful motif that not only held sway over the colonial imagination, but would come to embody the significance of America itself:

> The ruling motive of the good shepherd, leading figure of the classic, Virgilian mode, was to withdraw from the great world and begin a new life in a fresh, green landscape. And now here was a virgin continent! Inevitably the European mind was dazzled by the prospect. With an unspoiled hemisphere in view it seemed that mankind actually might realize what had been thought a poetic fantasy. (Marx, 3)

The pastoral essentially represented a natural, but unthreatening and harmonious, alternative to wilderness and city that was capable, in the words of Irwin, of "feeding and clothing a growing and vibrant nation."

As the industrial revolution transformed nineteenth-century America, the pastoral was adapted to fit this new experience, forming what Marx termed a "middle landscape." At heart, the definition of "middle landscape" remained closely connected to the pastoral tradition in that it represented a landscape infused with human values and under human control, but over time the term has also come to include a more general merger of nature and civilization that may potentially include any type of place from a traditional agricultural environment to a modern office park. Consequently, it makes sense to talk about the middle landscape in gradations since it, in some cases, has remained highly influenced by the classic pastoral, and in others (as with the plants used in a shopping mall-type environment), is closer to a techno-industrial ideal.

The pastoralized middle landscape is of relevance to the present discussion since it is most often found in connection with modern tourist sites. The appeal of such designs is the sense of order and authenticity that is also associated with the pastoral and which seeks to counter the disorder and confusion associated with modern existence. Today that type of middle landscape is found all over America in amusement parks like Disneyland in California and Cedar Park in Ohio, and it is prevalent in a number of Tennessee Valley tourist sites. The Chattanooga Choo Choo, for example, displays an antique train in the midst of a flower garden. Similarly, when the Tennessee Valley

Authority (TVA) created the Norris Dam State Park in the 1930s, the new dam was placed in bucolic surroundings and an old grist mill was moved, reassembled, and subsequently displayed in the vicinity of the new power turbines, integrating the history and future of technology just as it merged culture and nature. Furthermore, at Dollywood, the region's foremost tourist attraction, the general emphasis is on traditional regional craftsmanship, music, and foods, and the park displays a replica of the small log cabin that Dolly Parton grew up in, as well as a replica of an old grist mill, an antique train that takes passengers on a ride through the foothills of the Smoky Mountains, and an old wooden church that reminds visitors of a time when religion was a more integrated part of life.

Marx's distinction between a simple and a complex pastoral structure is also relevant when talking about the middle landscape. Whereas the simple pastoral represents a sentimentalized retreat from modernity into a world of harmonious balance between nature and civilization, the complex pastoral essentially comments upon the futility of its own design by showing the retreat from modernity to be threatened or temporary. Interestingly, both the simple and the complex pastoral revolve around some idea of history, as when in the complex pastoral of Twain's *Adventures of Huckleberry Finn* the romantic European past and American progressivism collide in a young boy's mind. The same is true at the Museum of Appalachia, which adheres to the category of the simple pastoral by engaging in what Edward Relph has called a process of "museumization" (Relph, 101-103). In a sense, the museum becomes a kind of moral geography, a historical vacuum infused with meanings and values, where working demonstrations of crafts, agriculture, and other historic everyday functions are used to celebrate the past as an oppositional structure to the fleeting values of modernity.

This kind of critique of the modern has been central to the open-air museum as far back as Skansen. Created by Swedish scholar Artur Hazelius in Stockholm in 1891, Skansen was the first of its kind and was created as a response to the rapid industrialization of the late nineteenth century. Hazelius, who was interested in peasant culture, collected well over a hundred historic buildings and placed them in a bucolic setting as a tribute to the country's rural past. The museum eventually proved a major source of inspiration and, in the following decades, open-air museums began appearing all over Scandinavia, the rest of Europe, and America (Wallace, 11).

The anti-modernity theme also carried over into early twentieth-century American heritage centers such as Henry Ford's Greenfield Village and John D. Rockefeller's Colonial Williamsburg. Seeking to recreate a pre-industrial experience, Ford's and Rockefeller's museums essentially imitated the original purpose of the pastoral which, in its Virgilian sense, had served as a retreat from the politicized and morally ambiguous civilization into a simpler and more innocent landscape; an attempt to start anew in a landscape that was under human control. The irony of Ford's and Rockefeller's fascination with the pre-industrial nation is hard to miss, but it also testifies to the immense popularity of romanticized visions of the past.

The idea for Ford's museum originally came about when his childhood home in Dearborn, Michigan was threatened by road construction. To save the building, Ford bought it, moved it, and undertook to renovate it to match his childhood memories. In the following years he collected other old buildings, either because they matched his memories of the town he grew up in, or because they had belonged to historic Americans that he admired. Thus, it was clear from the start that Ford's vision was a highly personalized and idealized version of his hometown with added features relating to a more general American history. When completed, Greenfield Village had become an idealized small town complete with a white-spired church, a courthouse, a school, and an inn all set about a New England-style village green. Ford revived folk dances and traditional dresses. At his village people could see history played out before their eyes in the form of old buildings, demonstrations of crafts, and collections of Americana. They could see a stylized vision of a golden age America that became a symbol of what had been lost and what the nation might become again if people wanted it.

However, the pastoral is far from unproblematic from a historical point of view. Escapist at heart, places such as Greenfield Village, Colonial Williamsburg, and the Museum of Appalachia essentially offer visitors a chance to retreat from modernity into a simplified world that is everything modernity is not. Built around an idealized vision of a society void of unpleasant social and political issues, such sites offer a spatial and temporal vacuum, presenting an escape not only from the modern world but also from history itself. Visitors to the Museum of Appalachia encounter a nostalgically driven narrative of simplicity, harmony, and bliss that emphasizes the life and values of a white, independent farmer of, what is most likely, English or Scots-Irish background. Irwin cultivates what amounts to an idealized pre-capitalist, pre-industrial, and essentially pre-lapsarian America frozen in time, where the displayed buildings and tools all tell stories of the good-natured and self-reliant mountain folk that created the nation.

In some respects, the story that Irwin tells evokes a Jeffersonian vision of a non-industrial nation based on the virtues of the hardworking, democratic, and honest yeoman farmer whose knowledge of, and connection to, nature is at the very heart of the American narrative:

> [My grandfather] had known many of the old pioneers and he had learned from them the secret of the forests, the streams, and the countryside. He taught us the names of every plant, every vine, and every tree. He pointed out that the ash split easily and made good stovewood, that the hickory made good handles for axes and hammers, that the cedar and locust made good fence posts, and that the chestnut made good rails. He showed us, deep in the woods, where the muscadines grew, and where the huckleberries, the 'possum grapes, and fox grapes could be found. He took us to the top of Pine Mountain one time and showed us where the summer grapes grew. (Irwin, 14)

Irwin's childhood memories, combined with a larger national discourse of trailblazers and pioneer farmers who tamed the wilderness, inform the narrative structure of the Museum of Appalachia. The problem is, of course, that just as Jefferson's vision was not viable in early post-revolutionary America, so Irwin's idyllic portrayal of the Appalachian past hardly presents the complete story.

Visiting the Museum of Appalachia is a pleasant and informative experience and the work which Irwin has put into recreating the past as he remembers and has experienced it is both commendable and honest; certainly there is no guarantee that a trained historian would have structured a museum of the Appalachian past in any other way. But as with any exhibition, the Museum of Appalachia narrative inadvertently transforms history by what is included and what is left out, how the exhibits are spatially and visually composed, and by the kinds of information that the visitor is given. Thus, Irwin's museum remains an example of how memory is structured, and as a brief look at the history of the region reveals, the pastorally inspired narrative that is told at the Museum of Appalachia is a simplified version of a past that was as hard, brutal, and lonely as it was complex.

One example of the museum's narrative structure is the absence of the TVA from the exhibits. Although the agency was instrumental in creating much of the pastoral framework that is so central to Irwin's museum, and although the construction of Norris Dam flooded his family's ancestral lands forcing them to sell their farm to the government when Irwin was only four years old, the TVA is not a part of the history presented at the Museum of Appalachia. Large federal subsidies and government intervention, it seems, hardly fits a tale of independent mountaineers. Also, little or no emphasis is given to the many hardships of rural life in the nineteenth- and early twentieth-century southern Appalachians, and whereas some issues are completely left out of the narrative others are merely given a positive spin. Poverty, for example, seems mentioned only to accentuate the generosity of the mountain folk who willingly shared what little they had, and there are no displays pointing to the many diseases that were common to the area, such as pellagra and hookworm, which were caused by an inadequate diet, dirt floors, and going without shoes. Malaria also posed a serious threat to people at the time, affecting, according to CDC estimates, close to one third of the region's population prior to the arrival of the TVA.[2]

Irwin's museum does include exhibits that potentially could counter the dominant narrative, but these stories are never allowed to seriously question the validity of the pastoral structure. The displays of Native American history and culture focus on portraying the tribes of the area as parents and farmers rather than as violent stereotypes, and consist of everyday artifacts such as clothes, woven baskets, and tools. But although Irwin's attempt to humanize the tribes of the region is admirable and historically correct, there is nothing about the horrors of forced relocation necessitated by the process of creating a garden in the wilderness which the axe exhibit celebrates. Likewise, there are no displays pointing to the significant presence of African American

[2] Centers for Disease Control and Prevention. "The History of Malaria, an Ancient Disease." http://www.cdc.gov/malaria/history/index.htm.

workers in the iron industry of the East Tennessee Mountains, and the museum has only recently added a slave cabin to its collection. Displays pointing to the industrial interests of lumber and mining companies in the region are never truly associated with issues of politics, race, or class, and the stories of large-scale environmental destruction caused by various industrial projects that left in their wake a landscape of clear-cut hillsides, low soil fertility, and desert-type environments caused by the release of airborne contaminants are also absent from mining exhibits.

The museum seems to suggest an abundance of available land that needed only to be cleared of trees through hard work and a good axe, but one of many hindrances to settlement which the independent farmer experienced was the lack of good land. In the latter parts of the eighteenth century the colonial government had given tracts of farmland to the men who had served in the French and Indian War, and early in the nineteenth century much of the remaining good farmland had been purchased by land speculators seeking to profit on the increasing westward expansion. This left many independent farmers to settle on poorer land in the mountains, which was often located in flood prone areas. Consequently, when rivers and streams went over their banks, destroying crops and sometimes washing away entire fields, farmers were forced to either clear new land or move to other parts of the country.

Geographic isolation also took its toll on the people of the region making the transportation of goods into and out of the valley both difficult and costly. To the north, east, and south mountains complicated travels, and to the west, the wild and often unnavigable Tennessee River ensured that little trade was conducted via the waterways. The construction of the Saluda Turnpike constituted one early attempt at creating a passageway across the southern Appalachians, but travelling the toll road was expensive and effectively raised prices on goods. As a result, the region remained economically, technologically, and culturally backward compared with other parts of the country.

Travellers to the region noticed these hardships. During a trip in 1830 British born geologist George William Featherstonhaugh made a note in his journal of the primitive way of life in trans-Appalachian frontier Tennessee:

> The people here, especially the women, were scarce above the level of savages, either in manner or appearance [. . . .] they had neither butter, eggs, milk, nor vegetables of any kind. Salt pork, bad corn, bread mixed up with dirt, tobacco and whiskey, formed the whole list of their necessaries and luxuries.[3]

Some thirty years later, when nature enthusiast John Muir travelled the area on his way to the Gulf, little had changed. As Muir noted in his journal, the Kentucky, Tennessee, and North Carolina landscapes are "the most primitive country I have seen, primitive in everything. The remotest hidden parts of Wisconsin are far in advance of the mountain regions of Tennessee and North Carolina" (18). And when, in the 1890s, the American forester Gifford Pinchot travelled the region, his descriptions of mountain living also emphasized poverty, disease, and isolation, painting a picture of mountain farming that is hardly the material of pastoral idyll:

> [The Mountaineers] regarded this country as their country, their common. And that was not surprising, for they needed everything usable in it – pasture, fish, and game – to supplement the very meagre living they were able to scratch from the soil of their little clearings, which often were no clearings at all, but mere "deadenings," filled with the whitening skeletons of trees killed by girdling. (Pinchot, 61)

Pinchot goes on to note the ordinariness of diseases such as hookworm, typhoid fever, and tuberculosis, and concludes that "the lives of these mountain people were literally as cribbed, cabined, and confined as the country in which they had their being was spacious, rich, and beautiful" (61).

As recent studies of southern Appalachian history have shown, the isolation was never complete. Although many farmers in the area were independent or relied on local trade, there were still ties to regional and national markets. But subsistence farming in the mountains was difficult, and many had to take jobs in the growing lumber and mining industries. Others simply left for cheaper and more fertile lands in the Deep South or in the trans-Mississippi West. Consequently, one would be hard put to include the general history of

---

[3] Featherstonhaugh, 173. This text is available online through the Library of Congress.

Tennessee Valley in the pastoral discourse that was popular in much of the nation in the nineteenth century, and few at the time did. While scores of artists and tourists praised and memorialized Niagara Falls, the Catskills, and the Hudson River Valley, few tourists went to the Great Smoky Mountains and no nationally acclaimed artists painted its many natural wonders.

It was not until the TVA initiated the large-scale transformation of the region in the 1930s that the area first became included in the pastoral discourse. In fact, prior to the TVA only a few regional projects had relied on the pastoral as their central metaphor. In the 1880s the traditional romantic rhetoric of Mary Noailles Murfree's stories first brought the southern Appalachians into national awareness, and in the 1890s George Washington Vanderbilt's impressive Biltmore mansion used the pastoral ideal in its imitation of English estates. The pastoral was also an important theme in Benton MacKaye's visionary plan for an Appalachian Trail that would include socialist inspired self-sufficient communes where weary workers could relax and regain a connection to nature. However, these were exceptions.

Although the TVA's primary responsibility was the electrification and modernization of the valley, it seemed clear from the beginning that the creation of a controllable and multi-functional natural environment was central to the overall plans for the development of the region. With the help of the Civilian Conservation Corps several reforestation programs were begun to fight the widespread soil erosion. Harmful agricultural practices that exhausted the soil were changed through demonstration farms and cheap fertilizers, and thousands of acres of farmland were transformed into pastures as traditional farming was supplanted by fruit and nut crops as well as dairy and sheep farms.

With tourism on the rise, recreational initiatives were found to be one of the most efficient ways in which to improve the economy of the region. The once wild river, now a series of controlled lakes, was marketed for water-skiing, swimming, and fishing; forest areas promoted hunting and hiking; public parks sought to fully utilize the new landscape's tourist potential. During the first decade of its existence, the TVA even pursued a strict policy of buying not only the land that was to be flooded but also the land bordering the new lakes. There was some internal debate about this policy of heavy purchase.

The TVA's experts on agricultural and power issues wanted merely to purchase the land that was being flooded, leaving as much land as possible for farming while minimizing the costs. Those working with recreational issues, however, argued that such land "ought to be forever reserved for its scenic beauty, to be used for public parks and playgrounds and to prevent the growing up of private developments that may mar the beauty of the land and lead to speculative profits for a few on what should be a public benefit for all" (Lilienthal, 68). By purchasing the land adjoining the reservoirs the TVA ensured proper landscaping for the pastoralized shoreline and kept the areas free from the heavy industrial developments that might scare away the tourists (McDonald and Muldowny, 128).

The underlying pastoral ideals of TVA's new landscape popularized that type of narrative and in the post-World War II years a growing number of Tennessee Valley tourist attractions, like the Museum of Appalachia, began relying on similar spatial designs. In addition to the pastoral framework, and partly as a way of making the pastoral more vivid, Irwin created the sense of authenticity that many visitors are looking for, and indeed claims of authenticity are central to the appeal of the place. In his book about the museum, Irwin points out that if "an item is separated from its history, then its meaning and significance is greatly diminished, or lost altogether" (23). For that reason all of the exhibits are presented with detailed descriptions placing them in a very specific context, explaining what the item was used for, as well as who owned or made it. On the museum's website Irwin further explains that he has "aimed for the 'lived-in' look, striving for, above all else, authenticity [and] to make the Bunch House, the Arnwine Cabin, and all the other dwellings appear as though the family had just strolled down to the spring to fetch the daily supply of water."[4]

But the idea of historical authenticity nevertheless is problematic. Unable, or unwilling, to verify or question such claims of authenticity, the tourist often chooses to believe in the narrative that he is being told. And is it, for example, even possible to claim authenticity on behalf of exhibits such as the General Bunch House (an old log

[4] Museum of Appalachia.
http://www.museumofappalachia.com/About%20the%20Museum.htm.

home that became the first dwelling exhibited at the Museum of Appalachia) after it has been moved from its original location and furnished with unoriginal furniture collected from other homes? Or does the notion of authenticity merely relate to the experience of anti-modernity? In fact, one may ask if there is any room for authenticity in the modern world.

In 1961, eight years before Irwin opened his doors to a paying public, Daniel Boorstin wrote his critique of a society which he felt was becoming increasingly unable to distinguish reality from pseudo-events. In *The Image* Boorstin argued that the increasing focus on advertising along with the media's growing tendency to create news rather than to merely report it, had caused expectations of life to become so overpowering and unrealistic that people increasingly gravitated toward events that were planned and orchestrated to fulfil these exaggerated demands. Tourism, as an expression of this trend, was increasingly dictated by the growing demand "that we can make the exotic an everyday experience (without its ceasing to be exotic); and can somehow make commonplaceness itself disappear" (Boorstin, 77).

Boorstin's insightful and provocative analyses, which were forerunners of hyperrealist and postmodern thought, opened a debate regarding what constituted an authentic experience in the modern, or postmodern, world. The focus on the plasticity of modern society spread into other academic genres as well, as when in 1976 geographer Edward Relph used the existentialist notions of authenticity and inauthenticity to study place in modernity. "Authenticity," Relph writes, "[. . .] connotes that which is genuine, unadulterated, without hypocrisy, and honest to itself not just in terms of superficial characteristics, but at depth" (64). He goes on to show that inauthenticity in fact defines a majority of modern places, giving examples such as suburban landscapes, highways, and tourist attractions including theme parks and, interestingly, recreated or restored pioneer villages.

Of course, Relph is right in recognizing the restored and recreated experience at places like Colonial Williamsburg, Greenfield Village, or the Museum of Appalachia as less than authentic, but his insistence upon a clear-cut authenticity/inauthenticity dichotomy leaves little room for nuances. As Dean MacCannell's anthropological study

The Tourist points out, the very idea of authenticity and inauthenticity is problematic in regard to tourist experiences. Recognizing modern man's essential need for something genuine to counter the increasing feeling of being alienated from his own existence, MacCannell instead suggests the introduction of the term "staged authenticity":

> The touristic integration of society resembles a catalogue of displaced forms. In this regard it is empirically accurate. The differentiations of the modern world have the same structure as tourist attractions: elements dislodged from their original natural, historical and cultural contexts fit together with other such displaced or modernized things and people. The differentiations *are* the attractions. Modern battleships are berthed near *Old Ironsides*; highrise apartments stand next to restored eighteenth-century townhouses; "Old Faithful" geyser is surrounded by bleacher seats; all major cities contain wildlife and exotic plant collections; Egyptian obelisks stand at busy intersections in London and Paris and in Central Park in New York City. Modernization simultaneously separates these things from the people and places that made them, breaks up the solidarity of the groups in which they originally figured as cultural elements, and brings the people liberated from traditional attachments into the modern world where, as tourists, they may attempt to discover or reconstruct a cultural heritage or a social identity. (MacCannell, 13)

MacCannell's argument is sound. In a postmodern world where one would be hard put to identify an authentic environment and where displacement seems to be the norm, it is of little importance if the artifact is removed from its original location since it is essentially up to the tourist to reconstruct its meaning anyway. Based on a number of factors that may include social and political standing, ethnicity, cultural background, religion, gender, personality, and the knowledge, or lack thereof, brought to the encounter, the tourist essentially negotiates the significance of the experience.

What we are left with when visiting a place like the Museum of Appalachia is neither authentic nor inauthentic, but a staged attempt at presenting something that a visitor might recognize as genuine; and for many people that is enough. The museum exemplifies not only a type of narrative structure to which tourists obviously respond, but also shows how the boundaries of what is perceived as authentic are constantly negotiated by what people want to believe. Ultimately, at the heart of the tourist's willingness to escape the turmoil of everyday life by entering idealized worlds such as Disneyland's Main Street USA or the Museum of Appalachia is a critique of the political and social complexities of modernity. With his museum John Rice Irwin has created a world of simplicity and perceived authenticity that serves as a counter narrative to the modern, and, albeit an illusion, it is one which most visitors want to believe in because it signifies that there is something real left after all.

Bibliography

Boorstin, Daniel J. *The Image: A Guide to Pseudo-Events in America*. New York: Vintage Books, 1987.

Featherstonhaugh, George William. *A canoe voyage up the Minnay Sotor; with an account of the lead and copper deposits in Wisconsin; of the gold region in the Cherokee country; and sketches of popular manners; &c. &c. &c.* London: R. Bentley, 1847.

Irwin, John Rice. *The Museum of Appalachia Story*. Atglen, PA: Schiffer Publishing, 1987.

Lilienthal, David E. *TVA – Democracy on the March*. New York: Harper & Brothers Publishers, 1944.

MacCannell, Dean. *The Tourist: A New Theory of the Leisure Class*. New York: Schocken Books, 1989.

Marx, Leo. *The Machine in the Garden: Technology and the Pastoral Ideal in America*. New York: Oxford University Press, 1970.

McDonald, Michael J. and John Muldowny. *TVA and the Dispossessed: The Resettlement of Population in the Norris Dam Area*. Knoxville: University of Tennessee Press, 1982.

Muir, John. "A Thousand Mile Walk to the Gulf." *The Wilderness Journeys*. Edinburgh: Canongate Classics, 1999.

Pinchot, Gifford. *Breaking New Ground*. Seattle: University of Washington Press, 1972.

Relph, Edward. *Place and Placelessness*. London: Pion Limited, 1983.

Wallace, Mike. *Mickey Mouse History and Other Essays on American Memory*. Philadelphia: Temple University Press, 1996.

Fig. 1. The American Axe Exhibit, Museum of Appalachia. 2006. Photograph by author.

Fig. 2. Museum of Appalachia. 2006. Photograph by author.

Fig. 3. Museum of Appalachia. 2007. Photograph by author.

Constructing Main Street: Utopia and the Imagined Past

Miles Orvell

The American small town has taken many different spatial forms across the United States, but generally they all have a central thoroughfare known as Main Street. Main Street is not only an actual physical place but, more generally, a symbol of core values in American society: community, democracy, and family. The symbolic meaning of Main Street evolved gradually, from the late nineteenth century on, and took hold partly through such self-conscious and idealized versions of community as Williamsburg, Virginia and Greenfield Village. Designed as a way of holding on to values that seemed to be disappearing in American life in the face of industrial expansion and urbanization, the idealized image of Main Street reached its apotheosis in Disneyland in the post-World War II era and has, since then, expanded its influence through the ever-growing new urbanist movement. Main Street is both a nostalgic symbol and a pragmatic response to development, and it contains elements of both freedom and repression, technology and the pastoral.

Every town or settlement has a central artery running through it, and in the United States that road or avenue is known as "Main Street": it is the essence of the small town and synonymous with it. The concept of "Main Street" is generic, but the term encompasses an enormous range of particulars, from the village green of the New England town to the Southern courthouse square to the linear Main Street of the Mid-West (Francavigila, 65-129). As the commercial district of a small town, Main Street is usually lined with storefronts—starting with the ubiquitous drugstore and the general store or grocery—that provide the essential services required of a residential population; and depending on the density of the population and the breadth of the region pulled into the town, Main Street may contain basic entertainment venues as well—movie theater, restaurants, concert hall. Meanwhile, a Town Hall might house the essential governmental functions, while one or more houses of worship will serve the religious needs of the community.

While the American Main Street is not unlike any European town center or English High Street, Main Street has a peculiar resonance within American culture. It is not only "real" public space; it is also a mythical place, perhaps the most iconic (and sacred) of American places, fraught with nostalgia for a lost America, carrying the symbolism of political community, of democratic town meetings, of life in harmony with nature; not surprisingly, Main Street is also the imagined place of our American future, our utopian longings, as we rebuild our cities and expand beyond into exurbia under the embattled banner of the so-called smart growth, based on the ever-appealing model of the small town. ("Smart growth" presumably replaces the dumb growth of the post-World War II years, whose sign is Sprawl.) Yet contradictions abound, as one examines the enduring model of Main Street, its genesis and continuing influence, and it is the history of this model I want to explore.

One might say that the ideation of Main Street begins as a self-conscious remembering of place, a remembering that is a consequence of moving away from it. Main Street, in this mythical sense, is part of the deliberate creation of the idea of the "New England town" at the end of the nineteenth century. I am not concerned here with the actual New England town, but rather with the touristic destination that would serve the needs of an urban America seeking an escape to a Golden Age of simplicity, far from the madding crowds of the city and industry. As Donna Brown argues, the recovery of the putative origins of New England village culture began as early as the 1850s, when Portsmouth, New Hampshire sponsored a "hometown reunion" for townsfolk who had been dispersed to the far corners of the country. Not only was there a sense of nostalgia for what was left behind as people moved to areas of greater economic opportunity; there was also a retreat to a past that was socially less conflicted, less ethnic than, for example, Boston, which by 1890 had an immigrant population of 68%, including first and second generations (Brown, 139). In the last quarter of the nineteenth century, scores of New England small towns—Deerfield, Ipswich, Marblehead, Salem, as well as Nantucket Island—underwent transformation back to a colonial appearance that purified the accidents of history, positing an exaggerated homogeneity, with neat clapboard houses and steepled churches, that satisfied the tourist in search of "heritage" (Brown, 133).

To the extent that the idea of Main Street has embedded itself indelibly in the contemporary American mind, it is largely a function of *created memory*—not, we might say, first-hand memory, but a kind of second hand, ersatz memory, a collective memory, given form in the many materializations of the small town that marked twentieth-century culture and continue powerfully into our own time. These models—and I will examine just a few—constitute the adhesive glue, to the extent that there is such glue, holding together America's multicultural society, a mythos that evolved in the early twentieth century through a variety of material models.

One of the most intriguing of these models is *Roadside America*, a masterpiece of folk art, created as a lifelong project to construct a miniature of the built environment of the United States. Billed as "the World's Greatest Indoor Miniature Village," *Roadside America* first became a public exhibit in 1935 and occupied a succession of spaces in Eastern Pennsylvania, expanding in size, until it attained its present form and location in Shartlesville, Northwest of Philadelphia, in 1953.

The creator, Laurence Gieringer, supposedly had the notion of building a miniature town at the age of ten (in 1905), when he climbed a mountain near his home and looked down below at the city of Reading—with its seemingly miniaturized houses. One can suspect the usual intoxication of a god-like dream in the youth, but it was his astonishing perseverance over the rest of his life (he died in 1963) that actually brought this vision to fruition, resulting in a doll house world, 3/8 of an inch to the foot, that covers approximately 6,000 square feet. (On his frequent trips to New York City, Gierenger was fascinated by the miniatures at the American Museum of Natural History and the Metropolitan Museum of Art.) Gieringer's "America" comprises a variety of landscapes, from ranches to mining to agricultural; there is a railroad yard, an airport, a zoo, a circus, and a baseball field. But the space where people "live" in this America is an idealized version of the small town, with modest architecture and streets, with shops, gas stations, and a movie theater.

Yet what is missing from this otherwise complete representation of the American imaginary is as striking as what is present: the city and the suburbs, the whole of urban and exurban America. This absence is the sign, we might say, of its ideology, a world-view that places the small town at the center of American civilization. Gieringer's model embodies a cosmogony that moves from

Western frontier settlements, to mining operations, to the farm and the village, to the early twentieth-century town. The latter is portrayed as a contact point between the individual and mass culture—movies, department stores, etc. (Meanwhile, surveying the whole of this narrative space are life-size statues of Native Americans—posted at the corners of the exhibition hall—removed from the scene of action to a spectatorial position.) In Gieringer's teleology, the town is the end-point, the summit of civilization, the center of a world that, as we can see from our perspective, was being torn apart by two world wars and a great depression.

Gieringer may not be our national Virgil, but his miniature epic relates to the larger movement to preserve and recover the American mythos of the small town that was going on, at larger scale, in many other places in the first half of the twentieth century: Henry Ford's Greenfield Village, Old Sturbridge, Plymouth Plantation, and a couple of dozen more historical recreations. I want to explore two of these—Greenfield Village and Williamsburg—which have served, since the 1930s, as touchstones for the American imagination, visible symbols of public order and American energy, shrines to the perfected memory of the village origins of the national life, that occupy a space somewhere between the work of art and actuality. With Ford as the creator of Greenfield and John D. Rockefeller Jr. as the prime mover behind Williamsburg, these places are themselves the fulfillment of dreams not unlike Gieringer's, only they were infinitely better funded, supported, and promoted.

Ford and Rockefeller were part of a wave of historical restoration and preservation that began in the late nineteenth century and was gathering force in the early decades of the twentieth. The idea itself of separating space, isolating it from the everyday, is as old as the concept of "the sacred" itself, but such sacred space is timeless; the idea of preserving space with a *specifically historical meaning—whether as diorama, period room, house, town, or historical district—* begins in the nineteenth century and it begins in Scandinavia. Though there were efforts at preserving the American past in the 1876 Centennial (a "New England farmer's home" with resident guide), the major inspiration for these American villages was the open-air museum of Skansen outside of Stockholm, Sweden, the creation of Artur

Hazelius in 1891, who wanted to preserve what he perceived as dying folk traditions in the midst of industrial change.

The juggernaut of industry was flattening the craft world of the nineteenth century throughout the "Old World" and the "New," but it is always an irony worth savoring that the man who contributed more than anyone else to these industrial changes was himself the greatest preserver of the past: Henry Ford. Ford, with virtually unlimited resources, used his network of car dealers to amass a huge collection of American artifacts. Ford's whole approach to restoration was personal and idiosyncratic: he had already been restoring his boyhood home when in the late twenties he completed the restoration of the Wayside Inn—"Built in the old Colonial day, / When men lived in a grander way," as Longfellow wrote. Ford was the Janus of his age—looking ahead to the enormous changes wrought by the machine, but also looking backward to, as Longfellow put it in introducing *Tales of the Wayside Inn*, a "region of repose":

> A place of slumber and of dreams,
> Remote among the wooded hills!
> For there no noisy railway speeds,
> Its torch-race scattering smoke and gleeds;
> (2)

But just as the Model T was manufactured for the mass market, so were Ford's escapist fantasies to strike a responsive chord in the popular culture of the early twentieth century, as his great project of preservation—The Henry Ford Museum and Greenfield Village—was to prove.

Most accounts of the Ford Museum and Greenfield Village discuss the complementarity of the two, with the museum housing the history of technology within a narrative of progress, while the village, with its historical buildings, evokes the past.[1] Yet Greenfield is not quite as simple as that observation may suggest. Greenfield Village is at once a nostalgic representation of small town America that is at the same time a microcosm of America's material history, taking us from the early 1700s (in the Plympton Family House, imported from Sudbury, Massachusetts) to the late nineteenth century. The space of

[1] See Anderson; Kammen; Conn.

Greenfield Village is designed as a map of economic spheres, with separate areas distinctly conceived yet spatially contiguous that exemplify the social economy of pre-twentieth-century America: there is an area of working farms and related structures (e.g., cider mill, carriage barn); an area featuring craft production (e.g., pottery, weaving, glass, sawmills); and an area called "Main Street," which comprises shops and offices (e.g., post office, tavern, town hall, general store, jewelry store).

Ford's Village is obviously eclectic and strives for a broad representation of material culture and the topicalities of place, but the whole of it feels like a small town and is arranged with "Main Street" as its central axis and another leading avenue (Maple Lane) running off at right angles to Main Street. Ford himself loved to stroll around the Village for relaxation, enjoying an atmosphere that would be very far indeed from the industrial landscape he was building during these same years, the River Rouge automobile plant, located fairly close to Greenfield Village, yet at the opposite extreme of the industrial economy: begun in 1917 and expanded over the next decade, River Rouge was the epitome of advanced mass industrial production.

Though Ford put the material history of technology in the separate Ford Museum, technology is not absent from Greenfield Village, which pays tribute to the creator's own genesis (a scaled down version of the Ford factory is there) as well as to the genius of Ford's friend, Thomas Edison (in replica laboratories). But Ford was nothing if not eclectic, and the Village encompasses a Cotswold Cottage from the 1600s, a Swiss chalet, and a section of a London sweet shop (scaled down from five to two stories), not all of whose connections with the American town are immediately obvious. In short, Greenfield Village is strikingly promiscuous, mixing the authentic and the replica, in its representation of the great figures of the past. Still, it's not the buildings per se we come to see, it's contact with "history." History was not bunk (as Ford had famously said), as long as it was invested in objects. Yet as the creator of a culture of copies, Ford may have also felt that it was enough, in the end, to present a model of the original, just as the consumer was buying an authentic "Ford" that was at the same time one of a thousand or more.

What's striking about Greenfield Village is the contradictions: it is a celebration of a democratic small town ethos that focuses on great inventors. It is a paean to community that harbors within it the seeds of the industrial transformation that would destroy community. It is a hymn to the creative past without bringing it into conversation with the modern world that was erasing that past. In short, Greenfield Village induces a kind of intellectual vertigo.

Colonial Williamsburg, by contrast, is all sanity, all consistency, all perfection. It was conceived not as a collection of buildings associated with illustrious figures, but as a collection of buildings whose claim on our interest is precisely their authenticity. Moreover, Williamsburg is representing not some mythical place, but itself, the actual Virginia town that had formerly been the capitol of Virginia and that, by 1900, had been largely eroded by the forces of change. Nevertheless, it is "representing" itself, for it exists at one remove (at least) from the authenticity of historical being.

For Williamsburg is not simply a place; it is a moment—1790—chosen as the epitome of the symbolic tradition represented by the town itself. Certainly Williamsburg's national importance in the late eighteenth century as Virginia's state capital encouraged the selection of 1790, but that date reflected the growing consensus among elite Americans of the early twentieth century that the high point of American history, the magical moment, was the War of Independence and its Federalist aftermath, long before the immigrant hordes invaded America, long before the Bolsheviks began to over-run the Williamsburg section—of Brooklyn.

Though Rockefeller's enthusiasm for Williamsburg (Virginia, not Brooklyn) makes sense in this context—and the restoration became his lifelong passion—it was also an expression of a cultural allegiance that in some ways ran counter to the more habitual inclinations of greatly wealthy Americans, many of whom were busy living a dream of European aristocracy, building castles on Fifth Avenue furnished with the finest hand-crafted goods of Europe. Madame Olenska's caustic observation in Wharton's *Age of Innocence* (1920), is on the money: "It seems stupid to have discovered America only to make it into a copy of another country"—France, especially (Wharton, 242). What was needed, Rockefeller realized, was a copy of his own country—the United States.

Given the sacred moment of 1790, many hundreds of buildings had to be destroyed in Williamsburg to achieve the desired effect, while hundreds more had to be either restored or, in most cases, built anew on their original historical foundations. But it was the consistency and coherence of the effect, the homogeneity of the town in its desired state, that appealed to Rockefeller who, as Mike Wallace has astutely observed, "was interested in totalities" (Wallace, 15). What Williamsburg represented was, above all, a complete environment, a sealed universe, a dream town frozen in a moment in history. Also frozen in place were the various social strata that the town represented—from the largest structures of the ruling class to the humbler abodes of the artisans; all would have a place in Williamsburg, although it would be many years until the African American presence would be adequately acknowledged, and even then in a form that essentially cleansed the brutalities of slavery from the living record that Williamsburg was to be.

The contradictions of Main Street are nowhere more evident than in that most powerful icon of American culture—Disneyland, America's utopia, the greatest of great good places, the most desirable tourist destination in the world, the ultimate reward for work well done. "I'm going to Disney World"—the ad campaign begun in 1987 and featuring the greatest athletes of their time—became in the late twentieth century synonymous with victory of the rarest kind, whether Super Bowl or World Series. Disney represents life itself, the absolute alternative to a death sentence, and more than one person on trial for his life has probably responded, when asked where he would go if he were to be released, "to Disneyland."[2]

Disneyland fuses several elements not previously brought together in amusement parks: it was not just a single-theme park but an amalgamation of themes (fantasy, adventure, futurism); it was a fair with rides, but with an atmosphere friendly to families; and it exploited the Disney characters, moving them from the two dimensionality of the screen cartoon to the three dimensionality of the street. Above all, it placed at the center of the Disney universe an image of small town America that struck a deep chord with Americans after World War II,

---

[2] Spoken by Jason Baldwin, in the film, *Paradise Lost: the Child Murders at Robin Hood Hills* (1996).

an affirmation of cultural values that symbolized the anchor for a ship of state that had recently survived a cataclysmic war and was now in the midst of an equally challenging Cold War. In 1947, when workers and artists at the Disney studios were being organized, breaking the founder's dream of a work force that was "family," Disney was called before the House Un-American Activities Committee as a friendly witness. What is your personal opinion of the Communist Party, Mr. Disney?

> DISNEY: Well, I don't believe it is a political party. I believe it is an un-American thing. The thing that I resent the most is that they are able to get into these unions, take them over, and represent to the world that a group of people that are in my plant, that I know are good, 100 percent Americans, are trapped by this group, and they are represented to the world as supporting all of those ideologies, and it is not so, and I feel that they really ought to be smoked out and shown up for what they are, so that all of the good, free causes in this country, all the liberalisms that really are American, can go out without the taint of communism. That is my sincere feeling on it.

Disneyland, still a glint in the founder's eye at this moment, would be everything that Communism was not; it would be as authentically American as the House Un-American Activities Committee itself.

Yet from another angle, Disneyland is the essence of the inauthentic: it represents nothing but itself, its own factitious universe. Manufactured, it can be replicated, not only in Florida, but anywhere the population might support the requisite tourist activity—Japan, France, Hong Kong—for The Mouse has a global notoriety. Disneyland—in whatever version—is unique among theme parks not only because of the patented presence of Mickey and company, but on account of the peculiar Disney cosmogony, at the center of which is Main Street America, in whatever theme park one might enter.[3]

---

[3] Even in a *New York Times* article about the visible slippage of Disney's theme parks, written in 2005, the author concludes with a paean to the magic of Main Street, "Seeing the cobblestone streets and the rows of old-timey shops, I can't help but buy into the fantasy that is Walt Disney's Main Street, designed after the one in his boyhood home of Marceline, Mo. Some might call it synthetic, but that's missing the point: It never pretends to be real" (Charles Passy, "Some Ask if the Disney Magic Is Slipping," 31 July 2005).

Main Street embodies the very concept of safe public space in its material form, by employing the architectural types that we associate with a nostalgic and idealized representation of small town America. There are no amusements in the Main Street area comparable to the rides and shows in the other areas; instead, Main Street features the shops of a small town, reduced to 7/8 scale to mark a smaller, more intimate environment for the shopper. Indeed, the shops are the entertainment, for there are no shoe stores, no cleaners, no hardware stores, or other practical establishments; instead there are typically a magic store, specialty food shops, and above all souvenir shops with Disney products. (There is also a stand-up movie theater showing Disney cartoons.) Yet Main Street is teeming with activity—people walking, a tram passing by, a horse-drawn vehicle, Disney characters floating through the crowd, and at regular intervals, a parade. Disney's Main Street is a perpetually active and exciting place, a place that implies a harmonious community and a democratic society, it is what Americans (and the rest of the world) want to think America stands for.

Yet Disneyland is a preserved moment in time as well as space: it is a dream of the McKinley era at the turn of the last century, when McKinley could declare, with the Great Depression of the nineties and the labor unrest of the period over, that "Now every avenue of production is crowded with activity, labor is well employed, and American products find good markets at home and abroad" (Second Inaugural Address). And that feeling of confidence and comfort was doubtless augmented by the recent display of America's might in the Spanish American War. However perilous that moment in fact was—anarchists lay in wait for McKinley, the Lower East Side was seething with socialists, the seeds of a convulsive war in Europe were being sown—it was a time well chosen for at least an imagined point of stasis. We can never underestimate the power of nostalgia, and Disney's Main Street can hold its own against Tomorrowland, Frontierland, Fantasyland, and all the other lands that we inhabit imaginatively. It can even, we might say, hold its own against Las Vegas.

The actual town of Las Vegas, like so many Western towns, evolved organically, adding piece to piece over many years, beginning with the railroad cottages built by the Las Vegas Land and Water Co.

for employees of the San Pedro, Los Angeles & Salt Lake Railroad, in 1911. Though seemingly a yoking of opposites, Main Street and Las Vegas have a kind of symbiosis and even a consanguinity, since the revival of Main Street iconography in the new urbanism owes more than a little to the realization of Robert Venturi and Denise Scott Brown that one could indeed learn something, in terms of urbanism, from Las Vegas; and what one could learn is the power of the commercial strip—a linear Main Street, writ large and in neon signage—to invigorate urbanist design. The Las Vegas of today, however, is far from Venturi's Las Vegas: it is a collection of themed hotels, more Fantasyland than Main Street. And if Middle America goes to Las Vegas to escape the humdrum realities of "back home" and find adventure in the factitious collection of New York skyscrapers, the Eiffel Tower, Venice, Caesar's Palace, the Luxor Pyramid, and the other facsimiles of world culture, then we have to add one more piece to the mix, in some ways the most startling: Main Street Station Hotel, Casino, & Brewery. It is Main Street slouching back to Las Vegas, reborn as a rough beast.

What is Main Street Station Casino doing there? Is it for the world traveler (Japan? China? Europe? Africa?) who wants the "real America" in Las Vegas and is tired of Caesar's Palace? Or is it for the real American, who is xenophobic, even in fairy land, and wants a familiar piece of home, evoking the Victorian era, as the website declares, yet with "non-stop casino action and great dining," not to mention a collection that includes "Buffalo Bill Cody's private rail car, a fireplace from Scotland's Preswick Castle, chandeliers from the Figaro Opera House"—and more. (Why couldn't they stop at Buffalo Bill's private car? Whatever, it's there.)

Las Vegas, whose resorts comprise a living museum of historical places, is also a place where people live through the year. And while it has its healthy share of tract development outside the center city area, it also has a new "new urbanism development." Combining shopping and working within walking distance of home, it is being advertised online by Carpenter Seller Architects (CSA), as the "Village of Centennial Springs in Las Vegas." The entry corridor is Main Street, offered on the Centennial website as the distilled essence of utopia: "CSA researched historic main streets around the country to reconnect to the community friendly developments that are not found anywhere in Las Vegas."[4]

Centennial Springs—a Las Vegas version of a community type that is growing everywhere as the model of exurban smart growth development—represents an idea that is, of course, worlds apart from the Modernist ethos and the type of development represented by the typical condominium world of enclosed courtyards, gardens, and swimming pools. The latter excludes the public from its interior community. Centennial Springs invites the public in, via its Main Street. Yet this Las Vegas community, like the many other replicas of Seaside and Celebration that have spawned across the country, itself contains a contradiction that residents may only gradually come to understand: the perfection of the image of Main Street requires total obedience to the designers's dream of a public utopia. That is, no changes can be made in the original conception without compromising the whole: if wooden fences look right, look old-fashioned, in the desired way, then white plastic fences (more durable, but after all plastic and shiny) must be forbidden.[5]

In the new utopia of the planned small town, the price of community is conformity, and the laws of Town Center—far from being created by some democratic town meeting—are created by the local government officials, acting with the developer. Because these planned towns are appealing and attractive precisely because of their perfection, there is no room for imperfection. Because they were built at one moment, they have not had the luxury, one might say, of evolving a more layered, more organic appearance that would allow for variation. They are hyper-realities, not realities. (Levittown offers a counter-example of a homogeneous community, built all at once, that was allowed to grow and evolve along individualist lines and still kept its appeal, but Levittown was on a grander scale and never had the "town center" that defines these smart growth new towns.)

The hyper-reality of the new urbanist main street has come full circle from the historic beginnings of the small town, creating an image

---

[4] Rhapsody Partners. http://www.centennialsprings.com/index2.html.

[5] Jill P. Capuzzo, "A Town Divided Over a Fence," *New York Times*, 5 November 2006.

of utopia built on an imagined ideal of public space, a Platonic idea or composite, resulting from multiple forms of idealized touristic destinations (*Roadside America* to Greenfield Village to Williamsburg to Disneyland) that have embodied the ethos and icon of Main Street. The growing popularity of this model in new urbanist communities attests to the continuing appeal of Main Street as public space and as symbol of community, however self-consciously constructed it may be. Whether or not the new urbanist communities are more than symbolic representations of democratic public space is another question, but one that lies outside the scope of the present essay.

Bibliography

Anderson, Jay. *Time Machines: The World of Living History*. Nashville: American Association for State and Local History, 1984.

Brown, Donna. *Inventing New England: Regional Tourism in the Nineteenth Century*. Washington D.C.: Smithsonian Institution Press, 1995.

Conn, Stephen. *Museums and American Intellectual Life*. Chicago: University of Chicago Press, 2000.

Disney, Walter E. "Testimony to the House Un-American Activities Committee, October 24, 1947." CNN Interactive. http://www.cnn.com/SPECIALS/cold.war/episodes/06/documents/huac/disney.htm l.

Francaviglia, Richard V. *Main Street Revisited: Time, Space, and Image Building in Small-town America*. Iowa City: University of Iowa Press, 1996.

Kammen, Michael. *Mystic Chords of Memory: the Transformation of Tradition in American Culture*. New York: Vintage, 1993.

Longfellow, Henry Wadsworth. *Tales of a Wayside Inn*. Boston: Ticknor and Fields, 1863.

Paradise Lost: the Child Murders at Robin Hood Hills. Directed by Joe Berlinger and Bruce Sinofsky. HBO, 1996.

Wallace, Mike. *Mickey Mouse History*. Philadelphia: Temple University Press, 1996.

Wharton, Edith. *Age of Innocence*. New York: Random House, 1920.

Pasteboard Views: Idealizing Public Space in American Postcards, 1931–1953

Jeffrey L. Meikle

Cultural geographers distinguish between physical *spaces* and the concept of *place*, the latter often mediated by visual representations. The so-called *linen* variety of picture postcard offered popular interpretations of the American scene, ranging from natural landscapes to public works and urban views, during the Depression and the Second World War. These unique postcards, with extravagantly colorized images printed on textured surfaces, were developed by Curt Teich & Co. of Chicago and widely imitated. Unlike earlier view cards, which adhered closely to the details of the monochrome photographs on which they were based, linen postcards eliminated perceived imperfections, manipulated and fabricated details, offered a colorful but universal vision, and idealized American place as the streamlined product of an optimistic, upbeat, even utopian modernity. However, public spaces of linen postcards were filled with faceless crowds or, more often, oddly empty, even devoid of people. These visual representations interiorized America's traditional wide open spaces and miniaturized them for individual possession and consumption, thereby contributing to middle-class desires for secure public spaces and for a homogeneous sense of place.

Cultural geographers often suggest a problematic relationship between the concepts of *space* and *place*.[1] According to their distinction, *space* refers directly to the objective material reality of a particular physical location. *Place*, on the other hand, refers to a complex assemblage of human perceptions and associations only imprecisely or imaginatively related to a corresponding physical *space*. To further complicate this distinction, the perception of place, often a highly personal, even intimate matter, can be socially communicated only through representations, whether verbal or visual. Although processes of modernization over the past two centuries have brought continuous transformations in society's shared *spaces*, it is only through mediated representations of *place* that most people—whether capitalists, politicians, intellectuals, or ordinary citizens—have gained a measure

[1] For a convenient collection of views see Adams, Hoelscher, and Till.

of understanding and a sense of control over otherwise threatening changes to the spaces of their existence. These representations of place actively reflect or shape beliefs, confirm or motivate behaviors, and in turn influence ongoing transformations of actual physical space. In this essay I intend to focus attention on the picture postcard, an ephemeral category of material artifact whose visual images, however humble and disposable they may be, have reflected and reinforced popular concepts of place. For that reason alone they repay close attention—not to mention that they can be almost hypnotically fascinating in their own right.

This claim of fascination arises from personal experience with a collection of more than 5,000 postcards published in the United States between 1931 and 1953.[2] Those dates are not arbitrary but encompass the chronological range of a particular type of card, a uniquely American variant known as the *linen* postcard because its extravagantly colorful surface was embossed with a distinctive textured finish (fig. 1). The leading publisher of linen postcards claimed that the "striking note of smartness" of these "beautiful miniature paintings" marked them as the most "aristocratic of all post cards."[3] Introduced during the depths of the Depression, the optimistic images of linen view cards celebrated the vast extent of the American scene, almost as if following the words of the folksinger Woody Guthrie's spirited anthem, "This Land Is Your Land": "from California to the New York island, from the redwood forest to the Gulf Stream waters," moving along a "ribbon of highway" through "diamond deserts," over an "endless skyway" into a "golden valley."[4] Although natural landscapes predominated in the lyrics of Guthrie's song, written in 1940, linen postcards equally celebrated the grandeur and tempo of modernization during an era which journalists and social commentators often self-consciously referred to as a "machine age" (fig. 2). Newly constructed "ribbon[s] of highway" passed over dramatically engineered bridges, looped across the curving rims of hydroelectric dams that tamed raging rivers and provided millions of people with energy, passed along the productive infernos of

---

[2] All postcards cited are from the author's collection.

[3] *Sales Pointers*, 10; *Jobbers Profits*, 2; and *Sales Pointers*, 9.

[4] "This Land Is Your Land, This Land Is My Land." Transcription by Manfred Helfert of a Guthrie manuscript dated 23 February 1940, online at http://www.geocities.com/nashville/3448/thisl1.html.

steel mills, and offered panoramic views of expanding cities across the continent. While New York City best exemplified the machine age, retaining considerable emotional power even during the Depression, the smallest of American cities boasted a proud new skyscraper rendered as a marvel of modernity in its linen postcard image.

These celebratory representations of the American scene, often almost utopian in their intense rendering of place, issued from a relatively prosaic mass-production process developed by a German immigrant, Curt Teich, who had arrived in Chicago in 1895 at the age of 18.[5] Trained as a third-generation printer, he opened his own printing business two years later. In 1904 he briefly returned to Germany to investigate new lithographic techniques, which he then applied to take advantage of an ongoing postcard mania so widespread that for about a decade every major town boasted a shop dedicated to the sale of postcards—not only landscape view cards but also cards with holiday greetings, sentimental genre illustrations, painting and sculpture, and humorous novelties. Many local photographers produced batches of so-called *real-photo* views of prominent sights by printing directly from negatives onto chemically treated card stock. However, the real-photo process did not lend itself to mass production. The majority of view cards were printed in volume either by halftone processes imitating black-and-white photographs or by lithographic processes whose color inks, sometimes ten hues per image, were each printed separately over a halftone base, thereby simulating hand-tinted black-and-white photographs. Competing with printers in Germany who at that time produced most view cards of American scenes, Teich reduced the lithograph process to four color inks over black halftone without compromising the delicacy of the effect. His business prospered, especially after 1914, when war in Europe eliminated German imports, and throughout the 1920s. However, after the Depression seriously reduced the profits of Curt Teich & Co., he responded with the new "C. T. Art-Colortone" process, the source of the linen postcard.

As in Teich's earlier lithographic process, preparation of a linen view card began with a black-and-white photograph often

[5] Biographical information comes mostly from Teich Sr., *The Teich's* [sic] *Family Tree and History*.

provided by a local client—a stationer, druggist, hotel or restaurant proprietor, or the manager of a tourist attraction—but sometimes taken by a traveling sales agent. The agent also sent back to Chicago a tissue-paper overlay indicating by numbered code the appropriate colors, along with instructions for manipulating a scene to improve its attractiveness. In the company's art department, an airbrush artist retouched the photo by cropping, removing undesirable features, adding desirable ones, and in general sharpening and clarifying the image. Working with a halftone produced from this airbrushed photograph, a colorist then prepared a postcard-sized watercolor proof to be sent to the client for approval. At the plant in Chicago, printing plates for offset lithography were prepared in four colors—the primaries red, yellow, and blue plus an additional dark blue which contributed a dramatic sense of depth. The embossed linen finish, which seems never to have been previously used on postcards, enabled the Depression-stricken company to reduce production costs. Not only did the texturing disguise impurities in a less expensive pulp-paper card stock; it also allowed the firm to reduce the printing screen from the postcard industry standard of 175 lines per inch to 125 lines per inch. These savings in paper and printing plates almost inadvertently yielded the bright, bold, sometimes extravagant colors and forms of the linen postcard. Before abandoning the linen process early in the 1950s to print directly from color transparencies, Curt Teich & Co. produced more than 45,000 individual linen views.[6] These postcards were so successful that five major competitors and dozens of marginal printers imitated Teich's process, at least doubling the number of individual views. Total production of tens of millions of postcards brought an idealized sense of place to local residents, tourists, business travelers, construction workers on government projects, defense workers, and military recruits from the Depression through the Second World War.

Only recently have historians begun to address the significance of postcards as popular representations of place during the 1930s and

---

[6] The quantity of individual linen views created by Teich can be estimated by totaling the runs of annual serial numbers on a complete set of cards in the possession of the Curt Teich Postcard Archives, Lake County Discovery Museum, Wauconda, Illinois. I have adjusted the figure downward to make up for the approximate number of humorous and advertising cards (i.e., non-view cards).

1940s.[7] Greater attention has focused on the famous documentary photographs of the Farm Security Administration—nearly 200,000 images recording not only the farms and small towns of Depression America but also their inhabitants and those who were just passing through—in other words, humanity in all its diversity. The photographer Berenice Abbott, who documented scenes of New York City for the Works Progress Administration, expressed contempt for commercial imagery of the sort produced by Teich & Co. She believed that photography captured "the contrasts, the paradoxes, [and] the anomalies" of life, thereby offering "a reality superior to guidebook wisdom, an observation far more profound than the postcard view of the Empire State Building."[8] But the documentary photographer Walker Evans held the opposite opinion of postcards. At his death in 1975, his personal collection numbered more than 9,000 American view cards. Recent studies suggest that during the 1930s, while working for the Farm Security Administration, Evans often took inspiration from postcards and rephotographed the exact scenes they portrayed. Indeed, Jeff Rosenheim, a curator of photography at the Metropolitan Museum of Art, is currently preparing an exhibition and catalogue, scheduled to open in the winter of 2009, exploring what he defines as "the symbiotic relationship between Evans's own art and his interest in the style of the postcard."[9]

For Evans, early lithographic cards captured both the reality of physical space and the more nebulous spirit of place. In 1948 he contributed an article with color reproductions of eighteen postcards to *Fortune*, an upscale business magazine.[10] Most of them portrayed public spaces of the early twentieth century: busy streets, a courthouse, a square with promenading couples, a public beach. His brief text celebrated the "purity," "fidelity," and "restraint" of these "humble"

---

[7] See Ruby; Geary and Webb; Stevens; Meikle, "A Paper Atlantis"; Werther and Mott; Jakle; and Isenberg, 42-77.

[8] Berenice Abbott as quoted by Orvell, 29, 318n39.

[9] Rosenheim and Eklund, 202. For analysis of a single postcard view of Morgan City, LA, and a corresponding photograph by Evans, see Rosenheim and Eklund, 200-202, and Rosenheim, "'The Cruel Radiance of What Is,'" 66-67. Rosenheim discussed the upcoming exhibition in a telephone conversation with the author, February 2008.

[10] Evans, "Main Street Looking North from Courthouse Square," 102-106. See also his similar articles "When 'Downtown' Was a Beautiful Mess" and "Come on Down."

cards. Their "mood" was "quiet, innocent, and honest beyond words." These simple artifacts, which comprised for Evans "some of the truest visual records ever made of any period," conveyed a "feeling for actual appearance of street, of lived architecture." His commentary suggested nostalgia for a lost American scene and the lost medium which had captured it. "This is how the county courthouse rose from the pavement," he explained, "in sharp, endearing ugliness. These, precisely, are the downtown telegraph poles fretting the sky, looped and threaded from High Street to the depot and back again." By 1948, however, American postcards had entered "an aesthetic slump" from which Evans thought "they may never recover." They had become the "quintessence of gimcrack, [. . .] gaudy boasts that such and such a person visited such and such a place, and for some reason had a fine time." Although Evans blamed "current color-photography printers" for the postcard's demise as an accurate representation of American space and place, the true villain's identity remains uncertain. Was he thinking of the shiny new cards based directly on color photography which in 1948 were already cutting into sales of linen postcards? Or was it the linen view itself, often gaudy and extravagant, and definitely not suited to "the rendering of patina and the soft tones of town buildings and streets," which provoked Evans's nostalgia? Given that he reproduced no linen postcards in the article, it seems likely he considered them a degraded medium of representation. Why Evans might have thought so relates directly to the ways in which linen cards stylized the representation of otherwise diverse public spaces throughout the nation, thereby constructing a sense of place more congruent with machine-age modernity but lacking in the random messy vitality which appealed to the photographer.

Before the advent of the linen card, public space went relatively unmediated in postcard views. Although few photographs ever escape conscious shaping, many local real-photo cards from the early twentieth century exemplified the sort of "folk document" Walker Evans admired—in his words a "beautiful mess" in which "trolleys clang" and "the air is not entirely free of horse smells" (Evans, "When 'Downtown' Was," 101). A real-photo view of the main street of Sycamore, Ohio (fig. 3), dated 1907 in pencil on the back, depicts ramshackle commercial buildings, horse-drawn buggies clumped along uneven hitching rails, a dilapidated awning frame missing its canvas, two scruffy men loafing in a doorway, and rows of towering utility poles running up each side of the street, their intrusive modernity contrasting with a sea of mud in the foreground, crossed by trolley tracks and pockmarked by hoof prints. Whatever local pride the photographer intended to convey could not obscure the scene's messy physical reality. Another real-photo view, an undated postcard documenting a public parade in Galion, Ohio, though looking down from an upper-story window, does not begin to contain the chaotic force of its details, ranging from horse-drawn floats and intricate decorations to the various actions of hundreds of people in the crowd, recorded sharply enough to be individually perceived.

Although early color lithographic view cards imposed greater control over their images than was possible with real-photo cards, even today they retain a convincing verisimilitude. Two lithographic postcards published by Teich & Co. in 1911 might easily have appeared in Evans's *Fortune* portfolio. A view of the Union Depot Hotel by the railroad tracks in Marion, Ohio, reveals a nondescript structure with walls stained by mildew beneath the eaves. Loungers fill the benches, an unkempt industrial building is visible to the left, and to the right a fence is covered with advertising. Another view of Marion (fig. 4) portrays the main business street as a wide expanse of dirty brick paving with a few buggies and autos parked along it, pedestrians scattered up and down the sidewalks, a wealth of detail of commercial signage, and a wooden utility pole prominently displayed in the left foreground, the first in a series of poles running up the street, most leaning to one side or the other. These images reveal American public space of the early twentieth century as jumbled and ill-defined. Such views are typical of most pre-linen postcards, whether produced locally by the real-photo process or mass-produced in Chicago by color lithography, because they suggest little conscious effort to idealize the places they portray.

The linen postcard changed all that, not because Curt Teich possessed a coherent vision of an idealized American scene but because the medium's inherent aesthetic qualities coalesced with general cultural trends in representations of machine-age modernity. Given the coarser printing screen and textured surface, linen cards could not approach the nearly photographic precision of the previous generation of lithographic cards. Such details as individual human figures and small advertising signs became blurred and generalized, while fields of color became bolder and brighter. At the same time, images conveyed by linen cards reflected developments in the commercial aesthetic of the 1930s. Whether in retail architecture, in passenger trains and automobiles, in household appliances, or in graphic arts and advertising, designers employed the smooth, rounded forms of streamlining, a style which suppressed complex details beneath flowing, easily comprehended forms, thereby promoting faith in modern progress. Following streamlining's cultural logic and the technical aspects of the linen process, the urban public space of Teich's postcard views became smoother, cleaner, less cluttered, and to some extent emptied of messy humanity (fig. 5).[11]

While official guidelines of the Farm Security Administration prompted documentary photographers to record "fire hydrants, traffic signs, hitching posts[,] [. . .] watering troughs, [. . .] wagons and horses, [. . .] men loafing," and other specific details of everyday life,[12] linen postcards offered a vision of public place which often seemed, despite bold colors and forms, somehow visually and conceptually reduced, brought under control, universalized, and idealized. Everything was modern and up-to-date. Linen postcards erased the very remnants of pre-modern life which FSA photographers were instructed to record and eliminated any aspects of modernity—such as utility poles, power lines, delivery trucks, and beat-up Model T jalopies—which might detract from an overall impression of streamlined efficiency. Street surfaces, in reality composed of asphalt or concrete, seemed poured from an artificial substance so smooth that it voided all memory of dirty brick or deep mud. Not content with air-brushing old cars out of a source photograph, the artists in Chicago often inserted fanciful elongated automobiles bearing little resemblance to reality. Equally unreal were the many views of individual urban landmarks, ranging from railway stations to hotels, whose forms were abstracted against a neutral backdrop, with other buildings to left and right either removed entirely or rendered in a hazy, stylized manner.

The urban historian Alison Isenberg has used the surviving production files of Curt Teich & Co. to demonstrate that the air-

---

[11] On streamlining see Meikle, *Design in the USA*, 113-129.

[12] "Sample Items from a Small Town Shooting Script," reprinted by Orvell, 292.

brushing of main street linen views coincided, almost as if by intention, with the goals of city planners who sought in the 1920s and 1930s to beautify city streets by creating "a dignified and simplified retail corridor, as opposed to the existing chaotic hodgepodge of individualistic storefronts" (45). The result, at least in postcards if not so often in reality, was "a clear commercial vista in which the eye could travel, like a consumer, uninterrupted by the bodies of other shoppers and by the constant bother of hawkers and 'cheap' sidewalk sales tactics" (70)—in other words, a metaphorical main-street version of visual streamlining. Although some linen cards, depending in part on the original photographs from which they were adapted, did represent street scenes with lively crowds of people, the majority of such scenes were, in her words, "strangely somnolent and empty" (Isenberg, plate 4). For the most part, pedestrians were rendered as tiny figures indicating scale. Automobiles constituted the major medium of circulation, suggesting the virtual exclusion of pedestrians from future urban landscapes.

The manipulation of urban space practiced by Teich and other linen postcard publishers was nothing new in visual representations of American cities. Peter Bacon Hales, a historian of photography, has defined a tradition of "grand-style urban photography" which came to prominence during the Gilded Age of the late nineteenth century as upper-class Americans sought to invest rapidly growing, socially chaotic, perpetually unfinished cities with a veneer of European high culture. Typical of this genre was S. B. Frank's *Chicago Art Portfolio*, an album of halftone photographic views published in 1894, which organized the city according to a taxonomy still being followed forty or fifty years later, virtually unchanged, by the publishers of colorful, inexpensive, democratized penny postcards. Although linen cards existed by definition as scattered images, designed and published in no particular order and brought together by location only momentarily in the vending racks of souvenir stands, their representations, like Frank's *Portfolio*, encompassed a taxonomy of bird's-eye views, street scenes, public buildings, monuments, skyscrapers, boulevards, parks, harbor views, homes of the wealthy, churches, railway stations, and hotels. According to Hales, this standard visual packaging constructed "a rational, orderly, stable city, a place of perpetual presence rather than mutation and change." Whether conveyed by grand-style urban photographs or by the linen postcards which borrowed their iconography, these images "anchored" the "flux" of the modern city, "banished all traces of poverty, violence, class and racial hatred, and multicultural conflict," and ultimately offered "reassurance and respite" within a protective "*cordon sanitaire*."[13]

According to Hales, grand-style urban photography of the Gilded Age neutralized the supposedly corrupting influences of immigrants and ethnic minorities either by excluding them from representation or by rendering them as harmlessly picturesque. In the same manner, postcards from the 1930s and 1940s typically stereotyped Chinese Americans and Mexican Americans, marginalizing them as exotic fixtures of recreational destinations for middle-class whites seeking momentary respite from the pressures of modern life. While San Francisco's Chinatown did indeed function as a residential and commercial district in addition to offering an array of ethnic restaurants, the Olvera Street market (fig. 6) opened in Los Angeles in 1930 purely as a Mexican-themed tourist attraction. Its success stimulated the nearby construction of New Chinatown, a pedestrian zone of shops and restaurants whose first phase opened in 1937. Linen views of New Chinatown emphasized comfort for white visitors but also conveyed an aura of exotic mystery. The success of such permanent commercial projects, which capitalized on the popularity of temporary ethnic attractions at international expositions, stimulated development of similar semi-private, semi-public festival spaces into the late twentieth century as a means of creating an artificial sense of traditional place within the placelessness of an expanding automotive metropolis.

Representations of clearly distinguishable individuals were common in linen postcards depicting members of racial or ethnic groups whose social marginality and apparent difference marked them as available for and apparently worthy of display. That was the case not only with urban groups but also with such rural groups as southern African Americans, Native Americans of the southwestern deserts and southeastern swamps, and even white Appalachian mountaineers. However, distinct individuals from the white urban majority were only

[13] Hales, 186-188, 203; on the taxonomy of Frank's *Portfolio* see Hales, 126-130, 184-187, 200-202.

rarely represented on postcards. The general public tended to appear as a mass of people in a crowd with whom viewers of a card could easily identify, projecting themselves into a scene and merging with it. Most crowd shots involved leisure and recreation. Scenes of beach resorts were common, with postcard views celebrating unbelievable congestion. The boardwalk of Atlantic City, New Jersey, was portrayed as a site of mechanized pleasures little different from ordinary urban experience. Other cards depicted football stadiums, theaters, band concerts, the skating rink at Rockefeller Center, public benches in St. Petersburg, Florida, with retirees as tightly packed as neighboring parked cars, and a huge shuffleboard park with so many rows of courts that the players resembled assembly-line workers. Despite frequent images of pleasure-seeking crowds, however, the majority of representations of public space were nearly devoid of people, often lacking even tiny figures provided for scale.

This common emptiness suggested public space as a commodity to be viewed from a distance rather than directly inhabited and used by the public. A progression of cards depicting the Chicago lakeshore indicated a vast expanse of open space recorded as if from an imaginary vantage point hovering just above normal human reach. Across the country, public squares, monuments, fountains, and auditorium buildings in attractively landscaped settings were oddly unpeopled. Especially fascinating in terms of their relative emptiness were views of public parks—which had in fact been initially established to provide urban masses sufficient breathing room to maintain the health of the social organism. From New York City's Central Park to the Botanical Gardens of Brooklyn, from San Diego's Balboa Park (fig. 7) to Philadelphia's Fairmount Park, linen postcards portrayed public parks as colorful and visually compelling—and as deserted as America's vast natural landscapes. Although representations of public parks often depicted complex forms of rocks, trees, and bushes, dappled with sunlight and shadow, the eye passed smoothly over their surfaces, which afforded only momentary interruption to the frictionless technological and social functioning of the modern cities in which they were embedded.

Strangest of all was the relative emptiness of linen postcard views of such public buildings as transportation terminals and government buildings—in actual fact energetic hubs of the movement of people and power. For example, the plaza in front of Kansas City's Union Station (fig. 8) revealed a rank of taxis and some private autos picking up or discharging passengers but conveyed nothing of the hustle and bustle of a major railway station processing thousands of passengers each day. Nor did the interior appear busy in a view portraying a massive space with a handful of people leisurely standing around. The exterior of Union Station in Los Angeles resembled a park building with a few people lounging on benches while its cavernous interior was nearly empty. From city halls and courthouses to post offices and state and federal capitol buildings, the halls of government portrayed in linen cards typically revealed a complete lack of human presence. Although Louisiana's new state capitol, completed in 1932, boasted a nontraditional skyscraper exterior, Teich's representations of the interior were typical of the genre. Vast corridors and public halls led on to sweeping staircases and endless internal vistas, devoid of people but pregnant with an ominous sense of something about to happen. A series of cards devoted to Kansas City's Municipal Auditorium conveyed a similar effect. While a dramatic night view of the exterior conveyed a festive mood, the auditorium's interiors (fig. 9) were as empty as those of the Louisiana capitol. A card depicting the grand foyer of Radio City Music Hall in New York, to note yet another example, permitted a viewer to imagine moving unimpeded across a vista of richly ornamental carpet and then ascending a distant staircase whose wildly colorful mural promised some unprecedented revelation.

The relative absence of people from so many linen views (or their relegation to the status of insignificant ciphers) possibly afforded consumers of the cards greater latitude of interpretation. Without any central figures with which to identify, the purchaser or recipient of a card was free to assume a possessive attitude toward whatever scene a card represented. No obstructions prevented the adoption of a centralizing point of view. Bird's-eye urban views and wide-angled street scenes emphasized the importance of the city itself, beyond the influence of mere individuals, as a smoothly functioning, streamlined engine of modernity. In the case of government buildings, a viewer could interpret the absence of human individuals as conveying an implicit understanding that everything was running smoothly and efficiently. On the other hand, given the social chaos of the Depression, such representations may have left some viewers with a dismaying sense that no one was truly in control. And those who felt excluded from the benefits of modernity celebrated by linen postcards may have experienced the emptiness of these views as emblematic of their personal exclusion. The expansive vista of Radio City Music Hall's foyer might have invited some viewers, perhaps tourists gazing at a rack of postcards on the sidewalk, to aspire to enter into a grander reality. Or they might just as well have felt left out, standing on the outside, figuratively looking in through seductive pasteboard windows. Either way, postcards featuring unpopulated spaces invited viewers of all classes and backgrounds to project personal needs and desires onto them, with no obligation to evaluate, identify with, or relate to intervening representations of specific types of individuals.

Although the vast majority of linen views represented outdoor scenes that qualified in one way or another as public space, proprietors of hotels and restaurants often commissioned cards illustrating the interiors of their businesses. Perhaps because they feared that changing fashions of clothes would quickly date visual representations of an otherwise modern establishment, these interiors also almost always lacked people. This absence of people created a unique aesthetic for these interior views. Without people, interior views typically opened from a generous wide-angle perspective and narrowed into a long diagonal vista whose exaggerated depth seemed to extend infinitely. It was almost as if the wide open spaces of the American scene had been brought indoors. Whether depicting a hotel lobby or a Los Angeles lunch counter advertised as "the longest [. . .] in the world," these interior views often paralleled the layout of city streets as depicted in linen postcards. Interior arrangements of tables and counters took on architectural characteristics, mirroring commercial buildings arrayed in the grid pattern of city blocks, an effect rendered most obviously when two aisles in a restaurant met to form an intersection. In a striking example, the interior of Wolfie's restaurant in Miami Beach (fig. 10), with a sleek streamlined lunch counter, suggested a street corner with a curving moderne facade (fig. 11). Even more startling was the interior of a curio shop in San Antonio, with an island counter whose glass display cases, as rendered by Teich & Co., prefigured the glass office buildings of two decades later. Despite a clutter of antlers on the walls and native weavings for sale, the store's wide, empty aisles promised circulation as free and easy as that of the widest city streets.

Occasionally the designer of a commercial interior consciously employed this morphing of exterior-interior space, as in the decor of a Florida cafeteria (fig. 12) whose tables seemed to fill the vast space of an arcaded square somewhere in Spain, set off by red clay roof tiles, with fair weather eternally promised by a delicate sky-blue ceiling scattered with fluffy clouds.

This postcard may remind twenty-first century viewers of the private interior spaces of such faux-exterior pseudo-public streets and squares as those of the Forum Shops at Caesars Palace in Las Vegas—an observation leading to my final point about space and place in linen postcards of the 1930s and 1940s. Although Teich's delineators often brought the outside inside, treating interiors with the expansiveness of urban streets and the openness of public space, the primary cultural effect of linen postcards was the exact opposite. By removing people from the picture, by reducing even the largest city, the tallest skyscraper, and the grandest natural vista to a pasteboard card only 3 ½ by 5 ½ inches in size, and by printing these views in exaggerated artificial colors, Curt Teich and his competitors effectively privatized even the most public spaces for individual possession. These miniaturized approximations of the American scene—airbrushed, streamlined, and fundamentally under control—may eventually have contributed in some small way to the development of an apparent contemporary desire for safe, exclusionary public spaces cleansed of uncertainty and risk. However, such cultural effects obviously lay beyond the conscious intentions of publishers who successfully recorded thousands of individual views of the American scene, often with considerable accuracy, many of which would otherwise now be lost. Just as significantly, their work also gave expression to a popular mood or sense of place, unique to a particular historical moment, whose expectant optimism still emanates from many of these images.

Bibliography

Adams, Paul C., Steven Hoelscher, and Karen E. Till, eds. *Textures of Place: Exploring Humanist Geographies*. Minneapolis: University of Minnesota Press, 2001.

Evans, Walker. "Come on Down." *Architectural Forum* 117 (July 1962): 96-100.

—. "Main Street Looking North from Courthouse Square: A Portfolio of American Picture Postcards from the Trolley-car Period." *Fortune* 37 (May 1948): 102-106.

—. "When 'Downtown' Was a Beautiful Mess." *Fortune* 65 (Jan. 1962): 100-106.

Geary, Christraud M. and Virginia-Lee Webb, eds. *Delivering Views: Distant Cultures in Early Postcards*. Washington, D.C.: Smithsonian Institution Press, 1998.

Hales, Peter Bacon. *Silver Cities: Photographing American Urbanization, 1839–1939*. Albuquerque: University of New Mexico Press, 2005.

Isenberg, Alison. *Downtown America: A History of the Place and the People Who Made It*. Chicago: University of Chicago Press, 2004.

Jakle, John A. *Postcards of the Night: Views of American Cities*. Santa Fe: Museum of New Mexico Press, 2003.

Jobbers Profits. Chicago: Curt Teich & Co., 1935.

Meikle, Jeffrey L. *Design in the USA*. Oxford: Oxford University Press, 2005.

—. "A Paper Atlantis: Postcards, Mass Art, and the American Scene." *Journal of Design History* 13 (2000): 267-286.

Orvell, Miles, ed. *John Vachon's America: Photographs and Letters from the Depression to World War II*, by John Vachon. Berkeley: University of California Press, 2003.

Rosenheim, Jeff L. "'The Cruel Radiance of What Is': Walker Evans and the South." In *Walker Evans*, by Maria Morris Hambourg, Jeff L. Rosenheim, Douglas Eklund, and Mia Fineman, 54-105. New York: Metropolitan Museum of Art with Princeton University Press, 2000.

Rosenheim, Jeff L. and Douglas Eklund. *Unclassified: A Walker Evans Anthology*. Zurich: Scalo in association with Metropolitan Museum of Art, 2000.

Ruby, Jay. "Images of Rural America: View Photographs and Picture Postcards." *History of Photography* 12 (Oct.-Dec. 1988): 327-343.

Sales Pointers. Chicago: Curt Teich & Co., 1935.

Stevens, Norman D., ed. *Postcards in the Library*. New York: Haworth Press, 1995.

Teich Sr., Curt. *The Teich's* [sic] *Family Tree and History*. Chicago: privately printed, 1958.

Werther, Mark and Lorenzo Mott. *Linen Postcards: Images of the American Dream*. Wayne, PA: Sentinel, 2002.

Fig. 1. "Oranges and Snow in California." Linen postcard. 1931. Teich 1A-H412. This image probably derived from a montage of two or three black-and-white photographs. Personal collection of the author.

Fig. 2 . "Transcontinental Highway across Boulder Dam." Linen postcard. 1941. Teich 1B-H2433. Renamed the Hoover Dam in 1947, this site inspired dozens of linen postcards celebrating modern control over nature. Personal collection of the author.

Fig. 3. "N. Main St. Sycamore O." Real-photo postcard. Printed directly from photographic negatives, real-photo postcards afforded impressive verisimilitude but limited circulation. Personal collection of the author.

Fig. 4 . "Center Street, Marion, Ohio." Lithographic postcard. 1911. Teich A-28104. Although published for the F. W. Woolworth dime store chain, this postcard features the firm's competitor F. M. Kirby on the left, while the Woolworth name appears only faintly above the long striped awning to the right (Kirby and several other chains merged with Woolworth in 1912). A specialty shop devoted to "Post Cards" is located midway down the left side of the block. Personal collection of the author.

Fig. 5. "Essex Street, Main Business Section, Lawrence, Mass." Linen postcard. 1945. Teich 5B-H955. Hundreds of similar main street views suggest a striking unity in the face of superficial diversity. Personal collection of the author.

Fig. 6. "Typical of Early Los Angeles--Olvera Street, Los Angeles, California." Linen postcard. 1936. Teich 6A-H2613. A fashionably dressed Anglo woman and her tall male companion survey the picturesque attractions of the Olvera Market tourist zone. Personal collection of the author.

Fig. 7. "Looking across the Lagoon from Botanical Building, Balboa Park, San Diego, California." Linen postcard. 1933. Teich 3A-H1381. In a similar view of the Palace of Fine Arts in San Francisco (Teich 2B-H321, 1942), an artist updated an earlier card by adding two swans in imitation of a rival company's competing postcard but left the scene devoid of people. Personal collection of the author.

Fig. 8. "Union Station, Kansas City, Mo." Linen postcard. 1937. Teich 7A-H2914. Postcard artists often rendered colorized backgrounds with considerable realism but the addition of automobiles, usually in faulty perspective, often suggested an unconvincing diorama. Personal collection of the author.

Fig. 9. "The Orchestra Promenade, Municipal Auditorium, Kansas City, Mo." Linen postcard. 1937. Teich 7A-H745. The typical palette of linen postcards was well suited to representing the boldly colored forms of muralist Walter Alexander Bailey, a student of Thomas Hart Benton. Personal collection of the author.

Fig. 10. "Wolfie's Restaurant & Sandwich Shops, Miami Beach." Linen postcard. 1951. Teich 1C-H1566. Restaurants, hotels, and motels offered customers free postcards as an inexpensive method of advertising. Personal collection of the author.

Fig. 11. "Radio Center, Sunset Boulevard, Hollywood, California." Linen postcard. 1940. Teich 0B-H39. Because electric lighting tended to distort photographs made at night, postcard images of night views were based on photographs taken in daylight. Personal collection of the author.

Fig. 12. "Tramor Cafeteria The Finest Cafeteria in the South St. Petersburg, Fla." Linen postcard. 1939. Teich 9A-H511. Modernity in the guise of stainless-steel serving equipment intrudes into an otherwise traditional scene. Personal collection of the author.

Part Two: Contesting Public Space

“Terra Incognita” in the Heart of the City? Montreal and Mount Royal Around 1900

Nadine Klopfer

In the course of its history, Mount Royal, the mountain nowadays located in the midst of the city of Montreal, has been considered a part of the cityscape essential for the uniqueness of the city and constitutive of the so-called *montréalité* of Montreal. As a highly symbolic space, it became a focus of struggles for power in the city, especially since the late nineteenth century, when it was turned into a public park and thus no longer considered a simple topographical feature, part of the rural landscape, but a usable site for the city. This paper analyzes the “entry” of the mountain into the citizens’ consciousness as a precondition for the numerous highly emotionalized public controversies around the mountain’s usage in the late nineteenth and twentieth centuries. Focusing on the complex evolving relationship between the city and the mountain, the paper argues that the processes of Mount Royal’s physical and perceptual integration into the cityscape were dissociated. Even though the mountain was not fully integrated physically into the urban fabric until the 1930s, it was perceived as located right in the center of the metropolis well before. However, Montrealers viewed the mountain simultaneously as a close-by and a distant space in the sense of an antithesis to urban life, thereby describing the mountain and its park as “natural.” While this perceptual ambiguity in the relationship between city and mountain pervaded all discourses about Mount Royal, the controversies about the mountain can be grasped in terms of competing meanings of “natural” space, rather than as confrontations between proponents of a more urban and defenders of a somewhat wild Mount Royal. Ultimately, these different conceptions of an urban antithesis to urbanity were at the forefront of the underlying struggles to define the use of and access to Montreal’s most iconic public space.

A City and Its Mountain

“Hands off Mount Royal” was probably one of the most frequently recurring headlines in Montreal’s newspapers between the 1880s and the 1920s. The Montreal Street Railway Company periodically submitted plans to build a streetcar line through Montreal’s biggest park which was located on the southeastern part of the city’s mountain (fig. 1). Invariably, these projects aroused emotionally intense public debates. The proponents of the plans—among whom figured most prominently Montreal’s populist French-Canadian Mayor Médéric

Martin[1]—loudly denounced the elitist character of the park, pointing out that the cheap streetcar fare would finally enable the poorer classes to benefit from the mountain's recreational quality.[2] Opponents, however, insisted that the tramway would in fact ruin this recreational quality, depriving precisely those who needed it most "of a simple and healthy pleasure."[3] In particular, the largely Anglo-Protestant *Montreal Parks and Playgrounds Association* (MPPA) and the *Montreal Local Council of Women* (MLCW) vehemently denounced the plans to make the mountain more accessible via railway.

Although these groups claimed to be arguing on behalf of the poor, it quickly became evident that more was at stake. For the mainly Anglo-Protestant elites organizing the lobby against the tramway line, it was completely inconceivable that noisy crowds should invade the mountain, disturbing those engaged in contemplative promenades.[4] Although decades earlier, the park's landscape architect, the famous Frederick Law Olmsted (1822-1903), had already acknowledged the importance of the mountain park's accessibility, he had focused on people seeking recreation through pleasant carriage drives in a wilderness-like setting.[5] Indeed, a few years after the park had been opened, Olmsted felt prompted to point out:

> If it [the mountain] is to be strewn with lunch papers, beer bottles, sardine cans, and paper collars and if thousands of people are to seek their recreation upon it unrestrainedly, each according to his special taste, it is likely to lose whatever of natural charm you first saw in it. (Olmsted, 26)

Yet the intensity and emotionality of the recurring debates about the accessibility of Mount Royal were rooted not only in differing views of appropriate recreation, but also in the desire to define the purpose and

---

[1] Mayor from 1914 to 1924 and 1926 to 1928.

[2] "Mederic Takes Slam at Local Millionaires," *Montreal Daily Mail*, 21 April 1916. See also: Dagenais, 316.

[3] Petition to the Committee of the Mount Royal Park by the MLCW, YMCA et al, s.d., Archives de la ville de Montréal, Fonds de la Commission des parcs et traverses VM44, S4, SS2, SSS7, D13, 121-03-08-01.

[4] "Preserve the Park," *Montreal Daily Star*, 10 April 1922.

[5] For a focus on the pleasure driving see "Frederick Law Olmsted to Commissioners of Mount Royal Park," 21.11.1874, Archives de la ville de Montréal, Fonds de la Commission des parcs et traverses VM44, S4, SS2, SSS3, D1, 121-03-07-02.

usage of the mountain, to thereby mark it as one's own, and to determine access to it accordingly. In fact, the conflict about the streetcar line was not the only bitter fight over the definition of Mount Royal in the late nineteenth and early twentieth century. From discussions about the erection of the Royal Victoria Hospital on the mountain's slope in the late 1880s to conflicts over a site for the Université de Montréal and a gigantic cross on top of the mountain in the early 1920s, the usage of Mount Royal was a hotly debated topic that revealed the fragmentation of Montreal's population according to class, ethnicity, and religion. To many Montrealers, "the Mountain" that dominated the city's skyline represented not only a given natural space, a simple topographical phenomenon, but a symbolic place inextricably intertwined with the city's space and history.[6] Guidebooks, travel accounts, and advertisements of all sorts depicted Montreal as a city spatially defined by the St. Lawrence River and Mount Royal. The mountain, although not much more than a hill, was represented as the feature that gave Montreal its visual distinctiveness.[7] On the basis of its spatial elevation and in accordance with widespread perceptions of mountains as awe-inspiring and dominating places,[8] it was ascribed with protective power over the city. Mount Royal crowned Montreal—a "royal" mountain indeed (*Montreal 1907*, 21).

Additionally, around 1900, the very origin of the city as a French colonial town was seen as rooted in the mountain's physical presence. Popular narratives about Montreal's early history assigned Mount Royal a prominent position in the city's founding myth. Most of them included the ever recurring theme of explorers climbing up Mount Royal. According to these late nineteenth- and early twentieth-century stories, Jacques Cartier had christened the mountain Mount Royal to honor the King of France because he was stunned by the breathtaking view from its peak in 1535.[9] Retrospectively, the most memorable event from Samuel de Champlains visit to the Island of Montreal in 1611 seemed to be his ascent of Mount Royal (American Association

---

[6] Dagenais, 311; Jacobs, 11.

[7] *Souvenir of Montreal '93*, 11; Wolff and Erpicum, 3; American Association for the Advancement of Science, 46-47; See also: Schmidt, 14-15.

[8] MacLeod, 30-32; Cohn, 26-41.

[9] Phelps; *Murray's Illustrated Guide to Montreal and Vicinity*, 25. See also: Grenier and Bumbaru, 2 and Poirier, 43.

for the Advancement of Science, 20). About thirty years later, Paul de Chomedey, Sieur de Maisonneuve had founded the city at the mountain's base, soon climbing up to the summit with a wooden cross on his shoulder to thank the Lord for preserving the colony from flooding, a story that became an intrinsic part of Montreal's founding myth.[10] In the course of the eighteenth century then, the city's original name of Ville-Marie was gradually replaced by the designation derived from the mountain's name.[11] Without Mount Royal, so it seems in these popular narratives from around 1900, Montreal would never have been. At that time, the mountain was widely considered an essential element of the city's historical and spatial identity, a landmark from which Montreal derived much of its uniqueness, the *montréalité de Montréal*.[12] It was thus endowed with a very special meaning in relation to the city. Consequently, the power to define the character of that iconic site came to be viewed as an important symbolic factor in the competition for power in Montreal. Since roughly the mid-nineteenth century, various groups of citizens have tried to appropriate the mountain in different ways, ascribing changing meanings to this "artifact of nature" that were deeply rooted in their own cultural, social, and political backgrounds and interests.

Stepping back from these conflicts though, it is interesting to note that the emotional quarrels over the mountain only started around the 1870s, gaining public momentum especially after a large part of the mountain had been acquired by the city and transformed into a public park. Not until what had formerly been perceived as an awe-inspiring natural site was turned into a public, usable space did the citizens vigorously try to make Mount Royal a place of their own according to diverging ideals. I would like to argue that the collapse of physical space separating the mountain from the city due to urbanization, and even more, the reduction of the perceived distance between the two, represented the basic conditions for, and consequences of, the transformation of the mountain into a public space and thus into an object of contestation. Interestingly, the mountain was viewed as

---

[10] *St. Lawrence Hall Montreal*, 23; Lighthall, 48. For the importance of the mountain in Montreal's public memories see Gordon, 97-101.
[11] *Noms et lieux de Québec: Dictionnaire illustré* (1996), s.v. "Montréal, ville."
[12] For the notion of "montréalité" see Morisset, 158.

integrated into the city much earlier than it actually was, the processes of physical and imagined integration thus being dissociated. The topic of the mountain's "entry" into the city then raises the question of whether, in the course of its imagined integration, it evolved from a so-called natural or rural space to an urban space, as it might seem when considering physical urbanization processes. I will propose, though, that on a perceptional level, the disappearance of what could be called "imagined geographical distance" since the second half of the nineteenth century brought about an increased "imagined qualitative distance" between Montreal and Mount Royal resulting in a heightened sensitivity to the qualitative differences between the city and the mountain. Although the mountain remained for a long time at the city's periphery, hardly accessible, people thought of it as being right in its center. At the same time, they clung to its image as a distant place of retreat in the sense of an antithesis to urban space, hence applying the terms "natural" or "rural" to it. In this perspective then, diverging concepts of "natural" and "city" space were at the forefront of the trouble around Mount Royal, expressing the often strongly emotionalized attempts by different groups of citizens to define the character of the mountain according to their own values and interests.

Bird's-eye View: Urbanization Processes

Let us first take a quick look at urbanizing Montreal and its physical distance from the mountain. How did the relationship between the city and its landmark map out over the course of the centuries? In the seventeenth century, the French colonial town of Ville-Marie started as a small settlement huddled on the bank of the St. Lawrence. Viewed from the river, the mountain rose far in its background. The space between the town and the mountain was part of Montreal's agricultural hinterland, subdivided in so called *côtes*, clusters of long narrow pieces of farmland (Jacobs, 44). Throughout the eighteenth century, the mountain's slopes were home to farms and orchards, forming a rural landscape not unlike the rest of the island (Hanna, 39). By the early nineteenth century, several members of the North West Company—mostly of Scottish origin, like James McGill—who had made their fortunes in the fur trade, had acquired farms on the southern slope of Mount Royal in order to erect comfortable summer retreats in a rural setting (MacLeod, 32-34). A generation later, in the wake of a first construction and speculation boom in the 1840s, the slope of the mountain was more systematically settled. The new landowners—mainly upper-class wholesalers, retailers like the sugar baron John Redpath—came to live permanently in their suburban villas at the foot of the mountain. Additionally, they subdivided portions of their enormous properties into lots and opened streets thereby "planting the firm imprint of a grid pattern on the mountain slope."[13] At the same time, the city had also grown northward, urbanizing the mountain's eastern slopes by the early twentieth century. By the 1930s, urban settlement had almost totally surrounded Mount Royal (fig. 2).

The View From the City: Physical Access and Perceived Distance

If we now take a look at the mountain from the point of view of the city and its citizens—how was it perceived in terms of geographical distance and closeness, and what was actually done to reach it? In the days of the French Regime, the mountain was a day's journey from Montreal.[14] For the citizens, it was an important topographical feature in the background of their city, part of the rural space behind it (Jacobs, 13). Correspondingly a map from 1702 shows Mount Royal as an impressive, massive, almost hostile mountain range (fig. 3). Until the mid-nineteenth century, the mountain remained far away for the major part of the population. Only in the minds of very few, such as the sixteen different proprietors who owned land on the mountain at that time, was Mount Royal a place where one could actually go.[15]

This began to change when, in 1847, the Mount Royal Cemetery Corporation acquired land on the mountain to establish a Protestant cemetery, quickly followed by the Catholic Notre-Dame-des-Neiges Cemetery Company (Watkins, 53). (See fig. 4.) With the city expanding so quickly, valuable inner city land was needed for residential and commercial purposes. Moreover, the relocation of the cemeteries to a safe place away from the populous urban districts

[13] Hanna, 41. See also: Jacobs, 45; MacLeod, 41-42.
[14] "It's no Everest . . . but it is a mountain!" *The Montreal Downtowner*, 4 May 1983.
[15] As the writer Hugh MacLennan pointed out in 1952: "Even a hundred years ago Montrealers could look north over a wide expanse of rising ground covered with farms and orchards and see Mount Royal in the distant background" (MacLennan, n.p.).

seemed necessary in view of recurring cholera epidemics (Jacobs, 11). Apparently, the mountain answered these needs, which suggests that it was perceived as space far outside the city. Yet, the creation of the cemeteries also indicates that Mount Royal was clearly thought of as a usable space for the growing metropolis. In a sense, by the mid-nineteenth century, Mount Royal represented a kind of transitory space between an urban and a rural landscape at Montreal's edge, a space already defined in relation to the city, but chosen to fulfill a semi-public, urban function precisely and paradoxically because of its physical and imagined remoteness.

In the second half of the nineteenth century, the boundaries of urbanization as summarized above were pushed forward by an extension of the public transportation system (MacLeod, 172). At the same time, the provision of a means of transport and the attempts to systematically settle the area indicate a certain interest in reaching the territory near the mountain, which reflects a decrease in the perceived space between the city and Mount Royal. In turn, the integration of the mountain as a usable space into the consciousness of some Montrealers—primarily the middle and upper classes eager to settle on its slopes and to develop the area—required not only a spreading settlement but also the actual possibility of reaching the mountain within less time from all parts of the city. From then on, these three closely intertwined factors influenced each other in defining the relationship between the city and the mountain: the advanced urbanization of the territories surrounding the Mount Royal, the possibility of accessing it easily, and its imagined geographical closeness. By the 1860s, the concept of the mountain as a place to be used by urban dwellers was obviously widespread among influential citizens, so that they convinced the City Council that it presented the perfect site for a public park. In 1869, the Provincial Legislature of Quebec allowed the city to borrow a certain sum of money in order to acquire land for this purpose (*Statutes of Quebec*). From 1869 to 1875, the city proceeded to expropriate several owners of land on the mountain. After severe legal battles, a considerable part of the mountain finally became public property and the landscape architect Frederick Law Olmsted was commissioned to design a public park.[16]

---

[16] Bellman; Schmidt, 26-32.

At that time, Olmsted had already firmly established his reputation as North America's foremost landscape architect. Renowned for his design of New York's Central Park, he had also worked on parks for Chicago, Washington, Philadelphia, and several other American cities (Murray, 163). In his design for the mountain park, Olmsted focused on keeping the mountain's "charming natural scenery," which he saw as the site's greatest advantage.[17] Therefore, only plants that grew naturally on the spot were to be planted, and the intrinsic qualities of every topographical section of the mountain was to be respected. The roads and walks he designed meandered through the park, following the topography, curving their way up the mountain, designed so as to occasionally reveal breathtaking views on the countryside (Murray, 166-67). (See fig. 5.) In short, man's interference on the mountain site was to remain barely discernible, while in a sense subtly perfecting nature's picturesqueness (fig. 6).[18] Apart from its extravagant site, Mount Royal Park was thus to be a typical nineteenth-century North American park in the tradition of the English landscape garden, conceived along the ideals of Victorian landscape architecture, intended to exert a soothing and invigorating effect on the visitor's health and morals.[19] Inaugurated ceremoniously on Victoria Day 1876, the park was depicted as a "pleasure ground" for Montrealers of all classes. Addressing an audience of several hundred people gathered on the grounds, Mayor William Hingston pointed out:

> It [the Mountain] is given as a pleasure ground to you, wealthy citizens, who can take an afternoon, riding in well cushioned carriages; to you, merchants, shopkeepers and mechanics, who long for a free breathing space; but more than all, to you, poor laboring

---

[17] Olmsted, 21; Murray, 166.

[18] Elaborating on his design, Olmsted frankly spoke about "improvement" of "landscape conditions," and one of his aims was: "without destroying the essential picturesqueness of its [the mountain's] natural features, to add a greater beauty of foliage." See "Frederick Law Olmsted to Commissioners of Mount Royal Park," 21 November1874, Archives de la ville de Montréal, Fonds de la Commission des parcs et traverses VM44, S4, SS2, SSS3, D1, 121-03-07-02.

[19] For an extensive analysis of Olmsted's design see Seline, esp. 52-124. For Olmsted's biography see Rybczynski. For an overview of the park movement in North America see Cranz.

> men, who, fatigued and worn with toil, will seek to reinvigorate yourselves with the free air of Heaven.[20]

Inserted into celebrations of the Queen's Birthday and symbolic affirmations of Canadian loyalty to the Empire, the opening of Mount Royal park clearly bore the imprint of the wealthy Anglo-Protestant elite that dominated the affairs of the city at that time. In the early years of the park, this was reflected in its limited accessibility. Situated close to the wealthy, primarily anglophone parts of town, on top of a mountain, the park was fairly out of reach of the "poor laboring men." Until 1885, when an incline railway was erected, it had been virtually impossible to reach the park without a carriage, which contributed to the image of the park as a garden for the wealthy.[21] These limitations were quick to spark calls for extended physical integration of Mount Royal into the city. "A park[,]" observed a reporter of the *Montreal Herald* in 1884, "should not be a place to be gazed at from a distance [. . .] A park should be easy of access, and not require a whole day to be set aside to visit it[.]" And he went on to wonder "whether it [the Mountain Park] was not a terra incognita to nine-tenths of the public [. . .]?"[22] Although the transformation of part of the mountain into a public park suggested a diminishing imagined distance between city and mountain, Mount Royal was obviously still not conceived of as being in the city proper, even by the end of the nineteenth century, due to the obvious lack of accessibility.

Interestingly, around the turn of the twentieth century, Montrealers started talking about Mount Royal as being right in the heart of the city[23]—which de facto was not the case, considering the urban layout of Montreal at that time. But apparently, since the opening of the mountain park, the distance between the city and Mount Royal had broken down more rapidly in the imagination of the citizens than it had in physical space. At the same time, however, the feeling that the mountain was still not wholly integrated into the city persisted. This

---

[20] "Her Majesty's Birthday," *Gazette*, 25 May 1876. See also: "L'inauguration du Parc Mont Royal," *L'Opinion Publique*, 8 June 1976.

[21] Dagenais, 312. The *Montreal Herald* even dubbed the park "aristocratic," in the 5 April 1886 article "Mount Royal Park."

[22] "The Mountain Park,"*Montreal Herald*, 14 May 1884.

[23] "Again Deferred," *Montreal Herald*, 9 October 1905.

emphasizes the slightly paradoxical position Mount Royal was occupying in the imagined geography of Montreal in the late nineteenth century: a felt closeness that was not matched by physical closeness. By the turn of the century, the mountain—distant or centrally located—had been propelled into the heart of urban discourses thanks to its conversion into public space. Over time, the calls for better physical accessibility that had been voiced sporadically before intensified. It seems as though all of Montreal's major urban planning projects in the early twentieth century were to some degree guided by the desire to physically connect the mountain and the city more closely. All sorts of connections were explored. The Province of Quebec Association of Architects, following the ideal of a *City Beautiful*, proposed in 1908 to embellish the city by linking Mount Royal park to Parc Lafontaine through a representative boulevard (Wolfe and Jacobs, 51). Others, more keen on catering to the masses, brought up the topic of a connection to the public transport system. Although an incline railway existed between 1885 and 1920 to reach the mountain summit, this means of transport was run by a different company than the streetcar system and consequently cost an extra fare (Dagenais, 312). The *Montreal Herald* led the way by claiming that "it shall be made possible to get to the summit from any part of the city for one street car fare [. . .] [with] the Incline Railway becoming part of the system."[24] A quite diverse coalition supported that approach in the early twentieth century. The Montreal Board of Trade, speaking for the city's anglophone business elite, joined the Park Association in its critique of a proposed tramline which would disturb "the beauty, repose, and safety" of the park. Nevertheless, they endorsed a compromise. They highlighted the importance of affordable access to the mountain park, arguing that it was vital especially for that part of the population "whose means are limited" to enjoy the healthy and also morally uplifting effect of a trip to Mount Royal; the Board of Trade spoke out in favor of the incline railway solution.[25] In its rejection of the tramline,

[24] "Mountain Park for the People," *Montreal Herald*, 1 May 1899. The anglophone daily *Herald* was a politically more liberal newspaper than the conservative *Gazette* and *Star*. See Westley, 132.

[25] "The Montreal Board of Trade to His Worship the Mayor and the Aldermen of the City of Montreal," 21 December1895, Archives de la ville de Montréal, Fonds de la Commission des parcs et traverses VM44, S4, SS2, SSS7, D13, 121-03-08-01.

the Board feared the monopoly of the Montreal Street Railway Company and, in a classic progressivist-reformer impulse, the possibly corrupt entanglement of the city's political machine and the corporation that obtained the franchise (Gauvin). The francophone business community though, as represented by the Chambre de Commerce, heartily supported plans for a tramway:

> The Park cost and still costs the municipality enormous sums for its maintenance, without tax-payers being able to enjoy it by reason of its distance to the city and its elevation above the level of the city.[26]

This view was shared by the majority of the City Council who claimed to be arguing on behalf of their local constituencies. As Michèle Dagenais has shown, new, mainly francophone political leaders à la Médéric Martin thus laid claim to the definition of the mountain's usage, thereby asserting symbolically their power within the city and dethroning a more traditional, anglophone elite that had been crucial in creating the park in the first place.[27] These fragmentations along ethno-cultural lines have to be differentiated and not overemphasized. Interestingly, several members of an older francophone professional elite such as the liberal Senator Frédérick-Liguori Béique joined their anglophone counterparts of the MPPA in the protest against the tram.[28]

---

[26] "Le Parc de la montagne a coûté et coûte encore à la municipalité, pour son entretien, des sommes énormes sans que les contribuables puissent en jouir en raison de son éloignement et de son élévation au dessus du niveau de la ville." (Unless otherwise noted, all translations are my own.) "Petition de la Chambre de Commerce de Montréal, 8 March1895, Archives de la ville de Montréal, Fonds de la Commission des parcs et traverses VM44, S4, SS2, SSS7, D13, 121-03-08-01.

[27] Dagenais, 314-317. The fact that several aldermen were accused of having personal financial interests in the Montreal Street Railway Company should not be neglected (ibid., 315). See also: "Pour le peuple," *La Presse* (1903), McGill University Archives, Montreal Parks and Playgrounds Association MG2079, 2147D, container 7, folder 259 "MPPA Scrapbook, 1902-1917." and The aldermen denied the charge, "Aldermen deny an insinuation", s.l. (1903), McGill University Archives, Montreal Parks and Playgrounds Association MG2079, 2147D, container 7, folder 259 "MPPA Scrapbook, 1902-1917."

[28] In 1903, a delegation uniting members of the anglo- and francophone *grande bourgeoisie* appeared in City Hall to lobby against the tramway. See: "No street cars on the mountain," s.l. (1903), McGill University Archives, Montreal Parks and Playgrounds Association MG2079, 2147D, container 7, folder 259 "MPPA Scrapbook, 1902-1917"; "They urge city to stand firm," s.l. (1903), ibid.; "Une imposante

It thus seems that the lines of conflict were drawn between older and newer leading classes, these lines corresponding mainly, but not wholly, with the ethno-cultural divisions in the city. All of these elites claimed to be arguing in the name of the masses. Quite a few francophone and anlgophone labor organizations though spoke out against the tram project. Their objection was grounded in the distrust of public franchises for private corporations; instead, they called for the city to own the tramline to the mountain.[29] Nevertheless, whether pro or contra tramway, and out of whatsoever reasons, Montrealers agreed that the mountain was still too far away for a site of that importance to the city.

Considered unsafe and too expensive, the existing incline railway was shut down in 1920 before having been integrated into the city's transport network, and the coalition of supporters of extended public transport, led by Mayor Martin, focused energetically on streetcar projects.[30] With regards to accelerating urbanization, plans for a tramline now seemed even more urgent than at the beginning of the century. The streetcar was to reduce the spatial distance between Mount Royal and the citizens, thereby resolving the paradox of an almost unreachable public place in the middle of the city that seemed so bewildering to many contemporaries. *La Presse* articulated this contradictory condition: "Strangers visiting our city must find it highly curious that a park like the mountain park, situated in the midst of the metropolis, is still almost inaccessible to the population today."[31] By becoming linked to the city's public transport system, accessible with a simple transfer ticket and connected to the urban electricity network, the mountain would finally be an integral part of Montreal's multilayered administrative and utility systems, integrated into the web

---

délégation devant le maire," s.l. (1903), ibid. Several anglophone aldermen, on the other hand, supported the tram project, see "Mountain Park for the People," *Montreal Herald*, 1 May 1899.

[29] "Petition of the Order of Knights of Labor," 8 January 1896, Archives de la ville de Montréal, Fonds de la Commission des parcs et traverses VM44, S4, SS2, SSS7, D13, 121-03-08-01. For other examples see Dagenais, 317.

[30] "Mederic takes slam at local millionaires,"*Montreal Daily Mail*, 21 April 1916.

[31] "Les étrangers qui visitent notre ville doivent trouver fort curieux qu'un parc comme celui de la montagne, situé en plein coeur de la métropole, soit aujourd'hui encore presque inaccessible à la population." "Des ascenseurs sont indispensables," *La Presse*, 13 May 1922.

of the modern networked city. In a project proposed in 1923, the railway was even intended to cut through the mountain's rock: modern technology allowed for a negation of the mountain's protuberant physical presence. Before, traffic had been "forced to go around the mountain," as the *Montreal Daily Star* criticized,[32] a problem that could now be resolved. From this perspective, the mountain not only represented a perfectly reachable part of the city, but a traffic connection between its eastern and western neighborhoods.[33] Mount Royal's "mountainness" was no longer considered an obstacle; in a sense, from the point of view of circulation, it became a mere *lieu de passage* within a city. A tunnel underpassing the mountain, completed by the Canadian Northern Railway (CNR) in 1917 to link the city to the new suburb of Mount Royal Town, likewise demonstrates that Mount Royal lost its character as an obstacle to urban traffic (McCann, 281). City maps show the tunnel as a straight black line cutting right through the mountain as if it wasn't even there (fig. 2). At least in cartography, modern planning had left its geometrical imprint on the mountain's representation. With its physical and imagined entry into the city, largely due to the creation of a park on its top as a public space, the spreading urban settlement and the integration of the mountain into the city's networks, Mount Royal was "domesticated" as contemporary newspapers put it.[34] In 1930, the tramline was finally realized.[35] The topos of the "mountain in the city" became an unquestioned fact in Montrealers' imagined geography of their city, underpinned by Mount Royal's actual location within the metropolis and its easy accessibility. By the end of the twentieth century, long after public buses had replaced the old streetcars, a book about the mountain could easily state: "Mount Royal is perceived as the center of the city of Montreal, geographically as well as symbolically."[36]

[32] "Another Plan to Scale Mount Royal," *Montreal Daily Star*, 23 January 1923.
[33] "Les ouvriers de Montréal en faveur d'une ligne de tramway sur la montagne," *La Presse*, 5 February 1926. See also: Dagenais, 318.
[34] "Les premiers à admirer un nouveau Montréal," *La Presse*, 14 July 1930; Dagenais, 319.
[35] "Inauguration de ce service de tramway," *La Patrie*, 14 July 1930.
[36] "Le Mont Royal est perçu comme le centre de la ville de Montréal, tant géographiquement que symboliquement." Jacobs, 11.

From Natural Space to Urban Space? Imagined Distances

Thus, the 1920s and the decision to build a streetcar line appear to mark a crucial turning point in the relationship between the mountain and the city:

> While until then, Mount Royal Park had been perceived as a pastoral place at the edge of the city's circulation and noises, it became an urban space. Its very signification changed. [. . .] Thus, the mountain was no longer located on the periphery, but in the heart of the urban space.[37]

From this perspective, the 1920s represent the culmination of a process that had started around the mid-nineteenth century, a process which led to the physical and imaginary integration of the mountain into the city, thereby turning it into an urban space. Nevertheless, it is interesting to note that this process was not as smooth and homogeneous as it might seem. Against the background of the spatial relationship between city and mountain as outlined above, it is apparent that even at times when the mountain was not yet easily accessible for the majority of the population, it was already widely perceived as being in the heart of the city. To be located in the center of Montreal's imagined geography, the mountain didn't need a tramline or urban settlement surrounding it. The perceived geographical distance between Montreal and Mount Royal had decreased ever since the cemeteries, and especially since the park, had been established on the mountain's slopes and summit. From a mere topographical feature somewhere in the rural hinterland, Mount Royal had evolved into a public urban space, which in turn had generated a more widespread consciousness of and feeling of closeness to the mountain. On a perceptional and discursive level, the mountain had thus entered the city long before its physical inclusion, and this perceived closeness had in turn intensified demands for its full physical connection to the city's manifold networks.

Such a change in the time frame of the mountain's entry into the city and the differentiation of several levels of integration—

[37] "Alors que le parc du mont Royal était perçu jusque-là comme un lieu champêtre, en marge de la circulation et des bruits de la ville, il est devenu un espace urbain. Sa signification même a changé. [. . .] De périphérique, la montagne se retrouve ainsi au coeur de l'espace urbain[.]" Dagenais, 318.

physical or imagined—though has effects on our perception of its supposedly simultaneous and gradual transformation from a rural into an urban space. To assume that Mount Royal (people would often identify the park with the mountain) was first seen as a rural retreat far from the city, only becoming urban with increased accessibility in the 1920s, presupposes a correlation between urbanity and accessibility. This ignores the fact that contemporaries perceived Mount Royal as located close by, in the heart of the city, well before and independently from available means of access; in their minds, it had entered the city long ago and thereby become urban. At this point though, a certain paradox emerges. The feeling of closeness to the mountain, after all, had been boosted through the establishment of a public park. Consequently, following this logic, the very creation of the park would have caused the mountain to be considered an urban space. However, the landscape park erected on the summit of Mount Royal was generally viewed as a rural or natural space. Contemporaries could simultaneously think of the park—and the mountain—as a nearby space in the midst of their city, and a space very far away.[38] It is noteworthy that this ambiguity can be spotted in both the streetcar opponents' and its advocates' rhetoric. Tram opponents, for example, tended to represent the mountain as a distant place located far beyond anything urban, in fact, as the anti-urban space par excellence. To quote Sir George Drummond, president of the Parks and Playgrounds Association, Mount Royal was a place "out of the hurly-burly and heat and dust of the crowded city streets."[39] The city's women's clubs emphasized its "purely rural character"[40] while the chairman of the MPPA's parks committee pointed out: "The value of Mount Royal Park to the public depends on its quiet, its safety, its natural beauty, its rural charm, and its freedom from the danger of our noisy city streets."[41]

---

[38] This ambiguity pervades most of the Victorian park rhetoric. See also: Scobey, 229-230; Machor.

[39] "Again Deferred," *Montreal Herald*, 9 October 1905.

[40] "Women's Protest," s.d., McGill University Archives, Montreal Parks and Playgrounds Association MG2079, 2147D, container 7, folder 259 "MPPA Scrapbook, 1902-1917."

[41] "Peter Bercovitch to A. D. Braithwaite," 28 November 1922, McGill University Archives, Montreal Parks and Playgrounds Association MG2079, 2147D, container 9, folder 354 "MPPA Correspondence, 1917-1924."

Even more than other parks, the mountainous site contributed to the perception of being not only far away from urban life, but equally "lifted above"[42] the city. Olmsted himself had considered the Mountain Park to be the exact antithesis to urbanity. Its landscape was supposed to be "standing in winning competition against the sordid and corrupting temptations of the town" (Olmsted, 63). At the same time the opponents of a tramway also thought of the mountain as centrally located. For example, the *Montreal Star*, one of the most vehement protestors against the tram, declared:

> The mountain forms one of the most beautiful natural parks possessed by any city in the world. It is wild nature at its loveliest, almost in the centre of a great city, a place of beauty and relaxation within the reach of the poorest.[43]

Spatially, this ambiguity was reflected in the design of Frederick Law Olmsted; well connected to the city by different roads and entries, the park was nevertheless clearly set off spatially against the linearity of the city's grid (fig. 7). Equally, contemporaries who on their part wanted the masses to easily reach the mountain via public transport demanded its increased physical integration into the city precisely to allow everybody, including the poorer classes, an escape from the city.[44] Thus, while both parties in the tram conflict felt the mountain was a space in and of the city, they simultaneously viewed it as a distant and natural place.

This paradox resolves itself by differentiating two modes of perceiving distance. Speaking in terms of imagined geography, of quantitative distance, Mount Royal was nearby. Qualitatively, however, people constructed it as distant in the sense of different. Within both modes of perception, the opposite signifiers of "closeness" and "remoteness" were filled with meaning regarding the character of the mountain, "close" signifying urban, and "remote" implying natural or rural, with these terms often being used interchangeably. Therefore, in

---

[42] "Hospital Sites II," Letter to the Editor, *Gazette*, 24 July 1889. In the opening ceremony, the park was compared directly with New York's Central Park and its natural advantages as a site for a park were highlighted, "Her Majesty's Birthday," *Gazette*, 25 May 1876.
[43] "Keep the Mountain for the People," *Montreal Daily Star*, 23 November 1916.
[44] "Mountain Park for the People," *Montreal Herald*, 1 May 1899.

contemporaries' perception of Montreal's relation to Mount Royal, geographic closeness and qualitative remoteness (i.e., difference from the city) did not exclude each other, and neither did urbanity and naturalness. Ever since the mountain was thought of as an escape from the city in the heart of the city in the late nineteenth century, this ambiguity pervaded discourses about Mount Royal. Independently of the creation of a public means of transport for the masses, the mountain was considered an urban as well as a natural space.

Of course, streetcar opponents feared that Mount Royal's natural peacefulness would be endangered by the tramway, whereas its advocates did not. However, this does not mean that the former considered the mountain a natural space, while the latter viewed it as urban. Rather, the difference lay in the understanding of what 'natural' and its opposite, 'urban,' meant. Streetcar opponents, for example, argued that the tram would ruin the mountain's quality as "a lovely and healthful retreat from the city's noise and dirt, free from the vulgarizing, if not debasing, incidental accompaniments of electric cars and electric lights."[45] In this perspective, "nature" clearly implied distinctive ideas about recreation based upon specific concepts of physical health and morals. For the mainly middle or upper class WASPs united in the fight against a tramway, urbanity was closely related to an unhealthy, immoral lifestyle that would intrude upon the park along with commercial amusements, which, in turn, they obviously associated with the masses that a tram would bring to the mountain. Streetcar supporters like Martin and the vast majority of the City Council, on the other hand, felt that poorer people in particular should profit from Mount Royal's beneficial influence. Less elitist, they did not associate negatively perceived aspects of urban life with the masses. In fact, for them, a recreative site away from urbanity did not necessarily require solitude and contemplativeness; the "natural" spot in this sense meant a space of retreat from day-to-day labor. This last, more democratic vision won over the former when the tramline was finally authorized in the 1920s. The mountain did not become urban at that time, rather a different interpretation of what its natural

[45] "Letter to the Editor," s.l., s.d., McGill University Archives, Montreal Parks and Playgrounds Association, MG2079, 2147D, container 7, folder 259 "MPPA Scrapbook, 1902-1917."

character implied superseded the older one, testifying to the large context of changing power relations in Montreal: the new political, mainly francophone elite had clearly stated its claim to leadership in the city, challenging the power of the older, predominantly anglophone Victorian establishment.[46]

Yet one could argue that the streetcar advocates' vision of the mountain and its park was not one of a truly natural or rural landscape. Wasn't their concept far more utilitarian than that of the defenders of a classical landscaped park? After all, their vision of a natural retreat embraced picnicking families, playing children, and ice-cream parlors. One is tempted to agree with their critics: such a vision might depict a recreational escape from the strains of day-to-day life, but certainly not an antithesis to urbanity. The *Montreal Daily Star* put it quite bluntly:

> One of the most attractive features of Mount Royal is that it enables the citizens to attain sylvan solitudes without going out of the city. You can get as far from the noise and reek of humanity and as near to Nature on the Mountain [capital N and M!] as you can in the depths of the Laurentians. But if commercialized amusements were once permitted to obtain a foothold, goodbye to solitude and nature too.[47]

Yet, in the end, this was just one normative idea of what a natural space could be, no more or less "natural" than its competing vision. In fact, even if contemporaries imagined the park, and with it the mountain, as "rural" or "natural" in character, the moment they started thinking about the mountain in relation to the city, transforming it first into a cemetery and then into a public pleasure ground for urban dwellers, the mountain was thought of as an urban function, linked dialectically to the city. As such, no matter who used the Olmstedian landscape park, and what for, the variation lay more or less on the surface. Since roughly the 1850s, with increasing urbanization, the mountain evolved from a natural space in the sense of a mere topographical phenomenon[48] to a space perceived by Montrealers as a usable space,[49]

---

[46] See Linteau.

[47] "Preserve the Park," *Montreal Daily Star*, 10 May 1922.

[48] See Kelman, 9.

[49] Shields grasps these kind of processes as the "territorialisation of geographic space: transformation from undifferentiated 'natural' space into the coded topography of

taking the shape of a public space, a park, thereby entering the imagined geography of the city. Once present in the citizens' minds, symbolic meaning was ascribed to the mountain, which was considered a defining factor in Montreal's *montréalité*. From then on, the mountain and its park evolved within the larger category of public urban space, as a space that was frequently labeled "natural." Here, though, "natural space" is to be understood in the sense of a historically contingent, changing construct (Kelman, 8-9), used in different ways by different people to define and occupy the highly symbolic mountain—some of their understandings of "natural" being closer to the concept of an "untouched topographical phenomenon" than others. Particularly for a place as iconic as Mount Royal, the power to define its use and its users amounted to a certain power over the city. It is no surprise then, that the opponents of the streetcar fought for "hands off Mount Royal." Had they succeeded, the mountain would have remained a kind of elite domestic garden, far from those who might have soiled the park with "lunch papers, beer bottles and sardine cans"—a public space for the few, right in the heart of the city and yet far away.

Bibliography

American Association for the Advancement of Science, ed. *Hand-book for the City of Montreal and Its Environs*. Montreal: American Associaton for the Advancement of Science Local Committee, 1882.

Archives de la ville de Montréal, Dossiers de presse, bobine 258-2.101.

—, Dossiers de presse, bobine 258-3.64.

—, Fonds de la Commission des parcs et traverses VM44.

Bellman, David. "Frederick Law Olmsted et un plan pour l'aménagement du Mont-Royal/Frederick Law Olmsted and a Plan for Mount Royal Park." In *Mont-Royal, Montréal/Mount Royal, Montreal*, edited by David Bellman, 31-43. Montreal: McCord Museum, 1977.

Cohn, Robert L. *The Shape of Sacred Space: Four Biblical Studies*. Chico, CA: Scholars Press, 1981.

'civilized' territory" (11-12). The perception as "undifferentiated 'natural' space" might also be considered a cultural codification in itself. In this perspective, the transformation of the mountain's perception might rather be seen as a fundamental shift in the character of the "coded topography of 'civilized' territory": from topographical background to usable city space.

Cranz, Galen. *The Politics of Park Design: A History of Urban Parks in America.* Cambridge, MA: The MIT Press, 1982

Dagenais, Michèle. "Entre tradition et modernité: Espaces et temps de loisirs à Montréal et Toronto au XXe siècle." *Canadian Historical Review* 82, no. 2 (June 2001): 308-330.

Gauvin, Michel. "The Reformer and the Machine: Montreal Civic Politics from Raymond Préfontaine to Médéric Martin." *Journal of Canadian Studies/Revue d'études canadiennes* 13, no. 2 (Summer 1978): 16-26.

Gordon, Alan. *Making Public Pasts: The Contested Terrain of Montreal's Public Memories, 1891-1930.* Montreal: McGill-Queen's University Press, 2001.

Grenier, Cécile and Dinu Bumbaru. "Le Mont Royal." *La Presse Plus* (1985).

Hanna, David B. "Creation of an Early Victorian Suburb." *Urban History Review/Revue d'histoire urbaine* 2 (Oct. 1980): 38-64.

Jacobs, Peter. "La Montagne magique." In *La Montagne en question*, edited by Groupe d'intervention urbaine de Montréal. Montréal: BANQ, 1988.

Kelman, Ari. *A River and Its City: The Nature of Landscape in New Orleans.* Berkeley: University of California Press, 2003.

Lighthall, William D. *Montreal after 250 Years.* Montreal: F. E. Grafton and Sons, 1892.

Linteau, Paul-André. "Le personnel politique de Montréal, 1880-1914 : évolution d'une élite municipale." *Revue d'histoire de l'Amérique française* 52, no. 2 (automne 1998): 189-215.

Machor, James L. *Pastoral Cities: Urban Ideals and the Symbolic Landscape of America.* Madison: University of Wisconsin Press, 1987.

MacLennan, Hugh. "Montreal: The Mountain in the City." *Mayfair* (June 1952).

MacLeod, Roderick. "Salubrious Settings and Fortunate Families: The Making of Montreal's Golden Square Mile, 1840-1895." PhD diss., McGill University, 1998.

McCann, Larry. "Planning and building the corporate suburb of Mount Royal, 1910-1925." *Planning Perspectives* 11 (1996): 259-301.

McGill University Archives, Montreal Parks and Playgrounds Association MG2070.

Montreal 1907. Montreal: s.l., 1907.

Morisset, Lucie K. and Luc Noppen. "Entre identité métropolitaine et identité urbaine: Montréal." In *Les identités urbaines* edited by Lucie K. Morisset and Luc Noppen, 157-179. Quebec: Nota Bene, 2003.

Murray, Alex L. "Frederick Law Olmsted and the Design of Mount Royal Park, Montreal." *Journal of the Society of Architectural Historians* 26, no. 3 (Oct. 1967): 163-171.

Murray's Illustrated Guide to Montreal and Vicinity. Montreal: Norman Murray Publisher, 1893.

Olmsted, Frederick Law. *Mount Royal, Montreal.* New York: Putnam, 1881.

Phelps, Henry P. *Montreal for Tourists.* New York: s.l., 1904.

Poirier, Jean. "Origine du nom de la ville de Montréal." *Revue d'histoire de l'Amérique française* 46, no. 1 (été 1992): 37-44.

Rybczynski, Witold. *A Clearing in the Distance: Frederick Law Olmsted and America in the 19th Century.* New York: Scribner, 1999.

Schmidt, Sarah. "Domesticating Parks and Mastering Playgrounds: Sexuality, Power and Place in Montreal, 1870-1930." Master's thesis, McGill University, 1996.
Scobey, David M. *Empire City: The Making and Meaning of the New York City Landscape*. Philadelphia: Temple University Press, 2002.
Seline, Janice E. "Frederick Law Olmsted's Mount Royal Park, Montreal: Design and Context." Master's thesis, Concordia University, 1983.
Shields, Robert. *Places on the Margin: Alternative Geographies of Modernity*. London: Routledge, 1991.
Souvenir of Montreal '93. Toronto: The Endeavor Herald Publishing Company, 1893.
St. Lawrence Hall Montreal: Tourists Guide 1911. Montreal: A. J. Higgins, 1911.
Statutes of Quebec, 32 Vict. (1869) ch. 70 sect. 20.
Watkins, Meredith G. "The Cemetery and Cultural Memory: Montreal, 1860–1900." *Urban History Review/Revue d'histoire urbaine* 31, no. 1 (Fall 2002): 52-62.
Westley, Margaret. *Remembrance of Grandeur: The Anglo-Protestant Elite of Montreal, 1900–1950*. Montreal: Libre Expression, 1990.
Wolfe, Jeanne M. and Peter Jacobs. "Urbanisme et embellissement urbain." In *L'architecture d'Edward et William S. Maxwell,* edited by Robert Lemire, 50-55. Montreal: Musée des Beaux-Arts, 1991.
Wolff, F. and A. Erpicum, eds. *Guide of Montreal 1908-1909*. Montreal: Mercantile Print, 1909.

Fig. 1. *City of Montreal*. Map. 1890. Bibliothèque et Archives nationales du Québec, Collection de Cartes et Plans.

Fig. 2. *Plan de la cité de Montréal et de ses environs, partie ouest*. Map, detail. 1931. E.P.J. Courval. Bibliothèque et Archives nationales du Québec, Collection de Cartes et Plans.

Fig. 3. *L'isle de Montréal divisée par costes*. Map, detail. 1702. François Vachon de Belmont. Bibliothèque de la Compagnie de Saint-Sulpice, Paris.

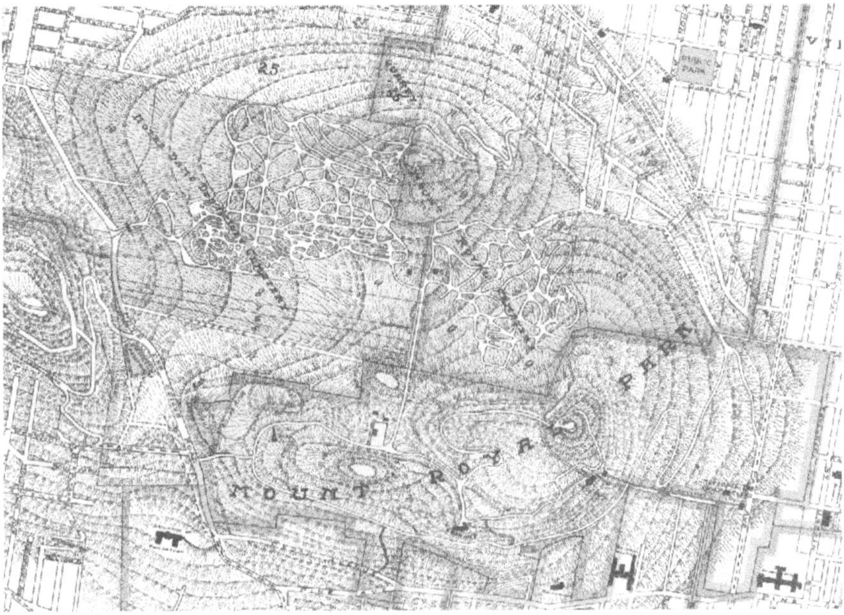

Fig. 4. *Atlas of the Island and City of Montreal and Ile Bizard*. Plate 25, "Around the Mountain," detail. 1907. A. R. Pinsoneault. Bibliothèque et Archives nationales du Québec, Collection de Cartes et Plans.

Fig. 5. *Glimpses in Mount Royal Park, Montreal*. Photograph, silver salts on paper mounted on paper, albumen process, 10x8 cm. ca. 1878. Notman & Sandham. McCord Museum, Notman Photographic Archives.

Fig. 6. *The Drive, Mt. Royal Park, Montreal*. Photograph, silver salts on paper mounted on paper, albumen process, 10x8 cm. ca. 1878. Notman & Sandham. McCord Museum, Notman Photographic Archives.

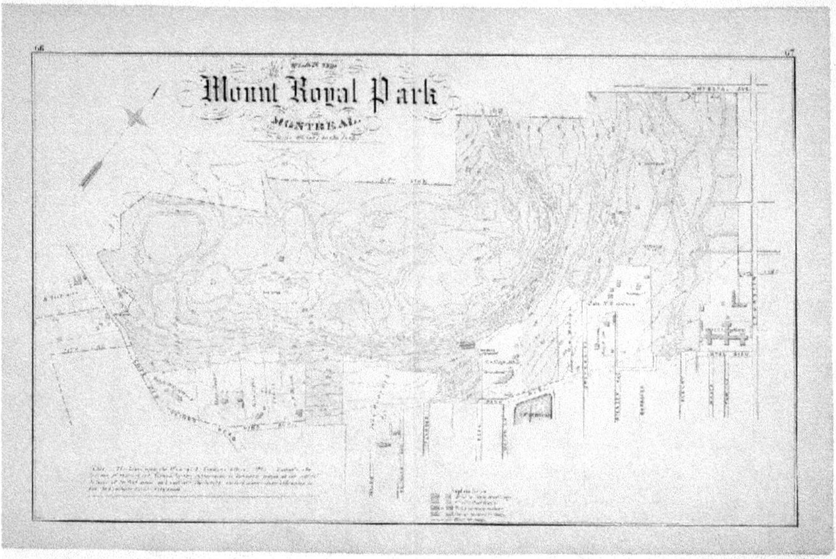

Fig. 7. *Plan of Mount Royal Park, Montreal*. Map. Ink on paper, photolithography, 38x61 cm. 1877. Attributed to Frederick Law Olmsted. McCord Museum, Notman Photographic Archives.

Grid, Regulation, Desire Line: Contests over Civic Space in Chicago

Peter Bacon Hales

Occupation and reoccupation of public spaces are usually accompanied by forms of conflict over ownership, rights, propriety, and opportunity. The built environment of Chicago stands in for that of many global metropolitan centers as a site for investigating the dynamics of public space. In these modern urban centers a particularly intensive, and correspondingly oblique, form of contestation over cultural ownership occurs, based upon the geography of urban occupation. At one level, this contest operates in a dynamic between regimented, regulated spatial formation and designation, and informal, intuitive, counter-regulatory actions. Those who regulate have the power of institutions at their behest; those who resist have other advantages—the fluidity and flexibility to form or reject alliances; the capacity to locate and relocate the sites of contest rapidly and pragmatically; the ability to act alone while aggregation of many of these individual acts can result in profound and lasting changes in the dynamic of public life.

We tend to think of public space as a staging-ground for ideal acts by citizens; as a site for public arts that uplift, ennoble, and rejuvenate the citizenry; as a natural refuge from urban enervation. But it is also, and perhaps always, the site of contests over the very definition of the telling words that surround it: *citizen*, *civic*, and *city*, but also *urban* and *public*. Occupation and reoccupation of public spaces are usually accompanied by forms of conflict over ownership, rights, propriety, and opportunity. It's not my place in a brief essay to develop a field theory that can unite these elements into some complex causal relationship that can then be seen to sustain itself over time. Rather I hope to introduce a series of moments in a particular historical locus, so that some of these tensions can be played out.

Behind my thinking is a theorem of sorts: that in urban centers a particularly intensive, and correspondingly oblique, form of contestation occurs, based upon the geography of urban occupation. At one level, this contest operates in a dynamic between regimented, regulated spatial formation and designation, and informal, intuitive, counter-regulatory actions. Those who regulate have the power of institutions at their behest; those who resist have other advantages—the fluidity and flexibility to form or reject alliances, the capacity to locate and relocate the sites of contest rapidly and pragmatically; the ability to act alone while aggregation of many of these individual acts can result in profound and lasting changes in the dynamic of public life. What follows are a series of anecdotal moments in the history of Chicago, a city where the tensions concerning the nature of the ideal civic culture are particularly strong and longstanding.[1]

Plat and Public Square

At the center of Chicago's downtown there is an imaginary vacant block, bounded by Washington, Dearborn, Randolph, and Clark Streets. It is a vacant lot because, when the city was first mapped out in 1830, its platter, James Thompson, designated that space a "Public Square." Thompson was no doubt thinking of the "greens" and "commons" of older towns and cities to the east; the Public Square was to provide a place of assembly, a parade-ground for citizen-militias, a grazing common for residents and visitors, and a parklike greensward in the midst of what would be a hectic, even frantic, process of citybuilding (fig. 1).

Thompson's grid plan was a means of inserting the new city into existing forms: within the larger template of national identity and growth, Chicago could be developed at a distance, with land purchased sight unseen. Chicago was rendered distinctive from its terrain and its citizens would be masters over it.

The Public Square was to serve as a catalyst to civil engagement by the citizenry. Around its boundaries were to develop

---

[1] This essay is the result of a two-year project on public buildings and spaces in Chicago; a history of the public face of Chicago's architecture will be published by Acanthus Press in 2009. My thanks to Barry Cenower of Acanthus for proposing the project, and to Sarah Dreller, my research assistant for part of the project, for her assiduousness and critical acumen. Research in Chicago's history is dependent upon the kindness of archivists at two stellar collections: the Ryerson-Burnham Libraries of the Art Institute of Chicago, where special thanks go to Mary Woolever and Jack Brown. At the Chicago History Museum (long known as the Chicago Historical Society), a string of archivists and librarians have facilitated this project; special thanks goes to Rob Medina.

the principal forms of civic life: the town hall, market, bank, and courthouse. It was a counterfoil to the life of the grid, whose purposes were speculative and commercial. Open to the eye, it would, in a sense, *be* an eye, not simply the center of the vortex of activity surrounding it, but a place for vigilant observation of all that surrounded it. It was to be the fulcrum of the new city.

By the 1850s, the Public Square was a warren of desire lines. Muddy and unkempt, it was a shortcut site in contrast to the grid in whose hypothetical center it was located. Its failure was its success; offering an alternative to the regulated patterns of movement within the built-up blocks surrounding it, the open square increased the efficiency of those who moved through it, while losing its place as a greensward. In 1851, it was redefined as the site for a common city-county government building which, over the decades before and after the Great Fire of 1871, swelled to occupy most of the block.

The story of the grid, the Public Square, and the fate of public space in Chicago's center recapitulates the tragedy of the commons in an urban-industrial form (Hardin, 1243-1248). When John Van Osdel designed the first iteration of the City-County Building, he conceived of it as a sort of symbolic hub from which radiated the lines of civic purpose and to which the forms of civic engagement, appeal, and responsibility might be drawn. Its rectilinearity echoed the grid surrounding it, but its overall design modulated and transformed the grid; to look carefully at the building's outlines is to see it as hiding a multiaxis form, a punctured octagon, from which the eight radiating pedestrian walks acted as rays. The Public Square became a public eye, panoptic, at the center—and defining the center—of a city. Over the decades, that would change, and dramatically (fig. 2).

Public Charter

The Public Charter of 1837 declared in legal terms the boundaries of a public sphere, the administration of that sphere by elected and appointed government, the funding and protection necessary to its continued existence, and an oblique template for the expansion of the city's boundaries, both physical and cultural.

Two passages delineate the right and extent of constituted government to direct individual citizens in the public sphere. The first declares the obligations of citizens to contribute time or money to the maintenance of those civic spaces essential to efficient commerce—streets and thoroughfares. The second cedes to city government not only the right to "regulate the burial of the dead," but also "to prevent the rolling of hoops, playing of ball or flying of kites or any other amusement or any other amusement or practice having a tendency to annoy persons passing in the streets and on the side walks[.]"

At its inception *as a city*, distinctive from its previous incarnation as a settlement and then a town, the very document that brings the civic into being also seeks to span and order space, to define the boundaries of decorum, and to mediate between individual behavior and the greater good. And already the authorities declare their right to suppress "amusement" and play in favor of the efficient motion of traffic—commercial traffic.

Enclosing the Public Square: The City County Buildings, 1874-1911

The Great Fire of 1871 destroyed Van Osdel's city-county structure and wiped clean the Public Square, even as it dramatically reinforced the need for an organized civic government, capable of regulating the public sphere, willing and able to determine when private interests encroached on the public good. A string of laws, ordinances, and regulations in the rebuilding years required making wide service alleys, splitting blocks, and providing access to buildings. They also dictated that buildings within the city be constructed in a "fireproof" manner. One effect was to change the scale and tone of the downtown, rendering its landscape more monumental, more rationalized, and more permanent.

The rapidly expanded city and county governmental bureaucracies emboldened by the string of post-Fire activity demanded an equivalent governmental space within which to operate this expanded regulatory system. Once again, the Public Square was capped, this time by an ornate double-building that sprawled out, sidewalk-to-sidewalk, eliminating Van Osdel's spoke-and-hub design, with its invitations to centripetal movement and its spaces for public congress and respite.

This City-County Building introduced a new conception of public use: to replace access with awe. With its complex of ornamental forms meant to symbolize the grandeur and weight of civic government, the building proposed a new relationship between individual and institution; dwarfed by its scale, overwhelmed by its looming presence, the citizen took on the role of supplicant (fig. 3).

But the building itself wasn't without irony; so elaborate was it that its foundations could not support its symbolic program: from its opening in 1885 it settled alarmingly, and stopgap measures failed to stem the trend. By 1905, when a particularly violent shift caused gas pipes in the basement to rupture, resulting in an explosion and fire, the building had come to symbolize the overreaching of civic government and its *hubris*. A replacement, designed by the established architectural firm of Holabird and Roche, brought a restrained, monumental Neoclassicism to the seat of civic government (fig. 4). Now the Public Square was a warren of corridors and offices, a maze within which power was contained, but hidden (fig. 5).

Haymarket, May 1-4, 1886

Without designated or even transitory informal places for public discourse, the city's streets became increasingly essential for the forms of assembly that mark the public sphere. But these were inhospitable and crowded spaces, and informality, play, and "amusements," were forbidden by charter, and discouraged by climate. As the speed of commerce increased, communal activity began to shift out of legally designated public arenas and into quasi-public spaces, with sometimes-violent results (fig. 6).

The Haymarket was a marginally public space pressed into service by one body of citizens, and retaken by the forces of institutional and commercial order. But the tensions that exploded there can be extended to a string of sites where nearly identical battles over public turf were fought by these two distinctive forms of "public" and "civic" identity. One might well argue that the discursive space of Haymarket as it has come down to us encompasses multiple historical places. Moving chronologically, the first might be the McCormick factory on the near South Side, miles from the bombing site; there the corporate manufactory had locked out workers and replaced them with scabs. Guarded by a private army abetted by city police, the factory grounds could be understood variously as resolutely private or as dynamically liminal, for the presence of official police simultaneously enforced and rendered porous the boundary between public and private. Beyond the factory grounds on the near South Side, similarly embattled sites ranged along the industrial and transportation corridors abutting the Chicago River. Upon one such, a union-sponsored picnic devolved into a clash with police, who fired on unarmed women and children.

Just exactly who owned the land upon which the May 3 rally and picnic was being held is unclear. But that clash, and the many that preceded it during the days surrounding that May-Day of 1886, signaled a shift in the balance of power between labor and capital, working-class citizenry and the larger "forces of order." All these parties sought to influence the definitions of public and civic space, the forms of propriety that could and should be enforced, the limits of power granted by the citizenry to its organized government forces, and even the variance in the definition of "citizen." For men like McCormick, like architect Daniel Burnham, and their merchant-prince society, it was a given that the voices of capital should be heard more loudly than those of strikers, workers, and their families. That McCormick owned the most influential newspaper gave weight to that assumption.

Yet it is also striking that the mayor, Carter Harrison, moved through the crowd of protesters at the Haymarket on the night of May 4, as anarchists decried police actions and labor leaders called for renewed fortitude in an epic contest. His action suggests that the moment was fluid, definitions were changing; Harrison could be, simultaneously, the titular head of the forces of organized regulation, and a pedestrian, a citizen, a witness. When he left to go home, satisfied that an appropriate form of civil protest was underway, the forces of "order" and the forces of "anarchy" responded with unexpected violence.

Haymarket was long a liminal space, an ambiguous boundary between private and public, a city-held widening of a street sufficient to allow for an open-air market that spilled over from the larger nearby produce market, occupied by small-scale merchants, farmers seeking to sell their harvests—private capital squatting, with some measure of permission—on city streetscape. The decision of the protesters to use the Haymarket reflected a complex of issues surrounding the older "Public Square," now entirely enclosed in a corseted frump of a Beaux-Arts building collapsing under the weight of its own symbolic architecture.[2] Without a bonafide place of political assembly, anarchists, demonstrators, and political orators turned to the liminal space of the *agora*, where commerce, contact, and controversy were tolerated in fluid admixtures.

First Regiment Armory

Public militias, private armies, security forces for hire, and city police forces mixed in a volatile stew after Haymarket. One of the functions of the Public Square had been to provide an assembly ground for such militia, originally conceived to protect the civic boundaries from incursions by invaders. By the later 1880s, these militias increasingly served the forces of "order," that is to say, served at the side of the forces protecting the engines of industrial wealth and those who were its beneficiaries.

Few sites so clearly spoke of this shift from populism to plutocracy as the First Regiment Armory built at Michigan Avenue and 16th Street in 1891 (fig. 7). The First Illinois Regiment was a haven for the privileged, many of whom joined the brigade after the Haymarket incidents. Lavish subscriptions resulted in the commissioning of Daniel Burnham, their most prominent architect, and his design—a castellated Romanesque fortress of a building—loomed at the apex between the Prairie Avenue neighborhood, where so many had built their mansions, and the rabble-infested neighborhoods to the west. The exterior of Burnham's building defined a boundary line across which neither side was encouraged to step.

Within, the Armory was part fortress-retreat and part luxurious private club. It contained a gymnasium and locker rooms, a library and suite of meeting rooms and lounges, and a central courtyard where the Regiment was to practice its military maneuvers—out of sight of the larger population. The courtyard served other purposes as well; by the

[2] After a rigged "competition," the county's own architect, James J. Egan, received the commission; his difficulties required that Van Osdel return to patch up the result, to little avail.

middle of the first decade of the twentieth century, it was a popular site for charity balls and other entertainments by Chicago's equivalent of "The 400."

The Armory was only one of a number of new forms of architecture-as-ideology. Its public face, intimidating and enclosed, telegraphed a warning to those who had sought throughout the 1880s to retake the public cityscapes of streets and sidewalks, parks and public squares, for a civic class feared and abhorred by its opposite. The celebrations within its walls contrasted with those held in the Coliseum just two blocks away. In 1908, the Coliseum hosted a First Ward Ball requiring 200 waiters and 100 policemen to minister to a crowd of city workers, ward heelers, prostitutes, magnates of the underground economy, and politicians, each of whom averaged three quarts of beer, topped by a total of some 10,000 quarts of champagne (Kingsdale, 472-489).

The White City and the Art Institute, 1893

Even as Burnham and the dominant figures in Chicago's established power structure sought to protect themselves and warn others off, they simultaneously engaged in the construction of an alternative utopian city, meant to model their conceptions of civic propriety, in architecture, in spatial array, in behavior, in economy.

The World's Columbian Exposition of 1893 brought together the principal conservative players in the civic arena, emboldening them not only to "make no small plans," in Burnham's words, but to embed their large plans across the core of the city. While the Fair was being designed and constructed at the far South Side region of Jackson Park, the architects, designers, and funders of the Fair built a downtown Orchestra Hall, a new Art Institute of Chicago and, over the years immediately after the Fair, pressed successfully for construction of a gigantic public library of Baroque ornateness, occupying a city block once designated for a public park.

Each of these projects served as incursions on the public sphere, whether of landscape or behavior. The Columbian Exposition was a quasi-public extravaganza set on land that was alternately once-public (Jackson Park) and *terra incognito* (if the vast stretches of land "reclaimed" from Lake Michigan can be so named). With the exception of "Chicago Day" and a few other moments, the public lands became private, requiring admission fees to enter; even on "free" days, the Fair was awash in proprietary regulation of everything from costume and comportment to the taking of photographs or the pace of movement. The Columbian Guards, dressed in white uniforms that were part-military, part-police, watched over the crowd's activities; violators of even the most trivial of regulations could be held for hours in detention before being ejected or set free.

With the Exposition serving as a model utopia for a new sort of orderly, patriarchal urbanism, Burnham and his compatriots looked to the downtown to demonstrate the utility of expanding the proposition into the city proper. A new Art Institute of Chicago, to the east of Michigan Avenue on public parkland, appeared as the first major permanent structure on a stretch of public lakefront land designated from the city's founding but never fully treated as civic space—cluttered with squatter huts, makeshift commercial sites, railway repair facilities, and overflow grounds for the First Regiment Armory some blocks to the south.[3]

Contemporaneous visual documents would suggest that the monumental structure sat upon a narrow strip of parkland sandwiched between Michigan Avenue and the rail yards of the Illinois Central. What the illustrations can't show is the controversy surrounding that open land, notably after the city government of Mayor DeWitt Cregier proposed the construction of an elaborate Civic Center on the grounds, and after the response of mail-order magnate A. Montgomery Ward who filed legal action to prevent the process.

Ward's lawsuits continued for decades, forcing the issue of public lands in what might seem a dramatically different direction (Willie, 71-81). The trend had been toward the enclosure, appropriation, and regulation of public lands, their conversion from open space with indeterminate, fluid utility to closed spaces with fixed functions. Promenades in rigid geometric forms, open areas without benches or amenities for play or even rest, and imperial vistas emphasizing sublime scale rather than picturesque immediacy served to restrain exuberant recreation, spontaneous play, family socializing, and respite in picturesque natural surroundings—the very ideals espoused

[3] The standard work on this subject is Wille's *Forever Open, Clear and Free*.

in Central Park by Frederick Law Olmsted, father of the two brothers responsible for this design. By 1909, Grant Park would be but one part of a far more expansive imperial city-plan: Daniel Burnham and Henry H. Bennett's *Plan of Chicago*, which envisioned a massive city rebuilding to bring the city into congruence with the neoclassical order of the White City (fig. 8).

Marshall Field's, the Palmer House, the Berghoff Restaurant: Private Public Spaces at the Beginning of the Twentieth Century

West of the Lake Park controversies, a new sort of street life began to emerge, rendering far more fluid the boundaries between the private spaces of commerce and the public spaces of sidewalk and street.

One signal development concerned the display windows of the new Marshall Field's Store, which between 1892 and 1914 came to occupy the entire city block bounded by State Street, Randolph Street, Wabash Avenue, and Washington. Designed by Daniel Burnham's firm and its successors, in stages, the store was a triumph of Chicago School architecture applied to the specific problems of the department store. New building technologies allowed for wide expanses of glass; the street-level display windows were remarkably tall and wide, each serving as a sort of stage for the display of goods from different departments of the store. Photographs of the time show pedestrians clustered around its displays, and the magnetic attraction of the goods within, artistically arranged and beautifully lit by the indirect light afforded by State Street's width, was sufficiently dramatic to induce Theodore Dreiser, writing his masterwork of naturalistic fiction, *Sister Carrie*, to render the theme of commercial display, temptation, and desire the dominant one of his novel (fig. 9).[4]

The fluid relations between street and window display, the creation of dynamics of desire within those stage-set displays, and the repeated appearance of entry doors to punctuate the sequence of displays all served to invite the passerby into the store itself. There,

[4] Dreiser had Sister Carrie's first encounter with the display windows of the new department stores take place in front of the Fair, built two years after Burnham's excursus; he had been avidly watching this trend to a stage-set drama of consumer desire since first taking the city as his subject.

other novelties of architecture and consumption continued the process. Most notable was the high, Tiffany-mosaic atrium, a gleaming light well that drew the eye upward to the successive floors of goods that had their open balconies fronting thereon; the effect was to render the store as a sort of ideal mini-city, an exotic European or near-Eastern market-square.

While the goal of architecture and design may have been simply to draw the curious, tempt them, and turn them into customers, the complex, open-air quality of the interior complicated the relationship between public and private; citizens came to consider the interiors of these stores as if they really were public market places, enabling a phenomenon that Dreiser had already noted: window shopping. In time, this process required that the shop-clerks change their own behavior, moving out from behind showcases and display tables to mingle with the crowds. Pinkerton police found it necessary to modulate their behavior, becoming more like friendly guides and less like vigilant enforcers. The quasi-public space became not simply spatially dynamic but socially and economically fluid as well.

This appropriation of private commercial space for public purposes was in some sense a response to the steady shrinkage of genuine public assembly spaces. By contrast, the lobbies of the Palmer House, the Palace Hotel, and many other showcase hotels in the downtown area served as meeting places where travelers and residents could interact, though again under the surveillance of porters, waiters, and private cops.

Whereas hotels benefited only indirectly from affording access to public activity, bars and restaurants in particular charged a sort of indirect admission fee or toll for the use of their spaces. Some establishments were specifically designed, or creatively adapted, to provide assembly places—halls and rooms that could be reserved or, if sparsely occupied, taken over on the spot. The Berghoff Restaurant in the downtown was a model of this sort of establishment, with multiple rooms that might fill up entirely at lunch or dinnertime, but at off-hours afforded spots for informal and formal meetings.

The Berghoff's history is particularly telling, as it was originally located at the corner of State and Adams, where it beckoned the weary and weather-worn with the promise of warmth and conviviality. When it was priced out of its original site, the restaurant moved westward along Adams, taking over the basement and first floors of one of Chicago's long-standing "public-hall" buildings, designed in 1872 to provide a string of public rooms. There the city's "Town Meetings" had been held, until their functions were distributed: to the City Council, to the Merchants' Club, to the First Regiment Armory, to the unions.

Each of these "quasi-public" spaces represents both a continuous example of and a paradigm for the ways in which citizens and their needs for public assembly and a sense of public identity intersected with and transformed the dual conceptions of *the private* in an urban setting. It is a truism to consider the city to be the perfect place for those desiring anonymity, a privacy verging on invisibility. But the city also afforded, and continues to afford, opportunities for malleable communitarianism: the quiet communality of those who might share the upholstered respite of a hotel lobby; the brief community of a beer-hall bar; or the sense of commonality shared among those who flocked to the Christmas window displays of Marshall Field's. And while these pieces of real estate were and continue to be privately owned, their dependence upon the commercial and cultural embrace of a broad public constituency has made them malleable environments of public comfort.

Even today, these forms of public assembly-space form a significant counterweight to the official public monuments of City Hall and Federal Post Office, the benevolently resplendent greenswards of the urban parks movement, and the celebrated forms of "urban design" taught to graduate students and schoolchildren alike (at least in Chicago). They are places of far greater fluidity, adaptability, and immunity than they might seem in their legalistic definitions as places where public and individual rights can be routinely abrogated to the aim of private regulation.

Daley Center Plaza

Even with the City-County Building occupying the entire Public Square, by the early 1960s city government found itself crowded into inadequate quarters. Construction of a massive new Federal Center Plaza designed by a towering legend of modernist architecture, Ludwig Mies van der Rohe, impelled the city to expand its own facilities to the city block directly to the east of the old Public Square. The decision was a traumatic one, as it required a massive campaign of eminent-domain-seizure, condemnation, and demolition of an area long considered essential to the fabric of public life in the downtown. Designer Jacques Brownson, a Mies acolyte, envisioned a bi-level plaza topped by multiple tall office towers. The result, he imagined, would be to afford a common, flexible assembly ground, located at ground level during good weather and in the climate-controlled lower plaza during inclement weather. His was an attempt at a modernist *agora*, a marketplace of goods, services, ideas, and activities.

Budgetary restraints resulted in the scotching of the bi-level plaza, and the result was not entirely dissimilar to the sterner Miesien Federal Center. With the addition of a Picasso sculpture that generated immense controversy shading, over time, into affection and landmark status, the Plaza became not just a tourist spot but a sort of hardscape park (figs. 10 and 11).

Over the next few decades, city efforts to locate festivals and marketplaces—from a summertime "farmer's market" to a winter Holiday Village—succeeded to some extent in reaffirming Brownson's original notion of a fluid space of assembly, regulated more by the vigilant presence of the tall city skyscraper than by the coercive presence of police or the inhospitability of architectural hardspace rendered in rigorous geometric form.

Twenty years later, on a lot kitty corner to the Daley Center, Brownson's successor at C.F. Murphy, the postmodern planner-architect Helmut Jahn, would attempt a similar feat with his State of Illinois Building (renamed the James R. Thompson Center), just to the north of the Daley Plaza. Jahn simply built a huge atrium, whimsically asymmetrical, with a bi-level interior plaza for services, activities, and commerce, while the governmental offices clung to the walls, offices open to the atrium and the public, with no doors closing offices from public view or visit. Regulation of this postmodern *agora*, to limit unstable public assemblies, noisy protests "disturbing" state workers, and commercial ventures not fully certified by state agencies, killed the plaza floors; above, bureaucrats insisted upon doors to offices and barriers between the business areas and the public aisles and aerial sidewalks. After 9/11, guards were posted at each elevator stop, to check credentials, appointments, and discourage any spontaneous action.

Grant Park, August 28-29, 1968

The contests, sporadically violent, over the days surrounding the 1968 Democratic National Convention offer a well-documented case of a battle between the two forms of "public order" that I have been proposing, one direct, often spontaneous, finding its origins in the heritage of "town meeting" governance, the other legal and regulatory, bureaucratic, and more or less distant from the populace that grants its powers.

Certain features of the confrontation aren't commonly given attention. The photographs of the demonstrations most often seen were made by press crews—still or video—representing established networks or new-wire organizations; the photographers stayed behind the police lines, in areas set aside for their work, with the result that the demonstrations tended to reflect the views from the Hilton Hotel and the police lines.

Dave Nystrom was a photographer for one of Chicago's alternative newspapers; Terry Southern was a well-regarded New Journalist for the journal *Esquire*. Both of them ranged behind the lines, moving with the demonstrators between Grant Park and Lincoln Park; Southern, at least, was tear-gassed, roughed up by cops, and generally treated as one of the demonstrators despite his press credentials, his maturity, his relatively conservative dress, his drawl.

Southern's infraction was spatial: to cross the line between the two areas of occupation was to cross other lines as well. Yet he was able to maintain his character while *within* the boundaries that separated propriety from impropriety: after the worst night of police rioting, Southern was able to get into the bar of the Hilton, where he could see the events unfold as if watching a larger, more panoramic television—until "at the height of the slaughter five or six kids were pushed through a plate-glass window on one side of the bar," followed by cops who clubbed the trespassers mercilessly before dragging them back out into the public sphere.[5]

The police action—to uphold the boundary between officially sanctioned, decorous, quasi-public space (the hotel bar) and the unregulated sphere of the streets—was consistent with the evolution of public and quasi-public spaces in Chicago and, by extension, in the United States. Over four days, police and protesters fought to claim, occupy, and redefine the public park as a place either of silent decorum or noisy debate. On the side of the protesters, most notable was their early move to "take the hill" in the otherwise flat Grant Park, claiming the General Logan equestrian statue, with much fanfare.[6] The phrase is telling: it appears in reports from both sides of the conflict—undercover cop Pierson used it in his testimony at the Chicago Seven trial, but so did feminist-photographer-protester Jo Freeman (fig. 12).[7] In a city of flat geographical jagged cultural and political topography, the struggle to define and claim a symbolic "high ground," and a panoptic vantage point, was telling.

Millennium Park, 2003

The "public-private partnership" that engendered Millennium Park directly to the east of the long-overbuilt civic spaces of the Public Square and the Public Park represents the reemergence—or perhaps just the continuation, in a more stridently public form—of the civic patriarchy that had dominated the City Beautiful movement in Chicago. When Burnham's *Plan of Chicago* had emerged in 1909, it was the product not of an urban-planning genius and visionary, but rather of a consortium of businessmen, philanthropists, civic leaders, wealthy

---

[5] Southern, "Grooving in Chi." This article was originally published in *Esquire* and republished online with permission of the Southern Estate at http://www.pbs.org/newshour/convention96/retro/southern.html.

[6] "When the march was midpoint past the Logan statue, the crowd broke and ran up the statue screaming, 'Take the hill.' They climbed the statue and displayed the Viet Cong flag, the red flag and the black flag." Police informant Robert Pierson, testifying in the Chicago 7 Trial. Transcript online at http://www.law.umkc.edu/faculty/projects/ftrials/Chicago7/Pierson.html.

[7] Pierson, transcript, http://www.law.umkc.edu/faculty/projects/ftrials/Chicago7/Pierson.html.

scions of the city's founders, and the architects and planners who had flourished by serving this elite urban class. In the intervening century, individual wealth had to some extent been supplanted by corporate power, but the emergence of various features of Millennium Park revealed the continuing tradition of individual entrepreneurial sponsorship. As the grand plans for the park grew grander and their flaws—engineering, social, tactical—surfaced and resurfaced, the costs of the park ballooned, and the privatizing of this public sphere became, increasingly, the means of continuing support and deflecting citizen and taxpayer rage at the inevitable redirecting of government resources from grassroots and neighborhood amenities to a grand downtown showplace. The Daley administration, long committed to a program that made the city center an attractive magnet for tourism, empty-nester returns, and high-rise urbanity, considered Millennium Park to be an essential symbolic and spatial capstone to the longer-range redevelopment of downtown and its reemergence from a long decline.

Daley's proposal was prescient. Remaking the downtown into a festival city, his administration simultaneously reinvented the city as a destination for global tourism, middle-class resettlement, and corporate nesting. The relocation of aerospace giant Boeing from Seattle to downtown Chicago was only the most visible sign of a reversing trend that, just a decade before, had seen Sears, Roebuck abandon the fabled Sears Tower for a sprawling exurban "office campus" far from the city.

Boeing was a relative latecomer to the corporate sponsorship of Millennium Park; Boeing Galleries opened in 2005. By then, Boeing needed to join a roster that included the McDonald's Cycle Center; Exelon Pavilions; BP Bridge; Chase Promenade; Cloud Gate on AT&T Plaza; and others, as well as the features underwritten by the prominent commercial families of Chicago, most of them central to civic propriety since the Columbian Exposition or earlier: Harris, Crown, Lurie, Pritzker, and McCormick (figs. 13 and 14). Contradictory images of Millennium Park dominate: as a theme park divorced from the civic energies of the city, and far from its neighborhoods; as a triumphant cap to more than a century of public-private partnership to render a humane and diverse civic fabric; as a public space actually owned by private interests and dedicated to their hegemony.

Homeland Security, 2002-2007

One of Millennium Park's most celebrated features is its array of monumental public sculpture. Largest is the Frank Gehry-designed Pritzker Pavilion and BP Bridge ensemble, with Gehry's trademark titanium forming swooping curvilinear ribbons against the sky and along the pedestrian walkways. Most popular is Anish Kapoor's *Cloudgate*, better known as "The Bean," whose super-reflective surfaces offer a fun house distortion of visitors and downtown architecture, lake, horizon, and sky. Perhaps the most lively and diverse, however, is the Crown Fountain. This feature combines two massive towers of LED screens which project four sets of images of Chicago citizens at sublime scale; water flows out of and down these towers to replenish a shallow reflective pool that lies between them. On summer nights, the water forms a wading pool for hosts of children, from housing-project residents to foreign tourists, overseen by an equally polyglot sample of parents (fig. 15). Atop these towers, an observant viewer in 2006 would have noted obtrusive overhanging metal superstructures upon which were mounted pairs of surveillance cameras (fig. 16).

Part of a citywide "urban security initiative" funded with a $54 million grant by the federal Department of Homeland Security, these cameras represented perhaps the most intrusive and simultaneously least visible instantiations of a new tension among conceptions of the public, of public order and safety, and of civic rights and responsibilities. More prominent were the concrete blocks, orange fencing, steel I-beams, and—eventually, at the Miesian triumph, the Federal Center Plaza—granite obelisks, all of which were put in place to deter terrorist attacks in the wake of 9/11.

To say that such intrusions on the public spaces of the city's built environment generate far more terror than they allay is by now a cliché, a truism. Less noticeable in the changed calculus after the civic response to 9/11 is the way that the city's form became more spatially regularized as it was more prominently regulated. In and around the federal buildings, federal officers, both uniformed and "plainclothes," now patrolled with a somber regularity that converted buildings and plazas into simulacra of the urban zones under authoritarian regimes: North Korea, Iraq, Turkmenistan. Homeland Security grants funded dramatic expansion of the city's traffic monitors, now given uniforms and a new bureaucratic authority, as the Traffic Management Authority (TMA) of the Office of Emergency Management.

The effect of this re-regularization of the city was, in many cases, to reassert the grid over the long-standing emergence of the spiderweb of desire lines that had formed so much of the informal spatial culture of the city. Public plazas, open spaces surrounding high-rises, areas where temporary activities had emerged on temporarily vacant lots sited between demolition and development, even public parks, had been marked, sometimes physically, by the evidence of citizen redefinition of their geography and their function. The Park District of the city had long ago given up on the futile project of discouraging these lanes of movement, in some cases reseeding and replanting annually or semiannually, in others actually paving and thereby legitimizing these diagonals, ovoids, and often irregular looping stripes of hardened, grassless soil. Jaywalking, diagonal street-crossing, midblock runs—these, too, had become part of the dynamic dance of civic denizens.

Homeland Security returned to the coercive assertion of proper movement, relegitimizing the grid. Yellow-jacketed TMA figures with whistles and black, cudgel-shaped pointers stood at downtown corners, often backed up by uniformed police waiting to stop and cite offenders. In many of the public and quasi-public open spaces, uniformed federal, state, city, and private police stood guard or patrolled the plazas. In Millennium Park, a small army of SegWay-riding TMA guards rigorously enforced behavior, preventing children from touching the sculpted surfaces, visitors from lounging in the wrong spots, street musicians from setting up in unregulated areas, cyclists from wheeling their bikes. Above them, the cameras that stood atop the Plensa towers had been removed from their prominent place, but others had appeared more unobtrusively, in trees and on streetlights.

The effect of this regulatory vigilance is marked by its contradictory qualities of active and passive voice, to borrow from the language of grammar. On the one hand, these symbols of authority enforce behavior, reasserting the power of the policing agents over the resistance of the citizenry. More subtly, however, the pervasiveness of these forms of surveillance and control, and their array across a spectrum from uniformed, gun-toting policeman to the surveillance camera hidden in a sycamore tree, have broader and more oblique regulatory effects. Knowing that you're watched, on camera, in sight of the authorities, eviscerates spontaneous, unselfconscious action, performance, or display. In its place is a different type of performativity: even the act of resistance, the declaration of hostility, the refusal to cross where *they* tell you to, takes on a ritual manneredness.

Critical Mass, October, 2007

One of the regulations most vigilantly enforced in Millennium Park is the restriction on bicycles. Other forms of wheeled transport are more-or-less tolerated: wheelchairs, by law; strollers and baby carriages by dint of their sentimental connotations; the SegWays of the city's own police and the Park District's monitors by dint of efficiency of enforcement and novelty of practice. Bicycles, however, can only be transported across Millennium Park by carrying the bike on one's shoulder.

On the last Friday of every month, on, at, and around the Daley Center Plaza, near dusk, at rush hour, thousands of cyclists congregate. Some are bicycle messengers, some are commuters, some are recreational cyclists, and some are even professional racers. Some ride bikes comprised of Rube-Goldberg-esque flights of fancy—bikes 12 feet high and more, made by inserting one bike frame into another, and another, and another—while some ride titanium and carbon-fiber performance cycles costing more than a good car. Without a set route, the anarchy of cyclists begins to negotiate a path through the city. At some point, the densely packed group reaches sufficient critical mass that it has spilled over onto all the major downtown arteries and it begins to move.

The goal, if one can call it that, is not to have a goal. There are impromptu track-stand contests, in which riders test how long they can stay upright on an unmoving bike, simply by perfect balance, while hundreds of other cyclists swirl by. Periodically, a segment of the Ride pauses for the *Chicago HoldUp*: riders dismount and hold their bicycles aloft. Sometimes, this evolves into the *Mexican Wave*, in which the rise and fall of the bikes passes along the length of the route. At each intersection, a rider or group of riders blocks the cross traffic for minutes, sometimes an hour: the vernacular term for this is "corking." At some point, the police begin their ritual of arrest and removal. At some point, the group has chosen a sufficient number of differing routes that the density has dispersed, and the city returns to motion—with bicycles still outnumbering automobiles, transit riders, and pedestrians for some hours afterward.

What is striking about Critical Mass is its translation of civic engagement from centralization and regulation to dispersal and what has been termed a "rhizomal" system of organization.[8] While some urban Critical Mass formations (and there are Critical Mass events at most larger cities in the U.S. and E.U.) reject any form of organization, locality, or identity, the Chicago Critical Mass is unusual in that it has a lively and long-standing presence on the Web. There, one can find an elaborate protocol for uploading and sharing "flyers," which can range from proposals for routes to manifestos to posters advertising this event or others. There, also, one can log in, that is to say, one can join up by generating a username and password. On the Webpage, flyers, statements of philosophy, aesthetic critiques of specific "interventions," and the like are given voice within an organized framework—a Webpage (fig. 17).

Located between anarcho-syndicalism, a long political tradition in Chicago, and counterculture protest, Chicago Critical Mass declares itself dedicated to "the right of the people to assemble [as] guaranteed in the Constitution, and Critical Mass helps people remember that right."[9] Like many of the moments of contest in the defining and redefining of civic rights and public identity in an urban setting, Critical Mass bleeds between conscious resistance and assertion, and a spontaneous aggregation of civic actions. Like so much of the dynamic tradition of public expression, it is simultaneously conscious of itself as an assertion of rights, and sufficiently bedded in spontaneity and social dynamic that self-consciousness seems an inappropriate term for the

---

[8] The term comes from social theory, possibly originating with Carl Jung but more recently promulgated by the French post-structuralist philosophers Gilles Deleuze and Pierre-Félix Guattari, in which points of entry and exit are not set in a fixed determination, but entry to and exit from an activity—intellectual, social, physical—can occur in a near-infinite pattern of possibilities. This notion of rhizomal organization is asserted on the *Wikipedia* entry for "Critical Mass."

[9] About Critical Mass, Critical Mass Chicago, http://chicagocriticalmass.org/about.

push and pull of the contested public sphere it takes over or, perhaps more appropriately, generates.

That it takes place in the civic geography, that it reasserts the public sphere as a geographical one, that it does so within a narrative dynamic, that it skirts the line of manifesto or declaration: all this seems to me to extend a significant body of tradition within which the definition of the civic is honed and redefined under different conditions. Carving out civic identity within city spaces is particularly difficult; by their nature, dense urban cores are complex assertions of physical restriction—monuments, buildings, gridded street forms, stop lights, park benches situated just *here* and not *over there*. The city's geography enables some sorts of activity, while discouraging or even rendering impossible others. From these possibilities and restrictions, as the city's first charter made clear, the boundaries and definitions of propriety are then set forth.

What is striking about that first charter is the way it sets from the first a civic obligation to restrict the freedom of individuals and groups, regulating their ability to use the public sphere and public spaces in ways that may be deeply satisfying to them, but serve to impede the activities of a "greater civic good." But not just *any* or *all* or *ideal* civic goods are privileged. Rather, as the charter gave special weight to commerce, to business, to the efficient practice of exchange (usually commercial but sometimes also social and cultural exchange), it demarcated a sphere separate from, above, the "common good."

The history of Chicago's contests over civic space as I've anecdotally outlined it here indicates the resurgent power of resistance to that demarcation, just as it shows continued pressures from those so privileged—the group often called "the commercial classes"—to ensure that their special privileges are ensconced as *rights*, and that the boundaries defining who is to receive those rights, and who is not, are tightly defined and, wherever possible, increasingly restricted. By contrast, those outside that sphere have chosen a wide and imaginative set of strategies to blur or relocate the boundaries, and to decrease the distance that privilege offers. Rarely violent, almost as rarely direct and vocal (Haymarket and the 1968 protests serving as clear exceptions), these forms of renegotiation and re-appropriation have more commonly applied oblique strategies of "squatting" on cultural spaces designated "public" or "civic," reorganizing by action the ways these concepts are understood in everyday space.

In this complex dynamic of taking and retaking, adapting and adapting to, the actions of Chicago Critical Mass on a Friday like that of Halloween weekend, 2007, might be seen not as outlandish or anarchic, but rather as firmly embedded in the everyday dynamics of urban public invention. Flyers—distributed by hand, by mail, and by internet—traced the informal and formal paths by which information moves in an urban environment. The appearance of crowds at the core of Chicago's public landscape, an open plaza just east of the earliest designated public Common, reaffirmed the right of citizens to occupy, enjoy, and organize in a space designated for just that purpose. And the demarcation of sinuous desire lines across the grid of the city's organizing principles served to claim a city still fluid, still flexible, resisting the rigidity of its grid and the force of its traditions, the heritage of the charter, the power of the proper classes, to renegotiate the right of assembly in dynamic form, overlaying one imaginary city with another.

Bibliography

Bruegmann, Robert. *The Architects and the City: Holabird and Roche of Chicago, 1880–1918*. Chicago: The University of Chicago Press, 1997.

Condit, Carl. *Chicago 1919–1929: Building, Planning, and Urban Technology*. Chicago: The University of Chicago Press, 1973.

—. *Chicago 1930–1970: Building, Planning, and Urban Technology*. Chicago and London: The University of Chicago Press, 1974.

Duis, Perry R. *Challenging Chicago: Coping With Everyday Life, 1837–1920*. Urbana: University of Illinois Press, 1998.

Grossman, James R., ed. *The Encyclopedia of Chicago*. Chicago: The University of Chicago Press, 2004.

Hardin, Garrett. "The Tragedy of the Commons." *Science* 162, no. 3859 (13 Dec. 1968): 1243-1248.

Kingsdale, Jon M. "The 'Poor Man's Club': Social Functions of the Urban Working-Class Saloon." *American Quarterly* 25, no. 4 (Oct. 1973): 472-489.

Mayer, Harold and Richard C. Wade. *Chicago: Growth of a Metropolis*. Chicago: The University of Chicago Press, 1969.

Miller, Donald L. *City of the Century: The Epic of Chicago and the Making of America*. New York: Simon and Schuster, 1997.

Smith, Carl. *The Plan of Chicago: Daniel Burnham and the Remaking of the American City*. Chicago: The University of Chicago Press, 2007.

Southern, Terry. "Grooving in Chi." *Esquire*, Nov. 1968.
Wille, Lois. *Forever Open, Clear and Free: The Historic Struggle for Chicago's Lakefront*. Chicago: Henry Regnery Company, 1972.
Zukowsky, John, ed. *Chicago Architecture, 1872–1922: Birth of a Metropolis*. Munich and Chicago: Prestel-Verlag and the Art Institute of Chicago, 1987.
—, ed. *Chicago Architecture and Design, 1923–1993: Reconfiguration of an American Metropolis*. Munich and Chicago: Prestel and the Art Institute of Chicago, 1993.

Fig. 1. Reproduction of Thompson's 1830 plat. A.T. Andreas, *History of the City of Chicago from the Earliest Period to the Present Time*, 1884.
Courtesy of the Chicago Historical Society.

Fig. 2. *Funeral Cortege of Abraham Lincoln leaves Cook County Courthouse and Chicago City Hall*. Photograph. 1865. Courtesy of the Wisconsin Historical Society.

Fig. 3. *Chicago City Hall and Cook County Courthouse*. Photograph. 1882. J.W. Taylor. Courtesy of the Ryerson-Burnham Libraries, Art Institute of Chicago.

Fig. 4. *City Hall and Court House*. Photograph. 1911. J.W. Taylor. Courtesy of the Ryerson-Burnham Libraries, Art Institute of Chicago.

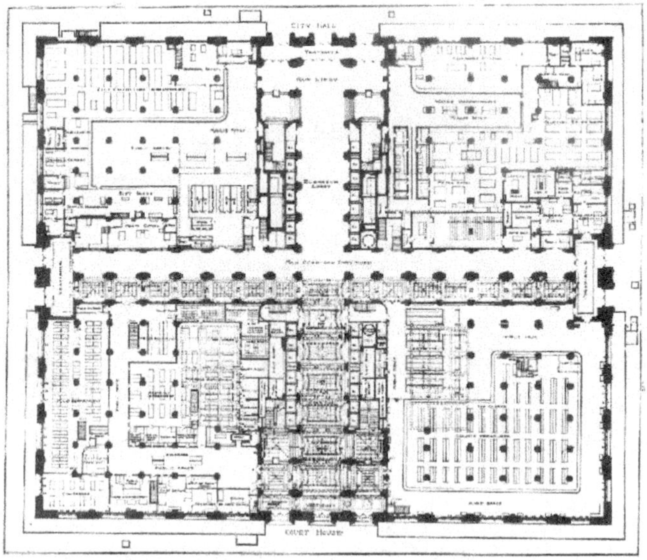

Fig. 5. City-County Building, Plan of First Floor. Blueprint.
Historic American Buildings Survey.

Fig. 6. *The Bomb Explodes, Haymarket Square, Chicago*. Lithograph. 1886.
Courtesy of the Wisconsin Historical Society.

Fig. 7. *First Regiment Armory, Chicago*. Photograph. 1891. J.W. Taylor [?]. Courtesy of the Ryerson-Burnham Libraries, Art Institute of Chicago.

Fig. 8. *Grant Park and Buckingham Fountain*. Photograph. 1929. Courtesy of the Ryerson-Burnham Libraries, Art Institute of Chicago.

Fig. 9. *Marshall Field's Display Windows*. Photograph. 1910. *Chicago Daily News*. Courtesy of the Chicago History Museum.

Fig. 10. *Chicago Civic Center Plaza with Picasso Sculpture*. Photograph. 1969. Hedrick-Blessing. Courtesy of the Chicago History Museum.

Fig. 11. Couple in front of Picasso Sculpture, Daley Center Plaza. Photograph. Courtesy of the Chicago History Museum.

Fig. 12. *Demonstrators Crowded Around General Logan Monument, Grant Park, 1968*. Photograph. Dave Nystrom [?]. Courtesy of the Chicago History Museum.

Fig. 13. *Pritzker Pavilion, Millennium Park.* 2007. Photograph by author.

Fig. 14. *Cloud Gate ("The Bean"), Millennium Park.* 2007.
Photograph by author.

Fig. 15. *Crown Fountain, Millennium Park.* 2007. Photograph by author.

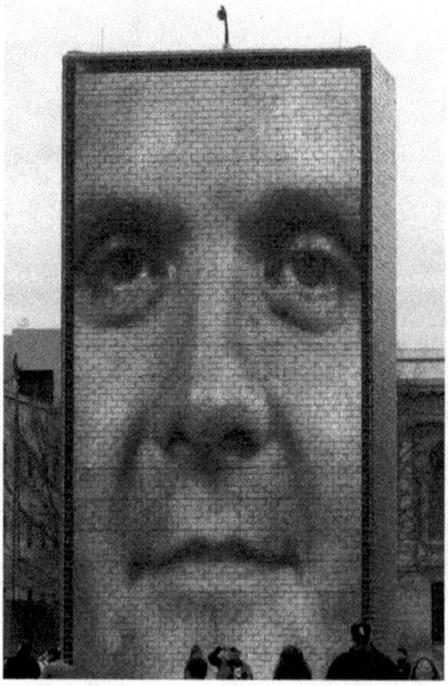

Fig. 16. *Security Cameras on Video Tower, Crown Fountain, Millennium Park.* 2006. Photograph by Devyn Caldwell.

Fig. 17. Poster/flyer for Critical Mass Halloween Ride. Lithograph. 2007. Spencer Thayer. Courtesy of Spencer Thayer and Critical Mass Chicago.

The Precarious Nature of Semi-Public Space: Community Garden Appeal, Complacency, and Implications for Sustaining User-Initiated Places

Laura Lawson

Since the 1890s, Americans have supported the development of community gardens as a local means to address personal, social, environmental, or economic concerns. Community gardens not only provide many tangible benefits, they also satisfy deeply engrained values associated with nature, individualism, and democracy; however, while generally appealing, community gardens are not validated as public resources. This essay suggests that the sustainability of community gardens requires unraveling their ambiguous role as semi-public resources. This ambivalence is partly due to attitudes about the role of gardening and partly due to the way most community gardens develop based on opportunities rather than long-range planning. Analysis suggests that permanence is a factor of both land-acquisition methods and sustained participation. The unclear status of community gardens has implications for a larger pool of user-initiated places that serve local needs yet remain ambiguously semi-public.

Introduction

Each year, seemingly in sync with the growing season, a crop of special-interest stories appears in local newspapers to praise volunteers who transform a derelict lot into a garden, to highlight a program that teaches children to garden, or to describe a community's effort to protect an existing community garden from pending demolition. Occasionally, community gardens make national news, as in 1999, when New York City officials announced that over 100 gardens in various New York neighborhoods were to be auctioned. In this case, articles, letters to the editor, and photographs described the battle between those who wanted to keep the gardens as community open space and those who considered the sites valuable for development. While garden advocates argued that community gardens provided local recreation, open space, and fresh food, Mayor Giuliani's administration countered that gardens on city-owned land were never meant to be permanent and that gardening was a privatized form of recreation that the public need not provide. Giuliani was reported to have taunted the gardeners, stating, "This is a free market economy. Welcome to the era after communism."[1] Fortunately for the gardeners, a Supreme Court injunction and a $4.2 million purchase stopped the gardens from being destroyed.[2] Seemingly solved, the public's attention subsided even though other community gardens in New York remained in tenuous circumstances. While 114 gardens were protected through purchase, others have since been lost.

A more recent flurry of articles highlighted another garden battle occurring in Los Angeles that brought out activists and actors in support of the primarily Latino gardeners at the South Central Farm. This 14-acre garden was established in the aftermath of the 1992 civil disturbances in Los Angeles as a way to heal the community.[3] Initially spearheaded by the Los Angeles Regional Food Bank, the Farm evolved and grew to become a self-managed community serving 350 primarily Latino families. However, even as the gardeners were busy planting corn, chayote, and other crops, the site became embroiled in a land struggle. The City of Los Angeles had initially acquired the site in the late 1980s through eminent domain for the purpose of building a trash incinerator plant, but, having abandoned this plan in face of public protest, failed to offer the original sellers "right of first refusal" that was required as part of the eminent domain proceedings.[4] Closed

---

[1] Michael Grunwald, "Mayor Giuliani Holds a Garden Sale," *Washington Post*, 12 May 1999.

[2] Stapleton, 7. Information on New York's community gardens was collected from materials provided by: Green Guerillas, the Council on the Environment in New York City, and Trust for Public Land; a presentation by involved activists and staff at the 2000 American Community Gardening Association Annual Conference; and interviews with participants.

[3] Initial information on the South Central Farm was collected in 1999 through a Food Bank press packet that included support letters and articles from the *New York Times, Los Angeles Times,* and *Tu Mundo*. The author visited the site in December 1999 and 2000 and June 2001 to informally speak with gardeners and take photographs. The fight to save the garden was covered by various newspaper and magazine articles. See: Hector Becerra, Megan Garvey, Steve Hymon, "LA Garden Shut Down: 40 Arrested," *Los Angeles Times*, 14 June 2006; Jessica Hoffman, "L.A. Urban Farmers Fight for Community Garden," *The News Standard,* 5 April 2006; Jessica Hoffman and Christine Petit, "14 Acres: Conversations Across Chasms in South Central Los Angeles," *Clamor* 36, Spring 2006, www.clamormagazine.org. A more complete case study is provided in Lawson, "The South Central Farm."

[4] For an account of the activism to stop the incinerator plant, see Di Chiro.

negotiations between the City and the original owners ultimately resulted in the repurchase of the property by one of the original owners for approximately $5 million, which is close to the price paid for the land seventeen years prior. The gardeners organized into the South Central Farmers Feeding Families to fight the pending site closure, arguing that the City process had not been public and that the publicly owned land was serving a public good worth protection. Efforts included letter campaigns, attendance at city council meetings, marches, protests, and ultimately site occupation that led to the arrest of famous activists and actors, including Joan Baez and Daryl Hannah. As the Mayor's office, foundations, and the Trust for Public Land put together purchasing packages, the original owners began planning for new warehouse space and other uses. The fight became increasingly contentious, to the point that some gardeners complained about feeling coerced into political action and the owner announced that he would never sell the land to the gardeners, even when offered his asking price of $16 million. Ultimately, the sheriff's department enforced the eviction on 13 June 2006 and the Farm was bulldozed. The City identified alternative gardening spaces under high-voltage power lines, to which some gardeners have since moved, reinvesting their sweat and energy into making new gardens on untilled land. Although the Farm was not saved, some of the gardeners and activists continue to work for community improvement through the recently formed South Central Farmers Health and Education Fund, a not-for-profit organization that promotes farmers' markets, gardens, and other community resources to assure healthier, culturally appropriate, and affordable food access (figs. 1 and 2).

Although public attention to community gardens may ebb and flow, the urge to provide land in cities for people to garden has been a persistent impulse. Indeed, communal garden programs have existed in U.S. cities since the 1890s. These include relief gardens run by philanthropic and public agencies during various economic depressions; children's gardens run by schools, clubs, and parks departments; war and victory gardens; neighborhood community gardens; and contemporary constituency-based programs serving homeless, at-risk youth, hospital patients, and others. Yet given the persistence of the idea, one wonders why community gardens are so rarely secured as public resources in the same way as parks, playgrounds, and other social and recreational public spaces. In this essay, I argue that the sustainability of community gardens requires unraveling their ambiguous role as semi-public resources. While the non-participating public tends to support the *idea* of gardening as a restorative and productive avocation, there is less clarity as to whether gardening, in light of the personal gains associated with it, belongs within the public sphere to be justified as public space. This ambiguity is influenced by vague regard for the capacity of gardening to achieve multiple goals and reinforced through opportunistic approaches to establishing community gardens that fall short of long-range planning to sustain public support. The conflicts inherent in community gardens pertain to other user-initiated spaces that evolve to address local needs, often as a result of social, ethnic, and lifestyle change in a neighborhood context, including informal and unofficial skateboard parks, dog runs, sidewalk vending, or even collections of chairs creating a semi-permanent social space at a corner or empty lot. The issues raised call into question the value, role, and permanence of user-initiated spaces.

The Appeal of Community Gardening in Times of Turmoil

At the most basic level, a community garden is land made available for people to grow food and flowers. Such a resource can be applied to a range of situations, from a neighborhood garden where local residents grow food for home consumption to institutionally run gardens that serve a particular constituency, such as recent immigrants, the homeless, or at-risk youth. The relatively straightforward land use is complicated, however, when one considers the many benefits associated with such ventures that result in a range of applications and programs. Whereas the allotment garden tradition in Europe grew out of policies to provide land to workers to compensate for lack of property or to supplement food resources, community gardens in the United States have traditionally been justified as serving a range of social, economic, or environmental agendas that reach far beyond land access and food production.[5] These justifications remain surprisingly consistent through various phases of community garden promotion over the past 100 years. During times of economic depression, inflation, or disinvestment, people turn to community gardens to subsidize household food expenses, offer job training, and help enterprising individuals earn income.[6] In times of civic unrest, the garden has been a nucleus for building morale, developing relationships with neighbors, and restoring the soul. Faced with environmental degradation, gardeners can take action to brighten up declining neighborhoods, restore habitat, provide environmental education opportunities, and encourage other kinds of community improvements. While one primary concern might instigate a community-garden campaign, other benefits are quickly added to further validate the effort (figs. 3 and 4).

Community gardens can serve an array of agendas because of their association with fundamental core values related to nature, individualism, and democracy. In a largely urban society that wears its agrarian ethos on its sleeve, community gardening provides a photogenic, nostalgic, and therapeutic representation of agriculture and its associated values. Community gardens serve as conduits to virtues attributed to country living, individualism, and simpler times. They also represent a return to "nature," particularly when juxtaposed to a confusing urban environment. In this vein, community gardens are often described as urban farms, refuges, or oases where individuals can remove themselves from urban life and current troubles, at least temporarily. At the same time, the hands-on, participatory quality of gardening also qualifies the community garden as an exemplar of individualism. An individual or a community can make use of an underutilized resource such as a vacant lot and, with a little organization and effort, grow food with minimal outside assistance. In the garden, one can see the results of labor, with harvests as proof of one's effort. Beyond feeding the gardeners, the act of gardening

[5] A historical account of Europe's allotment gardens is provided in Crouch and Ward. In addition to allotment gardens, there are also European urban garden programs that have social, environmental, and economic objectives.

[6] Historical examples cited in Lawson, *City Bountiful*. A summary of benefits is included in Malakoff. For contemporary descriptions of garden programs serving economic, therapeutic, or social roles, see: Bonham, Silka, and Rastorfer; Balmori and Norton; Hynes. For research on the restorative aspects of gardens, see Kaplan.

provides evidence of initiative—that the gardeners are taking matters into their own hands and doing everything possible to improve their situation. Along the same line, the garden provides an equalizing context that satisfies the "melting pot" view of the United States. A community garden provides space for everyone to garden, no matter what his background. In the garden, each person starts with the same resources and proves his worth through his yield.

That community gardening can claim such an array of benefits helps to explain why there has been a persistent if episodic activism to provide this resource in urban communities facing social, economic, or other concerns. Our gut instinct assures us that gardening in the city is a means to counter urban congestion as well as depopulation, to improve education and health, to ease labor relations, to curb social anomie, and to remedy environmental conditions. While satisfying such high ideals and larger aims, the community garden is also immediate and local. When global wars and economic systems wreak havoc on our daily life, the tangibility of the garden keeps us focused on what we *can* do to improve our situation. We may not be able to stop war, global warming, or a depersonalized global economy, but we can grow our own food, get to know our neighbors through gardening, and build ourselves a refuge and oasis. Community gardening has been and remains an appealing approach because it shows immediate results, is highly participatory, and is relatively cheap compared to addressing larger structural problems that manifest in a community.

Community Gardening: A "Nice" Idea Kept Vague and Opportunistic

While most people consider community gardening as a symbol of hope and positive action toward individual and social betterment, it is often a vague, benevolent approval that holds little weight when confronted with the realities of urban land politics and real estate. While many gardeners may consider their plot as a necessity, the general public tends to view gardens as "nice" but not necessarily as needed permanent resources. This is due to two issues: the vagueness given to the garden's role and the opportunism of investment.

First, the community garden has tended to serve as a means to other ends rather than an end in itself. The social, economic, environmental, and psychological benefits associated with gardening are largely anecdotal and descriptive and remain largely unaccountable. Some aspects can be measured, such as quantity of food produced, but other benefits are not measurable, such as community empowerment, psychological well-being, or environmental restoration. Indiscriminately grouping together such vague benefits avoids the exercise of clarifying and articulating a garden's primary or achievable objectives. As a result, urban gardens hold little proof of impact when pitted against development or more pressing funding concerns. The assumption that gardens can garner general support through their inherent goodness permits a lack of accountability about the good that gardens actually do. As a result, critics often state that community gardens only serve those who are growing and eating the food, ignoring the undocumented community and environmental benefits.

Second, the immediacy of gardening—transforming a vacant lot, getting the idle unemployed engaged in productive activities, or encouraging the social interaction of strangers—often spurs community garden projects to take advantage of whatever opportunities emerge, forgoing any long-range planning efforts. Impatient to start gardening, gardeners have made the most of volunteer work and donations. Historically, opportunism has allowed for waves of popularity and participation, such as the war gardens of World War I and victory gardens of World War II. These programs engaged millions of people in home and community gardening for a short period of time, but when the war ended so did most of the gardens. In contrast, however, most community gardeners today assume that they will be able to continue gardening, even if the garden is on borrowed land and tenuously funded.

Many community garden organizations are aware of the need to secure garden futures. The most recent national survey of community garden organizations, conducted by the American Community Gardening Association (ACGA) in 1996, found that permanency was a concern for most of the respondents (ACGA, 2). Although there had been a net gain in the number of gardens—1,851 new gardens created in the five years previous to the survey—there had also been 542 gardens lost. The survey found that the primary reasons given for garden losses were lack of interest by gardeners (49.4%), loss of land to a public agency (19.7%), and loss of land to private owners

(15.3%). These findings suggest that permanence is a factor of both land-acquisition methods and sustained participation (fig. 5).

Land Acquisition

Since the 1890s, community gardens have tended to rely on donated or otherwise underutilized land—vacant lots, rights-of-way, underused sections of parks, or other undeveloped land. Historically, gardeners use land through arrangements with public and private landowners who donate or rent land at nominal rates, with the agreement that gardeners, upon the owner's request, leave the site in good condition. Typically, as economic conditions stabilized and land regained value for development, gardeners acquiesced to the situation and either stopped gardening or moved to a different location. Much of the public attention subsided and programs that suffered from benign neglect ultimately closed. In some instances, efforts to sustain programs were successful, such as some World War II victory gardens in Washington D.C. and Boston that still exist as community gardens. In general, however, each subsequent phase of popularity has required that new land be found, usually a little further out on the city limits or discarded from some other use and in need of substantial clean-up and soil reclamation.

Today, even though most community gardeners intend their gardens to be permanent, the most common type of site procurement arrangement is the short-term lease. For neighbors who are eager to start gardening, a lease arrangement of one dollar per year for one to five years sounds very appealing. For the owner—whether a city agency, an institution, or an individual—such an arrangement allows the site to be used and maintained by others without jeopardizing the possibility for later development. Donating the use of land for a community garden is a gesture of good civic will that is made even more appealing when the gardeners' occupation also protects the site from vandalism or other damaging uses. The aura of permanence that full-grown trees and richly cultivated soil give to a ten-year old garden may belie the vulnerability of an annually renewed lease. Leasing public land is no more secure than leasing private land, as these sites are just as likely to be developed for other public purposes or sold.

Local development trends and community attitudes toward gardening influence city policies on land acquisition. For example, Boston's community garden movement thrived in the 1970s, so during a development slump in the 1980s, the Boston redevelopment agency worked with community-garden groups to assume ownership of garden sites.[7] In New York City, on the other hand, lease arrangements fluctuated according to development demand (in spite of the city's Operation Green Thumb program which actively promoted community gardening on city-owned vacant lots)—ultimately leaving gardens vulnerable to closure. Only a few cities, such as San Francisco and Seattle, include community gardens as open space in their city planning process.[8] When not acknowledged and protected through land use planning and zoning, the garden remains a vacant lot that can be taken away under the justification of "higher and better use."

One of the recurring solutions suggested by community garden advocates is the creation of neighborhood land trusts to purchase garden sites. As of 1996, only 5.3% of community gardens in thirty-eight cities were in land trust or on land owned by a gardening organization (ACGA, 2). Land trusts may reflect a larger trend toward privatized recreation and "friends of" organizations taking an increasingly prominent role in the upkeep of parks and other public spaces. However, while ownership assures site protection, it also avoids the more complex issue of public responsibility to facilitate what has been a persistently sought-after local activity, namely, gardening. Furthermore, this approach has environmental-justice implications. While wealthy neighborhoods may have the resources and connections to potentially buy their garden sites, poorer neighborhoods, even organized ones, may not have the necessary resources or social connections (fig. 6).

---

[7] Betsy Johnson, Director of Garden Futures, in an interview with the author, 13 April 1999, Boston.

[8] The American Community Gardening Association survey conducted in 1996 listed Madison, Philadelphia, Portland, Seattle, Somerville, and Austin as including community gardens in city plans. An unpublished study of municipal support for community gardens was conducted by Kendall Dunnigan of the Berkeley Community Gardening Collaborative in 1996 that listed Denver, Trenton, Asheville, Dayton, and Providence. For a discussion of various land tenure strategies, see Kirschbaum.

Participation

Opportunism has also applied to leadership and support for community gardens, particularly in times of national upheaval. Each time the idea to start a community garden program arises, a network of leadership emerges to advocate and provide technical assistance. In the past, this leadership was primarily volunteer and drawn from compatible interests in education, philanthropy, or civic improvement. Federal agencies, universities, and civic and women's clubs often provided the structure and guidelines that participants could plug into at the local level. Today, most people consider community gardens to be grassroots efforts that build up from local commitment and need. Since the 1970s, organizations have formed in cities around the nation to support community gardening efforts and provide technical assistance, including New York Green Guerillas, Garden Resources of Washington (GROW), and Boston Urban Gardeners (BUG). In addition, gardens are initiated and managed by city agencies, nonprofit organizations, neighborhood groups, and institutions serving specific constituencies including at-risk youth, disabled persons, the elderly, immigrants, and public housing residents. Successful programs tend to have both overseeing advisors and grassroots activism in place.

Because community gardens are built and maintained by their users, they are often considered models for community participation and empowerment.[9] This is not an inherent condition and the demands involved in sustained involvement should not be overlooked. In the past, for instance, didactic or economic objectives of various garden programs often minimized the decision-making role of participants. In the early 1980s when over-enthusiastic public agencies and organizations installed gardens without neighborhood involvement, some of the resulting gardens did not respond to the particular needs of the neighborhood and were eventually abandoned. Perhaps because of the wasted resources and negative publicity created by such failures, garden activists increasingly insist that gardens should not be "given" to a neighborhood but that assistance should be provided to enable

[9] Community gardening is cited to as an example of community activism and empowerment in Francis and Hester. Case studies also show this: see Mikalbrown.

neighbors to build their own gardens. In this way, the process nurtures and strengthens neighborhood social networks and organizing skills.

The internal governance structure of a community garden varies from anarchistic to dictatorial. Some gardens may have a citywide umbrella organization that they can turn to for support while other gardens are isolated and rely on internal resources. Quite often there is a "benevolent dictator" who organizes semi-regular meetings and oversees such matters as compost delivery, garden plot fees, and insurance. In fortunate cases, the volunteer leaders realize enough personal satisfactions to continue; however, sometimes these individuals burnout, leaving a gap in leadership. Gardens are volunteer efforts and the commitment required to maintain gardens—from cultivation of a plot to being a part of the gardening community's leadership—may tax otherwise occupied neighbors. Just like any group activity, interpersonal dynamics influence garden functioning. Some gardens may have clear rules that make involvement easy while others may struggle with opposing ideas about how to garden and how to share responsibility. Gardens in high demand may have waiting lists of people who want to garden while other gardens may go through periods of under-utilization. All these factors suggest the need for flexibility in garden planning and continued attention to nurturing participation and local leadership.

Perversely, the tenuous status of some gardens may unite participants and keep them prepared to exploit opportunities, work together, and fight to protect their efforts. As neighbors collectively work to establish a garden out of nothing, they develop social networks and learn organizational skills. Likewise, the effort to protect a garden has served as a catalyst for community activism. However, as in the case of the South Central Farm, this activism is no guarantee that the garden site itself will remain available. The energy and enthusiasm that goes into urban gardens and can lead toward other aspects of civic engagement can be lost if city officials do not consider the garden a legitimate resource in the community.

To encourage general public support, many garden organizations have developed policies to promote outreach and community involvement. Public events, educational programs, and collaborations with local institutions (churches, schools, half-way houses, etc.) are examples of such outreach. Currently, the American

Community Gardening Association is considering ways to increase its lobbying efforts at the local, state, and national levels. One of the main objectives is to transform the general appeal of community gardens into direct public support that reaches into city planning, legislation, and budgets (fig. 7).

Conclusion: Speculations on the Longevity of User-Initiated Places

While garden advocates, researchers, and laypeople may credit community gardens as resources for recreation, open space, dietary supplements, therapeutic relief, and more, there remains an underlying ambivalence regarding their ongoing public status. Although community gardens have been inserted into cities since the 1890s, permanent provision of land for this activity has not been universally validated in the same way as other once-innovative ventures, such as recreational parks and playgrounds. A broader issue here is not whether community gardens should become as standard to urban development as the public park or playground, but whether and how such user-determined insertions can be sustained as part of the public domain. When gardeners claim vacant land for gardening, when dog owners claim an empty part of a park for unleashed dog play, or when dirt-bikers or skateboarders create an obstacle course in an unused plaza or vacant lot, they all reclaim underutilized public space while also potentially withdrawing a piece of the neighborhood landscape from non-participants who may desire other uses for the land. Especially when proposed on public land, such specialized group activities may not be considered the "highest and best use" of a site. Under what conditions do special interest activities gain legitimization as a public function? General public good will and appeal to vague benefits is obviously not enough. More concretely, such ventures require simultaneous attention to land tenure and participation.

This suggests a need to retain some degree of flexibility to accommodate new and evolving types of recreation and social space, often the result of changing demographics in urban neighborhoods. The acknowledgment of social practices shaping public space is well established through concepts of everyday life as developed by Henri Lefebvre and others. Such user-initiated and maintained spaces have received increasing attention in the planning and design fields through the "everyday urbanism" approach to urban spaces that are shaped by use and social practice (Chase, Crawford, Kaliski). However, such places, while interesting to document and acknowledge, in practice tend to be tenuously held through appropriation and neglect. The same concerns that face community gardens—land tenure and ongoing participation—apply to most user-initiated public spaces. The meaningfulness of such spaces at the moment of analysis may not attend to the fact that such sites are community investments and resources that serve important social and personal functions. The shift in thinking about these spaces as semi-public to public suggests new opportunities for public investment in user-initiated spaces.

Bibliography

American Community Gardening Association. *National Community Gardening Survey*. Philadelphia: American Community Gardening Association, 1998.

Balmori, Diane and Margaret Norton. *Transitory Gardens, Uprooted Lives*. New Haven: Yale University Press, 1993.

Bonham, J. Blaine, Gerry Silka, and Daryl Rastorfer. *Old Cities/Green Cities: Communities Transform Unmanaged Land*. Chicago: American Planning Association, 2002.

Chase, John, Margaret Crawford, John Kaliski, eds. *Everyday Urbanism*. New York: Monacelli Press, 1999.

Cronon, William, ed. *Uncommon Ground: Rethinking the Human Place in Nature*. New York: W.W. Norton & Co., 1996.

Crouch, David and Colin Ward. *The Allotment: Its Landscape and Culture*. London: Faber and Faber, 1998.

Di Chiro, Giovani. "Nature as Community: The Convergence of Environment and Social Justice." In *Uncommon Ground: Toward Reinventing Nature*, edited by William Cronon, 298-320. New York: W.W. Norton & Co., 1995.

Francis, Mark and Randolph T. Hester. *The Meaning of Gardens*. Cambridge, MA: The MIT Press, 1990.

Hynes, H. Patricia. *A Patch of Eden: America's Inner-City Gardeners*. White River Junction, VT: Chelsea Green Publishing Company, 1996.

Kaplan, Rachel. "Some Psychological Benefits of Gardening." *Environment and Behavior* 5 (1973): 145-162.

Kirschbaum, Pamela. "Making policy in a crowded world: steps beyond the physical garden." *Community Greening Review* 10 (2000): 2-11.

Lawson, Laura. *City Bountiful: A Century of Community Gardening in America*. Berkeley: University of California Press, 2005.

—. "The South Central Farm: Dilemmas in Practicing the Public." *Cultural Geographies* 14, no. 4 (2007): 611-616.

Lefebvre, Henri. *Critiques of Everyday Life*. London: Verso, 1991.

Malakoff, David. "What Good Is Community Greening?" *Community Greening Review* 5 (1995): 4-11.

Mikalbrown, Kerstin. "Saving Esperanza Garden: The Struggle over Community Gardens in New York City." In *From Act Up to the WTO: Urban Protest and Community Building in an Age of Globalization,* edited by Benjamin Shepard and Ronald Hayduk, 229-233. London: Verso, 2002.

Shepard, Benjamin and Ronald Hayduk. *From ACT UP to the WTO: Urban Protest and Community Building in the Era of Globalization*. London: Verso, 2002.

Stapleton, Richard M. "Bringing Peace to the Garden of Tranquility." *Land and People* 11, no. 2 (Fall 1999): 2-7.

Fig. 1. Gardener at South Central Farm, Los Angeles. 2001. Photograph by author.

Fig. 2. A garden plot at South Central Farm that includes a small altar alongside nopales (pear cactus) and other crops. Photograph. 2001. Lewis Watts.

Fig. 3. A family collecting their harvest from their plot at Ocean View Farms, Los Angeles. Photograph. 2001. Lewis Watts.

Fig. 4. Man recalling the victory garden his family grew during World War II while resting in his plot at Ocean View Farms, Los Angeles. Photograph. 2001. Lewis Watts.

Fig. 5. The Interbay P-Patch in Seattle, which currently includes 132 community garden plots as well as other gardening resources and social amenities, is located on an old land fill and has had to move three times since its founding in 1973. 2005. Photograph by author.

Fig. 6. Seattle's Thistle P-Patch, located under power lines and adjacent to a light rail system, serves recent immigrants from Vietnam, Korea, and various Laotian tribal areas. 2005. Photograph by author.

Fig. 7. Liberty Community Garden in Battery Park City, New York, shown here as it appeared before the September 11, 2001 attack on the nearby World Trade Center, was nearly destroyed by the dust, debris, and aftermath of demolition and cleanup. At the garden's rededication ceremony, community gardeners from Seattle presented a cubic yard of compost made from flowers collected from a flower vigil in Seattle. 1999. Photograph by author.

Buy, Sell, Roam:
The Airport Calculus of Retail

Kay F. Edge

This essay examines and critiques design that is based purely on a calculus of demographic market studies and predictions of consumer behavior. The airport, as a public space and as a building type that incorporates generous amounts of retail space, provides a vehicle for such an examination. The airport aspires to civic completeness but succeeds perhaps only in being a fragmented facsimile of the city, filled with transient citizens and connected back to the urban center by thin strands of rail and freeway. These imposter cities contain a captive population of demographically correct citizen-consumers. At best they sit uncomfortably and indeterminately between public and private. With such a complete interpenetration and blurring of boundaries and with the "merging of public places with private markets," the role of the architect becomes problematic, yet crucial, negotiating between economic and aesthetic values.

Introduction

Eero Saarinen's desire to evoke an upward soaring quality in his TWA terminal at Kennedy Airport seems innocent now as design and cultural symbolism are threatened with irrelevance against a barrage of demographic information and market studies that pinpoint everything from the average yearly income of all U.S. airline passengers to expected income per square foot of airport retail space.[1] New types of

---

[1] The average annual income of the 23 million passengers traveling through Philadelphia International Airport is $75,000 (Becker, 26). The average yearly income for all U.S. airline passengers is $50,000 to $68,000 while the income of an average shopping mall customer is $29,000 annually and the average expenditure per departing passenger at the Pittsburgh Airmall is $9.50 (Al Chalabi, 103-104). Before the Airmall, passengers spent an average of $1.70 (Dunham-Potter). The average Japanese passenger leaving San Francisco International to go home spends an average of $200 (Milanese). At Heathrow's 600,000 square feet of retail space (Sudjic, 155), the average expenditure per passenger is $25 (Bruegmann, 205). Heathrow employs 55,000 workers who are also potential retail customers (Bosma, 64). A typical urban retail space earns about $600 per square foot; airport retail earns $1200 per square foot (Wolf). The average passenger at Chicago's O'Hare Airport is there for eighty minutes of buying time (Becker, 26). Travelers at the Detroit Airport spend an average of ninety

buildings and public spaces are produced and shaped by consumer culture as it intersects with design and demographics. The activities of shopping, leisure, and consumption have served to create a convergence of several types of buildings into few. Pure retail space (i.e., the shopping mall) begins to disappear as buying and entertainment space is incorporated into other building types, and goods, services, and experiences are made available to potential consumers in many different settings. There is a parallel in what has happened to farming and to the deregulated airline and banking industries—that is, the many have become the few and the small have joined to become the conglomerate. The behavioristic diagramming and computer animation of predicted consumer activity, marketing, and demographic studies of consumers, and retail design "rules of thumb" employed by the new hybrid marketing/design firms, have given us a sameness that sometimes makes it difficult to determine whether we are in a shopping mall in suburban New Jersey or an airport in Saudi Arabia. As public places are increasingly equated with and delineated by retail space, they become "privatized, corporatized," and made less permeable (Smithsimon). Spatial quality and choice are diminished and architectural form becomes unintelligible as more revenue-producing functions are compressed into fewer building types. What we once thought of as separate and distinct building types are now just different members of the same family in terms of form and now, more recently, in terms of function, that is, the function of retail, spectacle, and entertainment. Furthermore, what is considered public space is often delineated by private space and geared to private gain. This essay examines and critiques design that is based purely on market considerations and the analysis of consumer behavior. The airport as a building type provides a vehicle for such an examination.

Building Typology

Building types are not timeless but are located in history. They are of their time. The explosion of building types in the nineteenth century can be traced to the Industrial Revolution when new methods of

---

minutes of "dwell time" there and 70% of them pass through the 125,000 square feet of retail space (Morgan).

producing goods, and a wealthier population with more leisure time generated a need for new building types: arcades, department stores, exhibition halls, train stations, and winter gardens. New materials and new methods of construction with prefabricated, standardized elements facilitated the construction of these new types. While the nineteenth century may have been the time when the number of building types expanded explosively, it seems also to have been the time when the groundwork was placed for the reduction of the number of building types. The standardization begun in the nineteenth century continues in the twentieth and twenty-first centuries. Though computer technology is now allowing us to think in terms of mass customization, architecture is slow to change its direction. Because of real estate values there is an enormous premium on uniformity; the more homogeneous the architectural object, the more globalized and saleable it is. If the Eurodollar has its architectural equivalent, it is in these multi-use retail, entertainment, and travel complexes. Buildings are the neutral interchangeable units that facilitate global exchange.[2] With the present impetus to accommodate retail space at every opportunity, building typology, once based on Platonic ideals in the eighteenth century and used as a compositional device in the nineteenth century, now in the twentieth and twenty-first centuries, is based on computer models of consumer movement, demographic market studies, and projections of spending behavior.

The issue is brought into focus through an investigation of the airport as a building type that is particularly representative of our time. The new building types of the twentieth century, the parking garage, the shopping mall, and the entertainment complex have converged at the airport (Brawne, 341). Contemporary airport design represents historical continuity; the dream houses of the nineteenth century—the arcade, the train station, the exhibition hall, the museum, the department store, and the diorama—are now transformed and incorporated into a twenty-first century building type—the airport—to make the dream houses of the present.

As a place that contains, controls, and directs throngs of potential consumers, the airport offers the perfect scene for a theater of consumption. As Baudrillard has pointed out, consumption can never

[2] Richard Sennett (seminar, Yale University, 8 October 2000).

be a solitary activity. The airport as a building type presents architecture as a "medium" for shopping. The airport forms new retail geography away from the shopping mall and it houses waiting time that can be transformed into buying time—waiting for boarding, waiting for arrival, and waiting to be picked up. Within the interstices of this time, the traveler, the airport employee, and those greeting arriving passengers are offered food, drink, necessities, trinkets, and spectacle. Though scale and technology have changed, the current retail and entertainment landscape at the airport has a history that started with the airport building type in the early twentieth century.

A Brief History of Airport Retail

> The majority of airport visitors in all parts of the United States are evidently supposed to be satisfied with eating hot dogs, drinking soda pop and watching airplanes. This cannot last. The airport visiting public is due to rebel unless interesting attractions are scheduled for their entertainment while at the field.
>
> *Airports*, September, 1930[3]

By 1930, with commercial aviation barely a decade old, the editors at *Architectural Forum* were writing about the air terminal as a new building type, declaring that "it must be developed, independent of precedent, upon the bases of economic and social demands" (Sherman, 702). The first airports were businesses started and maintained by private companies. At the end of World War I, aviation changed almost seamlessly from a wartime operation to a commercial one. The aircraft manufacturing industries that stood to go out of business after the war branched out to provide commercial air service as a matter of survival.

The few airports that were making a profit in the 1920s and 1930s were doing so by way of "ample public attractions" for the "sight-seeing public" (Franzheim, 26). Airport administrators recognized from the beginning that it would be impossible to make a profit from flying and aviation activities alone. Designers and administrators sensed the opportunity to include in airport design, revenue-producing space not related to the business of aviation. While the shopping extravaganzas and multi-purpose entertainment venues of

[3] Brinkley, "Attracting the Public to the Airport and How to Do It" 35.

today's airports may seem to be an unprecedented phenomenon, airport retail and amusement have an extensive history. Research shows that it is impossible to look back to a time when the airport was a pure type, serving only the function of efficient aviation. Several examples illustrate that there was a convergence of types from the beginning. In 1930 it was possible to go to the Philadelphia Airport to swim in the new swimming pool or to play miniature golf—both concessions managed by airport administrators for the profit of the airport. The swimming pool was owned and operated by the airport but most of the concession and entertainment space was rented to the various concessionaires on an annual basis. Rent was payable in advance and the airport collected a percentage of sales income. The operating companies themselves were required to finance, install, and maintain the facilities. Of the revenue sources at the Philadelphia airport in 1930, concessions accounted for 36%, the swimming pool brought in 10%, and aviation income amounted to 6% (Green, "Concessions Are the Principle Source of Income," 43). The airport, the manager said, is a "logical amusement center" (Green, "Concessions Are," 53). In the airport master plan, the entire west side of the site was reserved for recreation and amusement. By the following year, June 1931, there were driving ranges, tennis courts, and shuffleboard alleys, and a full-sized eighteen-hole golf course was under construction.

Lockheed's Burbank Airport provides another example of the inclusion of non-aviation revenue space in early airport design. In 1945, it was possible to go to the Burbank Airport for a massage and a Turkish steam bath. A breakdown of their revenues for the year 1944 shows the major sources of income to be non-aviation related concessions and amenities.

Sources of Revenue (1944)	%
Skyroom restaurant and coffee shop	39.0
Airport ground rentals, landing fees	14.5
Cocktail lounges	12.6
Two garages (150 cars)	11.2
Parking lots (five in all)	4.7

Other buildings (offices)	4.6
Hangar rentals	4.3
Cigar and newsstand	4.0
Automotive service station	2.0
Electric power sales	1.9
Public telephones	0.3
Gift shop (sale of novelties)	0.2
Western Union telegraph	0.2
Barber shop	0.2
"Good Humor"	
(exclusive rights to sell frozen confections)	0.1
Baggage storage lockers	0.1
Steam bath and massage	0.1

The manager of the Lockheed Air Terminal had words of advice for other airport managers seeking to become profitable:

1. Merchandise the terminal—make it the community center
2. Develop all possible sources of revenue from the public
3. Make your services outstanding among those of the community (Taylor, 104).

For several decades of the early development of the airport as a new building type, a debate was carried on as to what the airport should be and what character it should have beyond its function as a transportation terminal. *Airports* magazine for the years 1928 to 1931 covered the debate and illustrated the preoccupation of airport administrators with making money in any possible way. The debate can be summed up by two opposing statements made at the National Airport Conference in 1930 and quoted in the January, 1931 issue of *Airports*. One airport manager at the conference expressed his disapproval of the inclusion of various money-making spaces at the airport: "I can't see why municipalities should have more or less a fair at its airport. I don't think it is going to be appreciated by transport companies and certainly not by the higher class of people who expect to use airports." Another airport manager asserted that the airport should make money in any way it could: "I think you are overlooking a good bet, gentlemen. We have four automobile gas stations, two restaurants, one lunch stand, a swimming pool and two miniature golf courses right now operating" (Green, "The Central Airport," 9). While some worried that a "carnival effect" at the airport would "cheapen" it, others advocated any means of making a profit (Brinkley, "Attracting the Public," 36). Most airport managers agreed that a restaurant and possibly even a hotel might be reasonable enterprises for an airport to engage in, but concessions and entertainment such as bowling, golf, tennis, and other amusements were a matter of disagreement. Airport managers who operated their airports as a kind of public utility or park rather than as a business were criticized by those who felt the airport should be run as a business. Airport designers were also criticized for not including enough revenue-producing space in their beautiful yet expensive airport designs (Green, "Concessions Are," 43). In a kind of early hint at the importance of demographics, one airport manager said in 1931, "Nearly every person who visits an airport has money to spend. It is up to the airport to get its share of that money." The manager went on to compare the airport visitor to any other visitor to fairs, amusement parks, and carnivals (Markey, 16). Attracting a crowd to the airport was the means to a profit, and a fascinated public in the early days of aviation was easily lured. Airport owners were aware, however, that the novelty of aviation would wear off and that the public would have to be lured to the airport by other means. A 1943 article published in *Pencil Points*, "Aviation as a Stimulus to Architecture," predicted that passenger demands would "grow more varied and complex" with the growth of the aviation business (43). A public that was blasé about flying could not be a dependable source of income unless other entertainments were provided and airport administrators came up with a variety of ways to attract a crowd to the airport. The suggestions made in the pages of *Airports* magazine were as varied as beauty contests, art exhibits, food shows, a Mardis Gras Ball, and dancing and costume competitions (Brinkley, "Draw Crowds to the Port With a Beauty Contest," 24).

The debate about the character of the airport continued into the 1940s and 1950s and the architecture journals of the day reflected this continued disagreement. An *Architectural Record* article published in 1945 asked the question "Are Airports for Planes or for People?" Another article in the same issue, "Up-to-Date Airports Involve

Passenger Amenities," answered the question (Goodman, 94). In a 1953 issue of *Progressive Architecture* devoted entirely to the airport as a new building type of the twentieth century, one article observed most presciently, "the airport terminal develops into a shopping center instead of a traffic station" (Arroyo et al., 71). The writers of the article urged airport administrators to "decide whether they are in the supermarket business or whether they want to provide terminals as a public service" (75). The new Pittsburgh Airport left no doubt on which side of the argument it fell. Pittsburgh became the first airport to incorporate a large amount of retail space. It offered such amenities as a dining terrace, a post office, a bank, observation lounges, a recreation center and movie theater, a 500-seat nightclub, a sixty-two-room hotel, a clothing store, a toy shop, and a florist. The November, 1952 issue of *Architectural Forum* proclaimed the Pittsburgh Airport unmatched in "profit-yielding facilities for passenger comfort." The article also foreshadowed the theme of airport as city by calling the terminal a "self-contained community" ("Pittsburgh," 138).

Despite the parallels between today's airport retail space and typical shopping mall space, there are some fundamental differences. Suburban malls represent a shopping destination while airport retail space is used by those passing through. Mall customers are free to come and go as they want; the passenger at the airport must contend with security, with flight schedules, and with navigating huge distances. Waiting is not an issue at the shopping mall while waiting is inevitable at the airport. The analysis of waiting time then becomes important in predicting the shopping behavior of an airport passenger.

While shopping malls, no matter what neighborhood they are in, represent a wider range of demographics and are less predictable in terms of spending behavior, there are several very specific and predictable target groups that airports can accommodate:

1. The largest and most important group in terms of profit is passengers who will buy necessary travel items, but who can also be persuaded through opportunity and display to make impulse purchases. The corresponding spaces for this market are duty-free stores selling liquor, tobacco, and luxury items as well as stores selling basics such as travel necessities, books, and newspapers. Other specialty or luxury stores can prompt

impulse buying. They also need a variety of restaurant types from hot dog stands to sit-down restaurants. Amusement facilities such as cinemas, discotheques, fitness centers, and casinos can cater to waiting passengers as well as non-passengers.

2. Airport employees, depending on the airport, can be a very large group of potential consumers. Heathrow employs 55,000 and the Frankfurt airport has a workforce of 40,000. They require the kinds of stores that offer the convenience of shopping for necessities after work. The Frankfurt airport in particular has provided for this group, offering stores such as supermarkets and hardware stores. It has paid off. One study showed that airport employees there spend about 15% of their net income at the airport.
3. Airline crews can also make up a large group of potential consumers; they need stores and services such as dry cleaners and beauty salons.
4. Meeters and greeters, though a very different group from passengers, can, like passengers, be persuaded to buy through opportunity and display.
5. Local residents provide a huge potential market, particularly if the airport is easy to get to and parking is readily available. Airport shopping facilities such as those offered in Pittsburgh are becoming destination shopping centers for local residents. Stores catering to these consumers could be any type of store typically found in a shopping mall.
6. Members of the business community: more than stores, these airport users need meeting facilities and services such as hotels, catering, fax machines, and computer hook-ups (Doganis, 114).

It is also possible for the airport retailers to very finely tune and quickly change their offerings to particular markets: at the Geneva, Switzerland airport, the drugstore sells crutches for skiers coming from the Alps. Jewelry stores at the Sharjah International Airport in the United Arab Emirates change their displays between twenty-two-carat gold and eighteen-carat gold depending on which charter group is coming through. For flights to and from Taiwan the jewelry counter is stocked with twenty-two-carat gold and twelve salespersons are available. For German charter flights the jewelry counter is staffed with only six salespersons and the jewelry on sale is eighteen-carat gold (Doganis, 145). Shopping malls have no such ability to fine tune their products and it is not possible for them to study their potential customers to such a degree.

Just as the large department store acts as an "anchor" in the shopping mall, the equivalent anchor for airport retail is the necessity of waiting. The suburban shopping mall as a building type becomes less used as retail space migrates to the airport and other multi-use buildings and entertainment venues. A recent design competition asked participants to "counter the trend towards the dereliction, abandonment, and 'death' of the regional mall and invit[ed] approaches to rethinking its urbanistic and architectural milieu (Techentin). Airports such as Pittsburgh that offer generous retail space have become shopping destinations for those who live in the vicinity just as the suburban shopping mall is/was a destination.[4]

Design, Demographics, and the Current Landscape of Airport Retail

Deregulation, the resultant lowering of landing fees, and the ensuing trend to privatization, then, have served to dramatically increase the amount of non-aviation, revenue-producing space included in airport design. Parking, retail, and entertainment income have been particularly important. The airport is still the "logical amusement center" described by the Philadelphia Airport manager in 1930.

James Ogilvy noted in *American Demographics* that our economy now is based more on the search by consumers for "vivid experiences" rather than the desire to acquire goods. Today's airports place themselves in a position to offer both goods and vivid experiences. Singapore's Changi Airport offers a karaoke lounge, a sauna, a swimming pool, and a putting green. Heathrow has four caviar shops. Frankfurt has the largest discotheque in Germany, twenty-six

---

[4] Staib and Hunt. They note: "The airport is a destination rather than transit point. The role played by services in airport shopping will dramatically increase. The relatively new airports in Asia and the Middle East demonstrate this trend. Large scale airports are becoming destinations themselves, the leisure hub Dubai is inseparably connected to the airport feeding customers into this tourism hot spot."

restaurants, a bowling alley, three cinemas (including an erotic cinema), a wedding chapel, a furrier, an antique shop, and a supermarket. Amsterdam's Schiphol offers a casino and tanning booths and recently applied for a prostitution permit. At Hong Kong's airport, Cathay Pacific offers first-class passengers a luxurious spa and bathtubs for two. Chicago has a full service medical clinic. The San Francisco airport has a library. Recent museum exhibitions at Philadelphia International have included: Totemic Sculpture, Political Memorabilia, Dance Photography, and Vitra Experimental Chair Designs.

The role of the airport as a business has dramatically affected the master plan, the building diagram, and the interior space planning diagram. The traditional airport diagram facilitated what was seen as the primary function of the airport—accommodating the needs of passengers, airlines, and freight handlers, and facilitating their efficient movement through the terminal. Impediments to this were kept to a minimum. The more current commercial airport diagram tries to accommodate efficient flow but at the same time provide places and opportunities for passengers to pause and buy. It attempts to process passengers as efficiently as possible, yet channel them profitably through the buying space. The design and placement of retail and concession spaces are crucial. New airports have the opportunity to design their retail spaces so that a majority of passengers are required to pass through them; older airports being renovated have to find the best possible way to insure maximum exposure to the new retail space they add.

The various airport plan diagrams and the forces behind their generation call to mind architectural theorist J. N. L. Durand's nineteenth-century declaration that "architecture is economy joined to convenience." These diagrams bear this out as they rationalize for modern use and a modern building type—Durand's philosophy of an architecture generated by economy and convenience. Only now, the generator of the airport diagram is retail space joined to crowd flow studies. The various terminal diagrams were developed around parking, walking distances, passenger flow, and, most recently, exposure to retail areas. Henri Lefebvre noted that increasingly, space is expected to pay for itself. The airport diagram serves to support that claim as it attempts to maximize consumer spending and to make the architectural space pay for itself. An analysis of airport retail and design serves also to illustrate Lefebvre's notion of abstract space, hyper-rationalized and diagrammed to arrange space and its contents in the most profitable way possible, always with an eye toward how space rules the available time of those who might buy.

Airport-specific demographics, market studies, and spending behavior studies have emerged in the attempt to make airport time and space as profitable as possible. In the same way that a biochemist might describe the human body as a collection of chemicals, the throngs of humanity passing through the airport are conceived by these studies in purely behavioristic terms—as a flow to be channeled, controlled, manipulated, and persuaded to spend. Consumer behavior is analyzed, diagrammed, and turned into statistical charts and tables, and these facts and figures are used to make predictions and design recommendations. The passengers amount to a tide of demographic facts spilling through the airport spaces. A 1943 *Pencil Points* article described the passenger as "a mobile unit, [which] must be controlled and guided for safety and operating efficiency, in his own interest" ("Aviation as a Stimulus," 43). In the current airport landscape, the "mobile unit" must not only be controlled and guided, but he must be induced to pause and spend his money; airport space and his "dwell time" must be manipulated and calculated to produce as much profit as possible.

The needs and buying behavior of various demographic groups that use airports have been studied in detail through surveys and market analyses. One study conducted by an airport official at the Brisbane airport in Australia considered the buying behavior of airport shoppers. The awkwardly titled paper, "The Effects of Emotion and Time to Shop on Shopping Behaviour in an International Airport," was presented at a 1999 Consumer Research Conference in the United States. The study draws the seemingly self-evident conclusion that the emotions of the shopper and the available time the shopper has are two factors which influence spending behavior at the airport. The study showed a correlation between shopping and available time: every minute of extra time spent at the airport increased the likelihood of shopping by a factor of 1.0114 (Bowes).

The purpose of such studies is to find a way of rationalizing, quantifying, and analyzing passenger behavior, and based on this analysis, to predict the factors involved in the propensity to consume. This analysis then is meant to generate design. There is a market niche now for firms specializing in airport retail design. These firms, which join design and market study, focus on passenger flow and shopping behavior, and have emerged to answer the call for more revenue-producing space at the airport. They use the tools of three dimensional modeling and animation to show passenger movement and anticipated exposure to retail areas, and they present their findings as factors that should heavily influence terminal design. They take into account passenger flow and shopping behavior, and their product consists of 3-D computer models and animations showing passenger movement and anticipated exposure to retail areas. The ancient philosophical issues of space and time are contemplated now in terms of how much income they yield up, and human movement through time and space is considered in terms of spending behavior.

One firm that combines passenger flow studies and shopping behavior studies with design recommendations for clients is Space Syntax, a research facility at UCL (London). Space syntax is described as "a set of techniques for the analysis of spatial configurations of all kinds, especially where spatial configuration seems to be a significant aspect of human affairs, as it is in buildings and cities." The firm has served as a consultant on a variety of mixed use urban projects and sports facilities, and recently applied their methods in the study of passenger/shopper flow and behavior in airports. They published an article titled "Passengers, Pedestrians and Shoppers" in the journal *Passenger Terminal World* which described their techniques. Their method is based on the idea that:

> [M]ovement and communication are essential to the social and economic success of public and private space and that it is the design of space, above all, which determines the movement and interaction of people in the built environment. (Major, Stonor, Penn)

Their method seems to combine sophisticated computer analysis and graphics, the old psychological theory of behaviorism, and the self-evident idea that space determines movement. They have developed their own software for this analysis. Another of their articles specifically addressed the area of airport retail design. Titled "Moving,

Browsing, Buying: Forecasting Passenger Behaviour," it was presented at the third Passenger Terminal World Conference in March, 2000. The difference between themselves and other passenger flow analysis models, Space Syntax says, is that while other models analyze "programmed activities" such as check-in, security check, and boarding, the researchers at Space Syntax also look at "informal" passenger activities in reaching their recommendations on airport retail design. They claim a 75% success rate in predicting passenger movement—such unprogrammed activities as "shopping, eating and waiting," asserting that the consideration of this kind of behavior is essential to the economic success of "mixed-use" facilities (Major, Stonor, Penn). Space Syntax begins by producing a graphic representation of the space in question and identifying the most prominent characteristics of it. This representation is then correlated with a study of passenger movement through the space and first-hand observations of how it is used. Using computer modeling, Space Syntax is able to examine a variety of possible layouts of retail space and predict how passengers will move through these layouts. They are able to calculate which routes are most accessible and offer the most exposure to retail offerings.

Another study on shopping behavior undertaken in 1994, "Consumers, Consumption and the Inanimate Environment: the Airport Setting," recognized the airport as a "sealed microcosm of consumption" and chose it as the setting for a study of the influence of design on shopping behavior. The study attempts to draw on the disciplines of marketing, psychology, environmental psychology, architecture, and anthropology in asserting the significance of "consumption surroundings" in convincing people to buy. The "mood inducing properties" of airports, the study says, "stem from their architectural form, which connotes life-style shopping and a stimulating atmosphere" ("Future Directions for Retail Design Management"). Factors which affect passengers' moods and the amount of time they have to shop are factors that will affect buying behavior. Such things as clarity of plan and "wayfinding" strategies, which may give the passenger more time to be a shopper, are shown to be important for a successful retail operation. The study does not recommend specific design strategies but points out that "design characteristics, i.e., the clarity of the layout or signage (orienting factors), can clearly assist with the accomplishment of set tasks (such as shopping)" and goes on to point out that "lighting, color, and texture" are the most memorable features of "consumption settings." With its behaviorist approach, the study identifies "approach-avoidance" qualities and attempts to quantify such things as "pleasure, arousal and dominance," demonstrating the effect of design factors (e.g., signage, layout, colors) on consumption. Not surprisingly, the study offers the observation that "positive outcomes may be induced by interaction with the physical setting."

These hybrid firms offer airport financial planning, passenger movement studies, including 3-D computer animations of passenger flows and exposure to retail areas, retail and concession planning, design and management, the development of "themes and design concepts," market studies in the form of passenger surveys, and surveys of other target "focus groups," and analysis of this market data. All seek to provide an "optimal consumer experience" and many, like Space Syntax, have developed their own software.

The foundation of contemporary airport design is thus a "calculus of objects" that includes space, commodity, and buyer and that carries Emile Zola's department store (in *Au Bonheur des Dames*), a masterpiece of retail design and crowd control, to new heights of architectural science and consumer pleasure. For example, bodiless negligees hang in the Victoria's Secret at Pittsburgh International, perhaps luring hurried passengers to pause and buy. At JFK International, a ghostly woman on a turning sign entreats passengers to "Buy, Sell, Roam." At any given airport, the front-porch domesticity of rocking chairs inserts the uncanny into odd niches. The Pittsburgh airport is a part of US Airways hub and spoke route system which airlines began employing as a result of deregulation. At the hub and spoke airport, passengers are gathered from many originating points and dispersed to their final destinations. Most passengers at a hub airport are connecting passengers with time to spend. The plan of the Pittsburgh airport itself works in much the same way as the route system, gathering passengers at its shopping mall center and dispersing them to their departure gates. The main terminal building is connected by means of an underground people mover to an X-shaped satellite departure terminal with pier concourses making up the "arms" of the X. From the main terminal building, passengers originating in Pittsburgh are taken, by way of the people mover, directly and inescapably into the Airmall, the 100,000 square foot retail center of the terminal. Any of the more than twenty million annual connecting passengers who must change concourses to change planes are also channeled through this buying space. In this shopping center the passenger can buy printer paper, a Big Mac, or a negligee. Body and mind can be soothed between flights—the body for a price at the gym or cosmetics counter, the mind at the chapel or reflection room, still free of charge. A chiropractor is also on the list of amenities at Pittsburgh International.

Philadelphia International Airport is another hub of the US Airways hub and spoke route system. The terminal is a long rectangular building with four "finger" concourses extending from it. Narrow connector corridors take passengers from parking and drop-off areas to ticket counters and security. Retail stores and concession areas stand as a second point of passage; beyond security, departing passengers must go through the buying space in order to get to their gates. Museum space mingles with retail space there; Plexiglas cases house sculptures and interchangeable displays. White rocking chairs and faux stone planters are placed in niches facing out onto the tarmac and along the mall-like spaces that lead to the individual concourses. Except for the airport signs, it is indistinguishable from a shopping mall corridor. The Philadelphia Marketplace is located between the B and C "finger" concourses. The original nineteenth-century shopping arcade, its commodities still there, now has a moving sidewalk; the *flaneurs* have less time to linger. These mall spaces and moving sidewalks serve to tie the "fingers" together so that anyone changing concourses to change planes is routed through the shopping mall. The sidewalks roll past commodity displays that provide just enough inaccessibility to create desire. In the glass store windows the consumer's gaze can take in simultaneously his own image and the *accouterments* of lifestyle creation. Neon light sculptures above replace the iron and glass of Benjamin's Paris Arcades. It is still a place of passage but now from concourse to concourse and airport to airport instead of from street to street. Those using the moving sidewalk as they make connections can look out on one side to the tarmac and the waiting and taxiing planes below; on the other side are the commodity windows.

The airport has the seemingly contradictory task of promoting efficient movement of ever-increasing numbers of passengers while seducing the hurried crowds to pause and buy. The moving sidewalk with its adjoining commodity windows may represent a massive compromise between the efficient flow necessary for a busy international airport and the pause necessary to choose and, ultimately, to buy. In the constricted space of the moving sidewalk, retailers "exploit the power of place to facilitate consumption"—the departing planes visible on one side, the articles of travel and luxury on the other (Lowe and Wrigley, 21). The untouchable objects, like the "look but don't touch" protocol of a world's fair or museum, are "valuable [in] that [they] resist our desire to possess them," occupying the space between pure desire and immediate enjoyment (Simmel, *The Philosophy of Money*, 67). The windows create a linear diorama of stuff beckoning the passenger, once off the moving sidewalk, to circle back around to the familiarity of the shopping mall and buy. Dioramas, once a substitute for travel, are now a prop and an invitation to buy in a space facilitating travel. These modern-day dioramas constitute the perfect ingredients for a dream world: movement, changing scenes, desire, visions of travel and commodities abundantly displayed. The space of flow joins with the stationary store windows full of what Walter Benjamin called "wish symbols." Neon light sculptures replace the iron and glass above. The *flaneurs* have changed; the objects have changed; but the calculus of objects remains.

Conclusion

In the design of these spaces the sheer weight of information begins to collapse traditional building typology, and design is replaced with information-gathering. Flow diagrams, parking distances, and exposure to retail areas begin and end a discussion of airport design. Simmel argued in 1903 that modern urban existence forced us into a calculative role. The more calculating tools at our disposal, the more fine-tuned and complex our calculations have become. In the design of urban public spaces, architectural qualities are reduced to statistical quantities, bearing out Simmel's complaint that we constantly reduce the qualitative to the quantitative. He says,

> The calculating exactness of practical life which has resulted from a money economy corresponds to the ideal of natural science,

> namely that of transforming the world into an arithmetical problem and of fixing every one of its parts in a mathematical formula. It has been a money economy which has thus filled the daily life of so many people with weighing, calculating, enumerating and the reduction of qualitative values to quantitative terms. ("The Metropolis and Mental Life," 71-72)

The latest generation of airports represents the first generation of market-researched architecture. The qualities of these buildings have been quantified as variables in an economics equation that measures in a behaviorist way the level of consumer pleasure or displeasure. Simmel's claim that qualitative features are quantified is illustrated in the intersection of market study and design in retail spaces at airports. The quantification of qualities and the homogenization of spaces are achieved by means of flow diagrams, retail exposure calculations, and retail design "rules of thumb." The same design techniques meant to increase consumption, used in a variety of building types, contributes to a universal sameness whether it is retail space at an airport in Pittsburgh or a shopping mall in Saudi Arabia.

The airport houses a multitude of spatial types and functions and it serves as a kind of filtered public space. The airport as a building type illustrates the intersection of design, demographics, and market study. Like a city the airport is likely to have its own security force and fire station, hotels, restaurants, shopping centers, heating and cooling facilities, banks, museums, and, in some instances, meditation chapels and sex shops (Bruegmann, 206). The airport aspires to civic completeness but succeeds perhaps only in being a fragmented facsimile of the city, filled with transient citizens and connected back to the urban center by thin strands of rail and freeway. These imposter cities contain a captive population of demographically correct citizen-consumers.

These settings make private space within what is considered by some to be public space. Through the control of crowd flow and the careful arrangement of retail offerings, the urban crowd is transformed into a collection of individual consumers. Through careful and meticulous use of demographics and market studies, airport managers use their "public space" to sell to a pre-selected segment of the population. To the consternation of many critics these spaces are considered by some to be the replacement for what was once civic public space. At best they sit uncomfortably and indeterminately between public and private. With such a complete interpenetration and blurring of boundaries between public and private, and with the "merging of public places with private markets," these categories as a description of space may no longer be useful (Zukin, 54).

The geographer Edward Relph enumerates the characteristics of placelessness in sports stadiums; but the characteristics serve to describe airports as well and they serve as a caveat for designers: (1) the standardized values inherent in internationalized, globally televised, synthetic entertainment, (2) the gigantism reflected in the formlessness and lack of human scale of megastadiums, (3) the uniformity of design in international styles of stadium architecture, and (4) the stadium as part of an entertainment district, resulting from its purpose of attracting outsiders (Bale, 41).

If we accept as unavoidable that economic forces inevitably and profoundly influence design, how is it possible to critique architecture as it is generated by a set of economic variables: parking, market and demographic study, crowd flow, retail exposure, and advertising facility? What is the architectural and urban answer to this kind of thinking—the rationale usually embraced by those who pay commissions? The agendas of those who pay commissions and those who design for commissions will always be different. In some sense, the architect must keep his/her aesthetic intentions "secret" as he/she works to produce something the client sees as cost-effective, efficient, and profitable. Market and demographic studies, efficiency, and cost-effectiveness may trump beauty for the client but they usually do not for the architect. Architecture as a long-standing part of human culture will remain irreducible to demographic facts and figures. The architect then, perhaps acting alone, is a social critic and his/her charge, whether retail space is included or not, is to refill a spatial experience emptied by market research and computer generated flow diagrams, and to resist what Tafuri called "the merciless commercialization of the human environment." The question to be asked is whether architecture can have any part in reversing the trend of consumerism. Can real spatial choice replace an apparent choice of consumer goods? Can consumerism be contained by offering a spatial alternative to material products? With late capital consumerism firmly in place the problem of diminished spatial choice and sameness of experience can only be addressed piecemeal. But architecture is one of the pieces.

Bibliography

Al Chalabi, Margery. "Airport Retailing Takes Wing." *Urban Land* (April 1998): 101-105.

Arroyo, Nicolas R., John Grisdale, Albert F. Heino, Francis R. Meisch, and Walther Prokosch. "Airport Terminal Buildings." *Progressive Architecture* (May 1953): 69-121.

"Aviation as a Stimulus to Architecture: Basic Requirements for Ground Facilities." *Pencil Points* (Nov. 1943): 41-45.

Bale, John. *Sport, Space and the City*. London and New York: Routledge, 1992.

Becker, T.J. "Long Overdue Changes in Store for Airport Retailing." *Urban Land* (Aug. 1996): 24-27.

Bosma, Koos. "European Airports, 1945-1995 Typology, Psychology, and Infrastructure." In *Building for Air Travel: Architecture and Design for Commercial Aviation*, edited by John Zukowsky. Munich and New York: The Art Institute of Chicago and Prestel-Verlag, 1996.

Bowes, Brad. "Research Highlights Unique Airport Retail Environment." http://www.bne.com.au/corp/media_releases/29_09_1999.html.

Brawne, Michael. "Airport Passenger Buildings." *Architectural Review* (Nov. 1962): 341-348.

Brinkley, Russell J. "Attracting the Public to the Airport and How to Do It." *Airports* (Sept. 1930): 35-36.

—."Draw Crowds to the Port With a Beauty Contest." *Airports* (Sept. 1931): 23-24.

Bruegmann, Robert. "Airport City." In *Building for Air Travel: Architecture and Design for Commercial Aviation*, edited by John Zukowsky. Munich and New York: The Art Institute of Chicago and Prestel-Verlag, 1996.

Doganis, Rigas. *The Airport Business*. New York and London: Routledge, 1992.

Dunham-Potter, Anita. "Airport Shopping Takes Off." *Smarter Living Travel*. wysiwyg://257/http://www.smarterlivingcom/columns/real/Airport19991209.1.html.

Franzheim, Kenneth. "The Public Point of View Comfort and Amusement Must Be Provided." *Airports* (April 1930): 26-27.

"Future Directions for Retail Design Management: Some Empirical Findings." http://www.textiles.umist.ac.uk/Research/TexOnLine/Paper99/Anewman1.htm.

Green, W. Sanger. "Concessions Are the Principle Source of Income."*Airports* (July 1930): 43-53.

—. "The Central Airport An Institution for Profit Sets the Pace." *Airports* (Jan. 1931): 24-30.

Goodman, Charles M. "Up-to-Date Airports Involve Passenger Amenities." *Architectural Record* (April 1945): 94-95.

Lowe, Michelle and Neil Wrigley. "Towards the New Retail Geography." In *Retailing, Consumption and Capital: Towards the New Retail Geography*. London: Longman Group Limited, 1996.
Major, M.D., T. Stonor, and A. Penn. "Passengers, Pedestrians, and Shoppers: Space Syntax in Design." *Passenger Terminal World*. Space Syntax Research Group. Barlett School of Graduate Studies. University College London. http://www.bartlett.ucl.ac.uk/spacesyntax/retail/retail.html.
Markey, Fred L. "Amusements." *Airports* (Mar. 1929): 16.
Milanese, Marisa. "SFO Is Set to Soar With Scores of Stores." San Francisco Bay Traveler. http//www.sfbaytraveler.com/sf/stories/sfoshop_20001013.htm.
Morgan, Amy. "Flying the Friendly Skies of Airport Retail." *Specialty Retail*. http://www.specialtyretail.net/issues/october99/flying.htm.
"Pittsburgh: Self-Contained Community." *Architectural Forum* (Nov. 1952): 138-139.
Techentin, Warren. "Dead Malls, Competition Brief." Los Angeles Forum for Architecture and Design. http://www.laforum.org/deadmalls/brief.php.
Sherman, Roger W. "Planning for Airport Buildings." *Architectural Forum* (Dec. 1930): 702-712.
Simmel, Georg. *The Philosophy of Money*. New York: Routledge, 1978.
—. "The Metropolis and Mental Life." In *Rethinking Architecture: A Reader in Cultural Theory*, edited by Neil Leach. New York: Routledge, 1997.
Smithsimon, Gregg. "The Technologies of Public Space and Alternatives to a Privatized New York." Department of Sociology. Columbia University. http://www.sociology.columbia.edu/home/greg/writing/techps.htm.
Staib, Daniel and Erin Hunt. "Airport Shopping 2025." *GDI Study* 24. GDI for Economic and Social Studies. Gottlieb Duttweiler Institute. http://www.gdi.ch/Single_News.1025.0.html?&L=1&tx_ttnews%5Btt_news%5D=25&tx_ttnews%5BbackPid%5D=20&cHash=9919fbc4dd.
Sudjic, Deyan. *The 100 Mile City*. New York: Harcourt Brace & Company, 1992.
Taylor, Don. "Records of Airport Profits Prove the Value of Extra Facilities." *Architectural Record* (April 1945): 102-106.
Wolf, Gary. "Exploring the Unmaterial World." *Wired News*. http://www.wirednews.com/wired/archive/8.06/koolhaas_pr.html.
Zukin, Sharon. *Landscapes of Power From Detroit to Disney World*. Berkeley: University of California Press, 1991.

Fig. 1. Buy Sell Roam: JFK Airport. 2000. Photograph by author.

Fig. 2. Rocking Chairs: Philadelphia International Airport. 2000. Photograph by author.

Fig. 3. Retail/Museum Space: Philadelphia International Airport. 2000. Photograph by author.

Fig. 4. Rocking Chairs and "Mall" Space: Philadelphia International Airport. 2000. Photograph by author.

Fig. 5. Express Walkway and Retail Windows: Philadelphia International Airport. Photograph by author.

Fig. 6. Victoria's Secret: Pittsburgh International Airport. 2000. Photograph by author.

Consuming Third Place: Starbucks and the Illusion of Public Space

Bryant Simon

Borrowing from Ray Oldenburg, Starbucks bills itself as a "third place"—a place between work and home for respite, socializing, and community building. Through participant observation and interviews, this essay looks at how Starbucks works on the ground. It explores how people use it and how the company creates the appearance—without the substance—of a public space.

Like a lot of people, *Boston Globe* columnist Alex Beam thinks Howard Schultz coined the phrase, "third place." But he didn't, retired University of South Florida sociology professor Ray Oldenburg came up with the term to describe places people gather other than at work or home. Still it is easy to see why Beam would make this mistake. Every chance he gets, Schultz uses this expression to depict his company's often busy and bustling stores. He calls them third places and links them not just to Oldenburg but also to the deep coffeehouse traditions of art, conversation, debate, community, and oppositional politics.[1]

Starbucks cafes are, Schultz tells talk-show hosts, journalists, university audiences, and stockholders, "places between work and home where people get a break that takes them far from the routines of their daily lives" (Schultz, 281). "Starbucks," he told a *New York Times* reporter, "has emerged as the third place [. . .] in America. It is the true extension of people's front porch. What people do inside Starbucks is talk, and coffee has been the centerpiece of conversation for hundreds of years."[2] Asked once about how his stores differed from McDonald's and Burger King, the one-time CEO, and now majority stockholder and company director of overseas operations, scoffed and said, "We're not

---

[1] Alex Beam, "He Gets a Buzz From Starbucks," *Boston Globe*, 20 February 2006. See also: Oldenburg, *The Great Good Place* and *Celebrating the Third Place*.

[2] Alex Kuczynski, "A New Magazine by Way of Starbucks," *New York Times*, 8 February 1999.

in the commodity business. We've created a third place."[3] In his autobiography, he boasts, "Almost everywhere we open a store we add value to the community. Our stores become an instant gathering space, a Third Place, that draws people together" (281).

But it is not just Schultz who talks the third place talk. He has drilled this language of civic service into his employees' heads, especially his public relations staff. A PR specialist announced, "Starbucks is that Third Space where people come to relax, be alone or meet friends and family over a great cup of coffee."[4] According to the company manual, a Starbucks employee needs to "[c]ommunicat[e] information to [the] manager so that the team can respond as necessary to create the 'Third Place' environment during each shift."[5] The word traveled within the company. Schultz, New England-based company spokesman, Jennifer Guebert, explained to a reporter, "loved the concept of the Italian coffeehouse." She added,

> It was a great way to provide a sense of community he felt had been lost in American culture. The original concept, from the first store at Pike Place in Seattle to the brand today, is to provide a 'third place' environment, between work and home, where you can sit and enjoy coffee, get a moment of peace and forget about the worries of the day.[6]

When Starbucks opened stores thousands of miles away in Australia, an official there declared, "We want to become the third place in people's lives, outside of home and work."[7]

Starbucks didn't start out creating third places. That happened over time. First it sold bulk coffee, and then it sold espresso. But people lingered. They didn't just pick up some beans or grab a cup of coffee and go—at least not always. So the company put a few stools, and then a few tables in its stores. Really, though, the building of Starbucks'

---

[3] Amy Wu, "Starbucks' World Won't Be Built in a Day," Forbes.com, 27 June 2003, http://www.forbes.com/2003/06/27/cx_aw_0627sbux.html.

[4] "Bagus." Singapore Tourism Board. http://app.stb.com.sg.

[5] Starbucks Coffee Company, *Learning Journey Guide* (2003), 16.

[6] Bill Kirk, "Starbucks vs. Dunkin Donuts: A Study in Contrasts," *Gloucester Daily Times*, 15 June 2007.

[7] Megan Doherty, "US Coffee Giant for Business in Manuka," *Canberra Times*, 22 April 2001.

trademark ambience was an accident of real estate, one coffee executive told me. When Starbucks tried to break into the New York area, it shifted its growth model. In Chicago and Washington, D.C., the company established itself downtown first, and then moved like the spokes on a wheel to the outlying suburbs. In New York, however, the company couldn't get into the downtown market right away, so it opened first in suburban Westchester and Summit. Executives noticed that profits in these areas were high and that customers stayed a while in the stores. Soon an employee got the idea of putting a couch in the corner and people sat there—and the people in line liked the looks of things and the promise of comfort and community connections. So Starbucks had a new template—it was not just a coffee dispenser anymore, it was a maker of "third places."

In this country where our schools, churches, and neighborhoods grow more segregated everyday, despite the gains of the Civil Rights Movement, Starbucks attracts a diverse range of people. In fact, the coffee shops are more diverse than most places. Of course not everyone goes there, but all kinds of people—old and young, African American and Latino, white and Asian, the very rich and the very middle-class—drink Frappuccinos and hang out on the comfy couches. But it is not just the presence of different people that makes the coffee shop look like a third place; it is also the energy of these places.

On a soupy July afternoon (not exactly coffee weather) in Washington, D.C., I stopped into a Starbucks on Dupont Circle—actually one of three or four stores in this immediate area. This one stands at the pointy edge where 21st Street meets N Street. The front of the store traces the contours of the intersection and is like the tip of a piece of pie; it gets wider as it moves away from the street corner. Still it is a small room; there is space for only two soft chairs, six round café tables, and one four-seat rectangular library table. When I walked in at three in the afternoon, the hum of the air conditioner and the buzz of conversation nearly drowned out the Starbucks soundtrack of Miles Davis and Bob Dylan. Every seat was taken.

A couple of months after I visited the D.C. store, Alfred Lubrano, a reporter from the *Philadelphia Inquirer*, called me. He wanted to do a story on my coffeehouse research. "Is Starbucks a third place?" he asked me when we sat down for an interview in a Starbucks store next to a Whole Foods outlet in Philadelphia's Art Museum area. "Yes," I answered, adding it is "a perfect third place." Feeling for a moment like a White House press agent, I even coined my own sound bite. "Starbucks," I told Lubrano with steady confidence, "is the corner tavern of the twenty-first century."[8]

But the longer I hung out at Starbucks, the less it seemed to me like a corner bar or a traditional coffeehouse or a fully operative third place. Lots of people, I quickly learned, use the coffee shops as a second place—for business meetings and as an out-of-the-office office. Students use it as a library, study lounge, and clubhouse. Even more people use it as a public fourth space, a place to get online (although you have to pay T-Mobile for the service) and talk with friends and strangers on MySpace, Facebook, and thousands of other virtual meeting rooms. But a third place? I don't see much of that going on at Starbucks. Different kinds of people definitely gather at the coffeehouse, but they only look at each other from afar. For Oldenburg, third places are supposed to work like Jürgen Habermas's public spaces. The community atmosphere of these sites should spill out into the streets and influence the way people think. Patrons should let their guard down and open themselves up to new music and new ideas and new people. Sharing a table or couch with, say an African American man, can encourage a white man—to imagine one kind of example laid out by sociologist Elijah Anderson—to rethink his ideas about race. Maybe through his third place interaction he learns that not all black men are criminals, so that the next time he passes a black man on the way to work he doesn't scamper across the street. But this change of heart and newfound tolerance requires not just observing the other, but talking—exchanging words, theories, and experiences.[9]

Over the last year and half, I have spent on average ten to fifteen hours a week at Starbucks. On only a dozen or so occasions have I spoken to someone I didn't already know. Several times my kids were involved. And I have seen this with others. With a sweet four-year old or a sleeping toddler by your side, you are marked as safe. But

---

[8] Lubrano, "Just the Place for People to Perk Up?" *Philadelphia Inquirer*, 16 December 2004.

[9] For another useful examination of how the public space functions and can, in some cases, bring people together, see Anderson.

somehow these conversations stall quickly after talk of parks and pre-schools gets exhausted. Twice outside the United States, I spoke with people—other Americans. But these interactions were one-off deals. The conversations never lasted long or got renewed the next day or the day after that at these stores.

One time, I sat in one of those comfy seats in a London Starbucks. Just as I settled in, a pack of German teens walked into the store. They crowded onto the couch next to me. I offered them my seat. They took it, but they didn't say much. Then a guy with matted hair and lace-less sneakers—he looked like a stand-in for the down-and-out Dan Akroyd in *Trading Places*—came up to me.

"People just don't appreciate kindness," he said.

I nodded and said without making full eye contact, "no they don't." He lingered for moment and then walked away.

Another time, I was sitting in a tiny Starbucks in Margate, New Jersey, a shore town just a couple of miles south of Atlantic City. A man started talking about his plans to build condos in Atlantic City. But he said—very loudly—he would have to sell them to New York Jews, not Philadelphia Jews, because Philly Jews, he bellowed, knew all about Atlantic City. Then he asked everyone in the coffee shop if they agreed. Two did and one wasn't sure.[10]

Judi Schmitt of northern Virginia went searching for a third place at Starbucks and didn't find it either. For the past three years, she told me in an e-mail, she and her friend Sharron have played weekly two-hour Scrabble games at their local Starbucks. "We kind of hoped to start something," she explained. But she regrets, "we have not [. . .] started a trend." Not a single person has ever asked to join them. A few other customers look up from their "babies, laptops, [and] school books," and occasionally will share "fond memories of playing Scrabble."[11] But that's it.

In search of that elusive third place, I went back to the pie-slice shaped Starbucks on Dupont Circle. This time it was a crisp, but still comfortable, January night. Again, it was packed with all kinds of different people and the crowd came and went. Students sat behind

---

[10] Conversation with the author, 29 August 2005, Margate, New Jersey; conversation with the author, 9 February 2006, Tottenham Court Road, London.

[11] Judi Schmitt, e-mail message to the author, 17 January 2006.

laptops and papers. Friends talked to friends. Businessmen barked instructions about orders and prices into cell phones. Lovers whispered to each other.[12] But no one talked with anyone they didn't know or anyone they hadn't come there to meet.

After an hour I packed up and went looking for something to eat. I spotted a bar on the other side of Dupont Circle with a college basketball game on the television. I pulled up a stool and ordered a burger and a beer. Immediately, the bartender started talking to me. The guy next to me started talking to me as well. He wanted to talk about politics. Before I could say anything, he volunteered that he hoped John McCain would run for president in 2008, that he respected Newt Gingrich, and that he had reluctantly voted for John Kerry. Another guy walked in wearing a University of North Carolina baseball cap. We started to talk about Tar Heel basketball. During a commercial, I found out that he had just got back from a trip to Spain, but that he had already broken up with the woman he had gone with on the trip. Two conversations in about ten minutes. The bartender offered me another beer on the house, but I said no. I wanted to hurry back to Starbucks and see if I could get anyone to talk with me. I sat on one of the comfy chairs in the back of the room. My knees just about touched the knees of the guy next to me. I made eye contact with him. But not a word; a nod, but not a word. Maybe we just know not to talk at Starbucks. We have been trained into silence. The next day, I went back to that bar and talked again with the guy in the Carolina hat—the politics guy wasn't there.[13] And then I went back to the Dupont Circle Starbucks again, and again, and never did I find a conversation.

Not long after my not-so-scientific Washington third place experiment, I went to a Starbucks in the bottom of an office tower in Philadelphia—one of those with the soft chairs in the front window—and I pulled out my copy of Oldenburg's *The Great Good Place: Cafes, Coffee Shops, Bookstores, Bars, Hair Salons, and Other Hangouts at the Heart of Community*. It is in this book that he introduces the term "third place."

---

[12] Journalists have also noticed the bustling nature of Starbucks stores, see, for example, Jonathan Lemire, "Starbucks Wacky World," *New York Daily World*, 20 March 2005 and Sandra Thompson, "Bringing Us Together – One Tall Latte at Time," *St. Petersburg Times*, 15 November 2003.

[13] Conversation with the author, 17-19 January 2006, Washington D.C.

Reading Oldenburg again, I realized that Schultz only cited—and only put into practice—parts of the third place idea. Third places certainly function, Oldenburg says, as spaces to hang out between work and home, but they remain, he insists, so much more. For starters, they are idiosyncratic and one-of-a-kind joints and haunts. They have their own feel and décor. Uniqueness makes them into agents of cohesion and community. But again, it is what happens in these quirky third places that matters. Oldenburg is more than just a romantic for a lost urban past, although he can come off this way at times. To him, third places serve not just as refuges or hideouts from the world. They perform a vital public service—they bring different people together, people who would not come in contact with one another in any other setting. And they do this, not just for commerce, but also for the larger social good, for democracy even.

Not long after I reread his book, I went to talk with Oldenburg. Thin and graying, with a bad back that made him move slower than he might have for his age, the retired professor blended into the early morning crowd at the Pensacola pancake house. Starbucks, he told me, had once asked him to work for the company. He turned down the offer only to have the executive lecture him in the back of a limo about the true nature of third places.

While Oldenburg admits that Starbucks has done some good things, he scoffs at the notion of Starbucks as a "third place." "It is an imitation," he said as he took a bite of his ham and eggs, adding "it's all about safety for them." Real third places, he insists, must have open doors and throw off a whiff of danger or a hint of unpredictability. In Oldenburg's mind, the owner plays a key part in creating the unique—and really transformative—character of a third place. Standing behind the bar or the counter day and night, he sets the tone for the place through his jokes, political commentary, wall hangings, jukebox choices, and gruff or gentle gestures. He welcomes strangers and brings them into the community by introducing them to the regulars. And he doesn't do this just for money; he does it for himself, out of a desire for social connections, and in service of his town or neighborhood. "Would Starbucks," he asks me, "give a guy who is down on his luck a job?" Essentially, he continues, third places are conversational zones, places to talk freely and openly, to sound off and entertain. With its "overriding concern for safety," Starbucks, Oldenburg concludes, as the check arrives, "can't achieve the kinds of connections I had in mind."[14]

Beau Weston is also skeptical about Starbuck's third place claims. Like Oldenburg, Weston is a sociologist—a Yale educated one. He teaches at Centre College, a tiny private school in rural Kentucky, perhaps most famous for hosting the 1988 Vice Presidential debate. Over the last couple of years, during the school's J-term (a short session of courses between fall and spring semesters) he has offered a class with readings from Oldenburg and the European theorist Jürgen Habermas on coffeehouse culture and public space.

When everything is working, Weston observes, the coffeehouse is "a place in which strangers can talk to one another" and debate the issues of the day. Over time, it became a gathering spot for all kinds of people, but also a sort of classroom. "Informed men, some educated and some not," Weston continues, "would come together and talk about stuff"—literature, plays, poems, the state of the economy, and politics. "Having a place to do that," Weston remarks, "enriches a culture. It takes us out of the cocoon of private life and into the public world. Cafes are important for creating a public life, particularly a democracy. It becomes a place where the town or, in the big city, where the neighborhood develops."[15]

In the early 1980s, marketing guru and business trends watcher, Faith Popcorn (that really is her name) noticed that middle-class Americans—Starbucks customers—were "hunkering down," "holing up," and "hiding out under covers." She called this trend of cocooning, "an impulse," she added "to go inside when it gets too tough and scary outside" (27). Everything from "rude waiters and noise pollution to crack-crime, recession, and AIDs," Popcorn maintained, led to this "heavy duty burrowing" (28-29). People, as a result, stayed home and avoided the few third places left.

The evidence seems to back up Popcorn's argument. In the 1990s, the mail order business in America tripled as it reached half a billion dollars a year. Over the same time period, the sales of Joe Boxer

---

[14] Conversation with author, 28 June 2005, Pensacola, Florida.

[15] Murray Evans, "Sociology Professor Takes Coffee Culture to the Classroom," *Washington Post*, 2 March 2005. On the coffeehouse and its history, see Ellis; Cowan.

pajamas increased by 500%. More and more families, meanwhile, moved into gated communities and fortified their homes with motion sensitive security systems, anti-snooping devices, and armed private guards.

But a strange thing happened just as Popcorn pointed her finger at the cocooning trend. Crime rates dropped just about everywhere and people started to come out of hiding. They backed their SUVs out of their steel-doored garages and went downtown for dinner and a show. Some joined book groups at Borders and saloons sponsored by the *Utne Reader*. For many, Starbucks became their favorite destination.

In the suburbs and the cities, women and men locked behind the walls of gated communities for years filled the tables and chairs of Starbucks stores. It was as if they were waking up from a long slumber and rubbing the crusties out of their eyes and looking around for the first time. Once they got there, they saw, much to their relief, that they were not alone. They sat and watched others. But they weren't quite ready, it seemed, to jump headlong into the loud, smoky, politically charged coffeehouses of old, or really even talk to anyone else. It is as if they had lost their third place skills. Maybe they didn't know how to talk to strangers anymore and surely they lacked the kinds of connectors—people who know other people and make the introductions—needed to facilitate the conversations. So people went to Starbucks to be by themselves and maybe watch others in public—to see at first if they really liked leaving home and maybe to dream about the possibilities of community. But quickly they learned that at Starbucks not only do you not have to talk, you don't talk; you keep your head down.

Starbucks has created a niche for itself by becoming a solo-friendly—and clearly controlled—environment. "Maybe," writes *New York Times* reporter, Anemona Hartocollis, "we only wish to drown our sorrows in a strong cup of coffee in cushy chairs surrounded by strangers who will grant us the illusion of community yet respect our privacy."[16]

Emerging cocooners might be alone at Starbucks, but it isn't clear that they want to be alone entirely. In many ways, I suspect, they

---

[16] Hartocollis, "Coping: Gazing Into a Coffee Shop and Seeing the World," *New York Times*, 29 September 2002.

want it both ways. They want to be safe and isolated from unknown others, but they also want—at least in principle or appearance—to share in the art and conversation of cafe culture. And Starbucks is there to fulfill and then consume that desire.

"Our coffeehouses," a Starbucks brochure promises, "add to the community life by serving as local gathering spots." Peter Maslen, the president of Starbucks Coffee International, told a reporter in 2000, "[W]hat our brand stands for is based on the European coffeehouse culture."[17] By invoking the romance and traditions of the cafe, Starbucks ties itself to a long and wild history, one that it, again, simultaneously embraces and rejects. In seventeenth-century England, shopkeepers and bankers, craftsmen and lawyers—really just about any man—met at the coffeehouse to hear the latest news. Literally someone would read aloud from the papers. When the reader finished, the noisy debate started. Intellectuals damned the government. Conservatives damned the intellectuals. And wits spread rumors and gossip and made fun of everyone. A century later in Paris, poets wrote in one corner while artists sipped absinthe in another. Out on the terrace, revolutionaries railed against the King and Queen. By the 1950s, Beatniks in baggy work pants and dark sunglasses took over the coffeehouse. As cool cats like Charles Bukowski read prose poems over a Charlie Parker soundtrack, beret-wearing hipsters clicked their fingers and sipped espresso from chipped porcelain cups. At least, that's the image popular culture has shown Americans.

"In the tradition of coffee houses everywhere," the company's web page proclaims, "Starbucks has always supported a good, healthy discussion." They got that right. The company recognizes that there is a desire out there for something beyond just a better cup of coffee than the diner serves, for, it seems, ideas, substantive exchanges, intellectual connections, and debate. So Starbucks has tried to link the brand to these impulses. A few years ago, the firm launched a magazine they called *Joe*. According to the managing editor, the glossy aimed to "replicate the ideas, conversations, and encounters in a coffeehouse." "Life is interesting" *Joe's* subtitle declared; "discuss" it offered. The magazine, however, didn't make it past a few issues. Perhaps there

---

[17] Catherine McLean, "Starbucks to Grind Through Europe," *Chicago Sun Times*, 8 October 2000.

weren't enough people in the stores willing to talk or enough ideas between the covers to engage them. But Starbucks didn't give up on creating the appearance of coffeehouse conversations.[18]

On a second try at getting the discussions going, Starbucks officials plastered quotes—now more than 200 of them—on the back of its white take-away cups. Dan Rapp, a Starbucks customer from Cincinnati says rather gravely on one of them, "I think every professional athlete should have to attend at least five kids' games every year, just so they remember what the sport is really about." That is quote number 73. In number 54, Morris Dees, the founder of the KKK-busting Southern Poverty Law Center, proclaims, "We are all brothers and sisters. Each face in the rainbow of colors that populates our world is precious and special. Each adds to the rich treasure of humanity." Number 79. The smoothed voice, model handsome singer, Michael Buble, whose tunes can be heard on Starbucks' Hear Music station on XM Satellite Radio, says, "Lips are the gateway to romance." In case anyone happens to disagree with the observations of Rapp, Dees, Buble, or any of the other coffee cup philosophers, the company denies any responsibility for the content. "This is the author's opinion," it says at the bottom of every cup, "not necessarily that of Starbucks."

Even as it distances itself from the cups and quotes, Starbucks, nodding in the direction of coffeehouse tradition, invites customers to join an online exchange about the views expressed on its containers, although it is hard to imagine who could quarrel with most of the lines. Who isn't in favor of the "rainbow colors" of the world or kids playing baseball or everyone loving everyone else? But, of course, when you are as big as Starbucks someone is going to be opposed.

"My only regret about being gay is that I repressed it for so long," novelist Armistead Maupin laments on a Starbucks cup, "I surrendered my youth to the people I feared when I could have been out there loving someone. Don't make that mistake yourself. Life's too damn short." For one Baylor University faculty member this quote was too long and too gay. In response to the professor's protests, the

[18] Alex Kuczynski, "A New Magazine by Way of Starbucks," *New York Times*, 8 February 1999. On the magazine's demise, see Susan Dominus, "Starbucks Aesthetic," *New York Times*, 22 October 2006.

Starbucks store on the campus of the Waco, Texas Baptist school stopped serving coffee in the Maupin cups. Linda Ricks, a university official, reported that the dining services agreed to ditch the offending cups out of respect for "Baylor culture." "There are different viewpoints on [. . .] campus," Ricks elaborated. "We pulled the cup to be sensitive." Starbucks, she told the press, supported the cup removal. "They aren't intended to generate conflict at all," Ricks said about the cup quotes. "Starbucks fully supported our decision because they understand our environment."[19]

Just like it cleaned up the quotes, Starbucks cleaned up coffeehouse culture. Almost every reference to intellectuals and powerful ideas is juxtaposed with a reference to safety and security. Whatever the lasting pull of the traditional coffeehouse, a darkish hue hung over these places. These were not the hangouts of moms and dads or Madison Avenue executives and Wall Street traders or honors students and cheerleaders—the people Starbucks was looking to get to its stores. The coffeehouse of the 1950s was a nocturnal place. Usually it sat in the middle of the block in a run-down part of the city or in a downtown basement. Often, then, you descended into it. After the walk down, you had to cut through a wall, not a veil, of smoke. Only a few wobbly mismatched floor lamps light these places.

At Starbucks, cleaning up the coffeehouse meant displaying order and transparency. The first Starbucks stood at the corner of Main and Main, on the best blocks in the best part of town. And there is no walking down. Saying that they have nothing to hide, many Starbucks stores are at street level and have floor-to-ceiling windows. You can see inside and know at a glance that this is a safe place. The lights are bright—not glaring like the ones at McDonalds or the diner, but not dimmed like at a roadhouse tavern or a beatnik coffeehouse.

Another quote about safety for cocooners: Starbucks bars smoking at all of its stores. Company officials insist that cigarettes interfere with coffee aromas. "Because coffee beans have a bad tendency to absorb odors," writes Schultz, "we banned smoking in our stores years before it became a national trend" (252). But this served

---

[19] Analiz Gonzalez, "BU Starbucks Pulls Cups Due to Homosexual Quote," *The Baylor Lariat*, 7 September 2005; Lornet Turnbull, "Tempest Brews Over Quotes on Starbucks Cups," *Seattle Times*, 30 August 2005.

another purpose. By the 1990s, smokers symbolized something else in health-crazed America. They had become pariahs. They were the undisciplined, the rough around the edges, the unclean. Before the city put a complete smoking ban into effect, each terminal at the Atlanta airport had designated smoking areas. These were bare, glassed-in rooms with no places to sit and no pictures on the wall, just a floor, a ceiling, and ashtrays. They looked like smoking cages filled with addictive freaks. You walked by and stared and felt better about yourself. By keeping smokers out, Starbucks marked itself as a safe, clean, and healthy middle-class universe—just the kind of place you could go to alone.

Starbucks designed its stores to enhance the alone-in-public feeling. Arthur Rubenstein worked as Starbucks' senior vice president for real estate. Yet he didn't just scout out locations, he also laid out the stores. Like everyone at Starbucks, Rubenstein spoke the language of third place, but this didn't stop him from building solo friendly enclaves. Round tables, rather than square tables, he advised, should be used in all seating areas. "A single person," he explained, "at a square table looks (and possibly feels) lonely." In *Built for Growth*, he adds, "A round table is less formal, has no empty seats, and the lack of right-angle edges makes the person seated at the table feel less isolated" (73). But these same tables, as Rubenstein understands, discourage people who don't already know each other from sitting together and talking. The furniture makes cocooning in public easy.

I know firsthand how this works. If I want to read a demanding book I don't go to my local coffee shop, I go to Starbucks—usually the one near the University of Pennsylvania campus or another one near a downtown trolley stop. I know I won't be bothered there. At first this sort of surprised me. I looked at the variety of people at Starbucks and thought I saw a third place. I told a friend about the lack of interaction at Starbucks and she said, "Hey, I know," adding, "I don't go to Starbucks to talk, I go to be alone."[20]

Now that I know the score, I have a Starbucks routine. Before ordering, I scope out an unoccupied corner table near a plug and spread out my stuff. After that, I get my drink—a tall coffee of the day. Back at my table, I pull out my computer, put my cell phone on the table, and

[20] Conversation with author, 21 January 2005, Philadelphia, Pennsylvania.

plug myself into my I-Pod. I have created my own virtual gated community. I couldn't be more alone in public and that's what I want. I never talk to anyone at these stores; I don't even recognize many people.

Idaho journalist Kathy Hedberg detected the same thing. When the company came to her neck of the woods, she went to explore its coffeehouse culture. "Caffeine," she observed, explaining a fact of cafe life from the beginning, "gives a jolt to your system that is like an electric current, and after a couple of cups people just start talking whether they have anything to say or not." The issue, then, is how to get the conversation started. After spending time at Starbucks, Hedberg thought its customers needed some help. "I have noticed," she wrote of the alone-in-public feel of the place, "that the folks who drink their coffee at Starbucks [. . .] are not big talkers. They're more an elite, standoffish group, not the gabby sort you run into at your neighborhood diner."[21]

Still Starbucks portrays itself as a maker of vibrant coffeehouse culture—the kind, for instance, that Weston talks about. To mark Ben Franklin's three-hundredth birthday, Starbucks launched the "Ben Franklin Coffeehouse Challenge." A press release called on the people of Philadelphia to rediscover the "civic generosity" of the city's famous Founding Father. "This is a town of unlimited ideas," an in-store poster proclaims, "Let's put them to use." "Join our fellow community members, Starbucks, and the Benjamin Franklin Tercentenary as we discuss the issues that face our neighborhood and find solutions that create a better community for us all."[22] Starbucks, in other words, promises more than coffee; it promises to solve or set up discussions about community problems.

Maybe this corporatizing of the conversation about social ills is inevitable with the retrenchment of government and deep cuts in funding over the last twenty years for libraries, parks, and schools. But still this trend limits rather than expands the debate and the range of participants. Only Starbucks customers are included and Starbucks will not fund any program that separates people from Starbucks. They will

---

[21] Hedberg, "Whoa: Starbucks Thinks Coffee Drinkers Need Help to Talk," *Lewiston Morning Tribune*, 3 January 2005.

[22] For more on this, go to: www.benfranklin300.org/chc.

support AIDs research, but not Queer Nation. The Ben Franklin program urges Starbucks customers to "plant more trees (and hug them)," but not attack polluters. "Why can't a vending machine," a hidden person asks on the back of a store handout, "sell locally made art?" But no one asks how standardized testing and George W. Bush's No Student Left Behind initiatives have narrowed curriculums everywhere or how federal budget cuts have choked off arts programs across the nation. Unlike the old coffeehouse where anyone could say just about anything, Starbucks stores are not places where all speech is free. Political parties, campaign meetings, and candidate fundraisers are not welcome, disturbing art never goes up on the walls, and workers are not allowed to talk about unions. "We are," explains a Starbucks spokesperson, clearly not grasping much about the coffeehouse tradition, "trying to promote respectful conversation, not incite controversial, taboo subjects."[23]

Opened in 1993, the Someday Café in Somerville, Massachusetts is a riff off the classic independent coffeehouse (and fills its own niche market). The owners have decorated the shop with mismatched furniture and they play a blistering soundtrack of alternative music. They leave graffiti unwashed (and unedited) in the bathrooms and the back wall plastered with fliers in a rainbow of colors announcing the shows of punk bands at cramped bars and alt-country acts appearing at reconverted theaters. Handwritten notes dot the community board, making it look like a paper patchwork quilt. Bands seek new guitar players. A vegan is looking for a "SUNNY ROOM IN HOME WITH VEGAN/MACRO KITCHEN." ACT UP announces an emergency meeting, anarchists urge coffee drinkers to attend a nearby protest against the death penalty, and the local Pagans invite anyone interested to a Wednesday night potluck.[24]

Tucked back in the corner of most Starbucks are the company's own version of community boards with phrases like "What's Happening" or "Starbucks Happenings" running along the top. But these are more simulations than the real thing. Like everything at

---

[23] Valerie Hwang, Starbucks spokesperson, quoted in Hedberg, "Whoa: Starbucks Thinks Coffee Drinkers Need Help to Talk," *Lewiston Morning Tribune*, 3 January 2005.

[24] Jason Gay, "Brewhaha," *The Boston Phoenix*, 24 June - 1 July 1999.

Starbucks, the company has a policy on the community boards. Over coffee at a store in Austin, Texas a former employee let me peak at the "Dos/Don'ts of Community Boards" from the late 1990s.[25] The list went like this:

> Do – showcase Starbucks' participation and involvement in the local community
>
> Do – post community events that Starbucks is involved in or sponsored
>
> Do – post photos of Starbucks partners' involvement in community events
>
> Do – post positive news article about your store's community involvement
>
> Do – post "thank you" letters and awards or certificates pertaining to our support within the community.

The rules instruct store managers not to post anything about politics or religion. Groups involved in recycling and conservation can use the boards, but not activists. Classified ads or calls for roommates are not allowed either. But generally, the company reminds employees, "Do Not post any information on any event not sponsored by Starbucks."

The manager of a Starbucks store in Washington, D.C. clearly followed the rules. When I stopped in for a tall misto, three things were stuck to the "Our Neighborhood" board. There was an advertisement for subscriptions to the official Starbucks paper, the *New York Times*. "Coffee," the ad says, "Makes News More Interesting and Vice Versa." Below this someone had put up a flier about, "Ethnos Water," Starbucks' own bottled water product and the clean water projects it supports in the developing world. Finally, a store manager had tacked up a copy of the company's social responsibility brochure.

Leslie Celeste manages a busy Starbucks outlet in Austin, Texas. When I asked about the community board in her store, she

---

[25] John Moore, in conversation with author, 10 January 2006, Austin, Texas.

chuckled. But she quickly settled back down and said, repeating company policy that she won't let any religious or political fliers go up. Occasionally, she said, she puts up calls for auditions at local theaters and notices about art openings at nearby galleries. When we went to look at the store's community board: it was empty. That's the way it usually is, she told me, and laughed again.[26]

I waited a long time to talk to George Ritzer about Starbucks and third places. On a soupy hot summer day, I finally had the chance to sit down for lunch with the author of a handful of probing books on McDonalds, everyday consumption, and the modern condition at the University of Maryland's new and rather formal cherry-wood paneled faculty club.[27] Ritzer wore shorts, sandals, and a blue-and-white Hawaiian shirt. His speech was as playful and provocative as his dress. To him, the Starbucks experience—the idea of a corporate-manufactured third place—was as empty as that community board in Austin. Bottom line, he said, the company strives to make money. "Everything else," he insisted, is "window-dressing." Coming back to his main point, Ritzer argued, that selling as much coffee and as many CDs, pastries, and sandwiches as fast as it can represents Starbucks only mission. When I tell him that grab and go customers make up 60 – 70% of the company's business, he lights up. "Right!" he calls out, adding that all the comfy chairs and coffee bean lights are nothing more than "dramatic devices," "props."

As he said this, I immediately thought of the cramped Starbucks in downtown Philadelphia I often go to. Office workers and sales people grabbing coffee on breaks and between meetings pack this place from early in the morning to late in the afternoon. Three wide soft chairs (the only ones in the store) sit in the front window of that Starbucks pointed toward the street. In London and New York, I saw this again and again. Company designers put their version of third places as comfortable community places as respites from the city, traffic, the kids, and daily life on display; they advertise this as much as they tout the coffee.

---

[26] Starbucks employee, in conversation with author, 10 January 2006, Austin, Texas.

[27] See Ritzer, *The McDonaldization of Society*, *Enchanting a Disenchanted World*, and "Islands of the Living Dead."

Even the people, Ritzer contends, are props just like the art, the newspapers, and even the smoking ban. "What attracts people most," the great city watcher, William Whyte, observed, "it would appear is other people."[28] That's what Ritzer tells me. The people in the "third place" let the customers in line rushing from the suburbs to the city, work to home, childcare to tumbling classes know that this is a popular place (making them popular), a safe place (suggesting that there is no risk), and possibly a community space (the desired notion of a third place). The theater also serves as a promise. Customers on the go, Ritzer speculates, get "warm fuzzies," watching other people relax, and they say to themselves "this is a really great place and one day I will sit down."[29] And some day, they will come to Starbucks in search of comfort, and maybe community, and they will most certainly find themselves sitting alone next to someone like me—sitting alone, barricaded behind a computer, and wired to a cell phone and an I-pod.

Bibliography

Anderson, Elijah. "The Cosmopolitan Canopy." *Annuals of the American Academy of Political and Social Science* (2004): 14-31.

Cowan, Brian. *The Social Life of Coffee: The Emergence of the British Coffeehouse*. New Haven: Yale University Press, 2005.

Ellis, Markman. *The Coffee-House*. London: Weidenfeld and Nicholson, 2004.

Habermas, Jürgen. *The Structural Transformation of the Public Sphere: An Inquiry into a Category of Bourgeois Society*. Cambridge, MA: The MIT Press, 1989.

Learning Journey Guide. Seattle, WA: Starbucks Coffee Company, 2003.

Oldenburg, Ray. *Celebrating the Third Place: Inspiring Stories about the "Great Good Places" at the Heart of Our Communities*. New York: Marlowe and Co., 2002.

—. *The Great Good Place: Cafes, Coffee Shops, Bookstores, Bars, Hair Salons, and Other Hangouts at the Heart of Community*. New York: Marlowe and Co., 1993.

Popcorn, Faith. *The Popcorn Report*. New York: Harper Business, 1991.

Ritzer, George. *Enchanting a Disenchanted World: Revolutionizing the Means of Consumption*. Thousand Oaks, CA: Pine Forge Press, 2005.

—. "Islands of the Living Dead: The Social Geography of McDonaldization." *American Behavioral Scientist* 47 (2003): 119-136.

—. *The McDonaldization of Society*. Thousand Oaks, CA: Pine Forge Press, 2000.

---

[28] Whyte quoted by Malcolm Galdwell, "The Science of Shopping," *The New Yorker*, 4 November 1996.

[29] Ritzer, in conversation with author, 29 June 2005, College Park, Maryland.

Rubenstein, Arthur and Collins Hemingway. *Built for Growth: Expanding Your Business Around the Corner or Across the Globe*. Philadelphia: Wharton School Publishing, 2005.

Schultz, Howard. *Pour Your Heart into It: How Starbucks Built a Company One Cup at a Time*. New York: Hyperion, 1997.

The Public Space of Urban Communities

Rickie Sanders

This paper explores multiple dimensions of urban public space—as organic spaces that emerge unplanned, as contested spaces, as spaces of exclusion, as sites of theater, and as spaces that provide entrée into understanding societal ideas and values. Its presence as a materiality and a conceptual framing of the world and ideas we hold makes it constitutive of processes that created it in the first place: race, class, gender, sexual orientation, religion, and so forth. Thus, public spaces are sites where meaning is made. Using three anecdotes, this essay argues that sometimes public space is not spatial, sometimes it is not public, and sometimes public space is not really public space. The question becomes, what happens if we lose those spaces we have come to value as public space?

Introduction

This essay on the public space of urban communities begins with two simple ideas. First, public space is important to the social and cultural life of a city and second, like the city itself, it is an exceedingly complicated and ambiguous concept. Dolores Hayden says of place, "it is one of the trickiest words in the English language, a suitcase so overfilled one can never shut the lid" (Hayden, 14). Such is the nature of public space. It has been used to refer to things as disparate as city parks, food courts in malls, sidewalks, and streets. It has also been used to capture the best intentions of community newspapers, internet sites, and literary projects/magazines. Although public space exists in every place and time, it remains surprising, misunderstood, and theoretically unexplored.

I want to explore public space both as a physical entity and also as an entrée into understanding societal ideas and values. On the one hand public spaces are the great formal plazas in the cities of Europe and Latin America. On the other hand public spaces are small, more organic sites that emerge out of everyday use and social appropriation. Both are important and both provide valuable insights into the values we hold and how we make our world make sense. The large, formal, designated spaces are places where protests and tensions are expected to take place. Groups (both official and unofficial) come together in an organized way to register their tensions and protests on a grand scale. In so doing they shape what is going on in the world around them, they give meaning to their times, their place in the world—and themselves.

Unlike the great public spaces, the smaller, more organic public spaces—streets, small neighborhood green spaces, and sidewalks—are also sites where meaning is made except on a smaller scale. They are theaters where neighbors stage ordinary (and spectacular) performances in front of an unsuspecting audience, where local everyday democracy is constructed, where individuals and small groups spontaneously engage in debate and contests (both physical and mental), and where communities decide who will be excluded and who will be included. The scale of these organic public spaces makes it possible to observe firsthand how people arrive at consensus, make sense of the larger questions impacting their lives, and negotiate the meaning of community. The small spaces are important staging areas for individuals and communities and may indeed be the only sites outside of the voting booth where they can give shape to their ideas and practice their notion of freedom. Because these spaces are ubiquitous and lack the presentation of the grand formal spaces, they go unregistered in our consciousness and, thus, are ignored and disregarded as important public spaces.

I became interested in this topic last year when a white colleague shared with me the story of jogging through North Philadelphia early one Sunday morning and being chased out of the neighborhood by a group of African American youth. His story captures so many of the complexities of public space. Often public space is not public; its "publicness" surfaces because societal tensions and contestations occur there. Often public space is not spatial; it is a site or location (e.g., virtual) where anyone who desires has access. And sometimes what appears to be public space is not public space at all; it is policed to such an extent that it loses all semblance of its original intent. One way to think of public space is that it is a space designated for the public to register protests but another way to think about public space is that it emerges because the people who use it daily claim it as a site to express their concerns. Public space belongs to everyone . . . and to no one.

For those not familiar with North Philadelphia where this incident occurred, it is not unlike Riverside Drive/Morningside Heights in New York City; the Dixwell neighborhood in New Haven, Connecticut; South Central Los Angeles; and several other African American communities where large, predominantly white universities are located. These places are regarded as the cultural hearth of their city's black community. In this case, North Philadelphia is home to the once famous Uptown Theatre, the Church of the Advocate, Freedom Theatre, and the many factories (now boarded up and shut down) that employed blacks in the initial wave of the Great Migration. Some locals remember when industries were still there and everybody had work. Back then, Saturday mornings were set aside to clean up the neighborhood. People would mop the sidewalks in front of their houses and afterward bring out a lawn chair or lean out of an upstairs window and "shoot the breeze." Even today, when some longtime residents who have moved away speak of "North Philly," there is a fondness and affection for the old neighborhood. For an increasing number of others, however, North Philadelphia is a place where it's good to have somebody "watching your back." Anything can happen and it may be "the cops or the robbers who get you."

As my colleague described what happened, he was jogging down one of the side streets, "not bothering anybody." Out of nowhere a group of five or six young African American boys ran behind him yelling for him to get off the "fucking street." Meanwhile, an elderly black woman on her way to church witnessed the incident as it was unfolding and told the youth to "leave him alone." She clearly held some sway over the youth (she might have been the grandmother or a neighbor), because my friend escaped having only been shocked and shaken up a bit.

What is intriguing about this seemingly ordinary event is not only its familiarity—as a matter of urban folklore—but its complexity. It suggests a simultaneous display of power, authority, social control, performance/playfulness, contestation, and disciplining. Even though my friend was in a public space—the street—it had been "claimed" by the five youth for the African American community. In their eyes he did not belong; and he could not use it. This incident provides an opportunity to address many of the knotty questions about public space that cannot be addressed by focusing on the grand, formal, designated spaces that occupy much of the literature about public space. What happened to my white friend in North Philadelphia has happened to many of my black friends in the public spaces of white neighborhoods or on the freeway; but is much less likely (nowadays at least) to happen in a public space like the National Mall in Washington, D.C., or Rittenhouse Square or Penn's Landing in Philadelphia. Organic, informal public spaces such as streets, neighborhood parks, and sidewalks that emerge out of everyday negotiations allow us to ask questions which cannot be asked if we limit our analysis to grand formal spaces or to grand theory. What is meant by "public space"? Who is the "public" in public space? What *rights* do individuals have in public spaces? Who can claim a space as theirs? What do these spaces tell us about American ideals and values? How do we make sense of what goes on in the seemingly trivial and overlooked public spaces—like sidewalks and streets? If public spaces are indeed sites of meaning and meaning making, what do they mean? Where does meaning lie? In the site itself? In its location? In the minds of the users? Or through its use? Finally, how is meaning made? By whom? In order to answer these questions, attention has to shift away from the formal, privileged sites and toward the smaller, ordinary, quotidian spaces that are ordinarily overlooked. Thinking about these other spaces and the questions they allow prompts us to move beyond the anecdotal observations garnered from formal public spaces and challenges us to expand our perspective and think outside the traditional ways of looking at the world.

I suggest here that because public spaces have material form we can interrogate them as objects, but we can also read and interpret them for what they might tell us about our social world and our place in it. Public spaces represent how we as a community live and what we value; they are also *representative* of how we live. This means that they provide insight into identity, place, power, and of course race, class, sexual orientation, religion, and gender. They portray ground-level realities and thus they are sites of "ground truthing" where practices and ideas formulated elsewhere can be tested for truthfulness on the landscape. In the incident described above, my friend was shaken up but, even more, his knowledge base was shaken. He did not question his right to jog through the back streets of North Philadelphia. What he was doing was fine. This was a public space. The youth however saw things differently. For them he was in *their* North Philadelphia—the cultural hearth of the black community. Here, he was the interloper. It would be the same if it were racially reversed; the interloper would be reminded that he was "out of his place."

Much of the discourse around urban public space connects to Michel Foucault's ideas around rights, space, discipline, knowledge, social control, and power. According to Foucault, space has trumped time as the great obsession of our era (239-256). *When* something happened is secondary, *where* it happened is primary. What matters is distance (how far, how near), juxtaposition, location, position, place, grids, flows, and networks. Public space is a special kind of space that requires that we ask more questions than we find easy to answer. Who has the power to decide what can take place in a public space? Or who can use a public space? Because it is public, does control reside in the state? Or does it rest in the hands of those who use it every day? If we get out of line in a public space, how are we disciplined; or, more interesting, how do we discipline ourselves in public spaces? The youths who challenged my colleague were "disciplined" by the elderly woman on her way to church. But what if she hadn't been there? What would have happened? It's possible they were just teenage boys being teenage boys performing for each other, but in the current economic and social climate in America fear, power, powerlessness, and consumerism have an enormous effect how we use, perceive, and behave in public spaces. An essential characteristic of public space is the idea that its use and meaning can be challenged, but do these challenges change the nature of public space for the worse and ultimately threaten its future? How do citizens negotiate the complex set of rules and codes around how to be seen and heard and, at the same time, protect their space in today's world when the moral code is ambiguous and there is a widely held belief that the state is capricious (at best) and preferential (at worst) in exercising its authority. The act of exclusion is critical for a space to function as a public space but, at the same time, it also contributes to its death knell. These complexities make urban public spaces a lightning rod for all of the post-9/11 issues—democracy, freedom, rights, identity, and citizenship. They also make safety measures, defense, protection, and precautions more pronounced. Other than the ballot box, public spaces are the only places where ideas of freedom and democracy are practiced and where people's desires and preferences are made known.

An Organizational Framework

As we ordinarily think about public spaces, they are formal, designated, clearly identifiable sites where people assemble and through congregating sort through issues of importance to them. The form these spaces take is often that of a town square or grounds punctuated by a marker that carries meaning, for example, the National Mall, Speaker's Corner, or a city park. In Philadelphia, examples are Penn's Landing, Rittenhouse Square, Love Park, and the steps of the Philadelphia Museum of Art. They are visible, recognizable places where clear and explicit clues regarding what is valued by the larger society are communicated. Public spaces are also sites of subconscious sensory and cognitive experience, where we hear, see, and process what is happening around us. Sometimes what is not present is more revealing than what is present.

The earliest example of public space is the square or commons—a place or space that is open to all (French, 48-49). In a classic, frequently referenced piece, "The Tragedy of the Commons," Garrett Hardin drew attention to the plight of common property that was shared and used by all. The commons that Hardin was concerned with were not urban public spaces, but his idea that a commons, if left unregulated, would be overused and misused is applicable to public spaces in cities where these concerns predominate. Hardin's solution was to restrict access through privatization and increased regulation. Forty years later looking back on the debate, all of the attention might have been displaced. Seldom does everyone have equal, unfettered access and rights to use a place. Even in places where in-group membership entitles everyone to equal access and rights, restrictions apply. Second, the focus on the "tragedy" may have deflected attention away from the bigger, more important issue of who has the right to use a space and who decides who can use a space.

In *Public Space* Carr et al. identify a hierarchy of freedoms or rights associated with public spaces: freedom of access, freedom to use a space in the manner desired, freedom to claim, freedom to change, add, remove or alter, and the right to own (137-187).

Freedom of access is the most basic of these. Public spaces are *intended* to be commons (accessible and open to all) but they are not always spatial. An event in Toronto, Canada, not long ago illustrates this very well. A U.S.-born African American basketball player married to a U.S.-born African American woman fathered a male child with a white Canadian woman. The wife sued for custody of the child, arguing that a white woman could not give a black male child what he needed to effectively participate in the larger society. The issue was one of national concern. An open forum was sponsored by the City of Toronto to discuss how Canada should deal with issues around parenting mixed race children. It was open to anyone who wanted to participate. The City provided the physical space but, more importantly, it provided an intellectual space where these ideas could be discussed and concerns could be recorded and catalogued. An important idea not often acknowledged in discussions of public space is that "public space" is not always spatial.

Next in the hierarchy is the *freedom to use a space in the manner desired* (for example, an early Sunday morning jog or a protest demonstration). Those who choose not to (or cannot) conform to the expectations of what is appropriate in terms of behavior (or appearance) are denied rights to use. The City of Atlanta recently proposed legislation to establish Free Speech Zones at public events. The legislation (Ordinance #07-0-1100) states that persons wishing to speak in public spaces must first obtain permission. If permission is granted, speakers must adhere to strict guidelines (e.g., fifteen-minute time limit) set forth by the City Council. In addition, everyone granted permission must wear a badge issued by the city indicating that they have been granted the right to speak. Atlanta's Free Speech Zone ordinance acknowledges free speech in public space as a constitutional right but it also acknowledges that rights may be taken away or limited in public spaces when the speech "interferes with the First Amendment rights of the organizer." They note in the proposed ordinance:

> [I]f a vegetarian society is holding an Outdoor Festival in a park that celebrates vegetarianism, and a person who is not part of the festival is handing out flyers to festival attendees advocating the

> health benefits of including chicken in one's diet, such leafleting will interfere with and dilute the message of the vegetarian society; [. . .] it is in the best interest of the City to designate Free Speech Zones for public property.

Demonstrations, rallies, distributing leaflets, speechmaking, and even taking a walk through one of Philadelphia's many urban parks require certain behaviors that must be sanctioned by the state.

The right to claim refers to the right to appropriate space for personal or community use. Here the Philadelphia incident is relevant. The five youth claimed the public space of the street as theirs. They saw it as a right based on continuous use and location.

Freedom to change, add, remove, or alter in order to make different statements about the place is the next level in the hierarchy of rights. Changes may be temporary or permanent; they may be constructive and sanctioned (building a swing set) or destructive and unsanctioned (untagged graffiti). They may occur over several years as particular groups gradually take over a space and restrict its use or they may occur all at once. Of all the freedoms, this may be the most complex since some activities, for example, unauthorized or unsanctioned and illegal activities, may ultimately result in changing how a place is perceived and thus how it is used.

Finally, there is the *right to own*. Ultimately it is the state that owns public space. Accordingly, it is the state that makes decisions about who can use it, when it can be used, and for what purpose. (Public) space is always contested; what we see on the landscape is simply who won the contest. Again, the Atlanta legislation is illustrative. The state has clear guidelines regarding how to obtain permission to speak, how far away the speaker has to be from another event, how large or small the Free Speech Zone is, who can exercise his or her First Amendment rights, and who is permitted to enter the zone. It owns the space. In belonging to everyone, it belongs to no one.

Carr's heuristic takes our thinking beyond Hardin's emphasis on the tragedies associated with the commons. However, it too falls short of asking the really important questions: Who is the public? Who is the state? What is the function of public space? Who makes the decisions? Who participates in the process? Who is excluded? Moreover, Carr's idea that there is a hierarchy of rights that can be ordered in a stepwise pattern may oversimplify what actually exists. In reality, as the incidents described above illustrate, individuals and groups exercise multiple rights simultaneously.

I suggest here that the best we have by way of public spaces that hearken to the original ideals of freedom and democracy where people challenge conventional notions are the ones that crop up organically. They are opportunistic spaces, claimed by neighborhood residents who see themselves as being unwelcome in other spaces and who often have retreated from participation in the public sphere. It is impossible to walk through any public space and not be reminded of one's race, class, and gender (Schein, 7). These small, informal, organic public spaces sit alongside the more formal, traditional spaces discussed earlier. Both are sites of tension/contests and theater where meaning is made.

Public Space as Sites of Tension and Contestation

Henaff and Strong suggest that public spaces are the spaces in which groups come together with the intention of determining how their lives in common should be lived (1). But who decides what can be included and what can be excluded from the life we are required to have in common with others. How are these decisions made? Where are they made? Who decides who can participate in the decision-making? On what basis are these decisions made? An agreed upon consensus can shift dramatically in a very short time as in the case of the discussions around the 9/11 memorial. Although the site itself has never been questioned, exactly what should be on the site, what it should represent, and the form it should take is still up for grabs. The important thing here is that when the memorial is completed, groups (official and unofficial) will assemble on the site to define it—and themselves.

Setha Low uses her work on public spaces in Costa Rica to uncover some unexamined issues about the intent and nature of "public" in "public space." She argues that the meanings attached to public spaces are actively manipulated by governments to facilitate their political and social objectives. While the social meanings of today's public spaces are fixed and static, at the same time they are the places where the conflicting and interacting dialogue associated with democracy takes place. The National Mall in Washington D.C. is an example. It is the place where protests around national issues are *supposed* to take place. Thus, it seems that the formal landscape of public spaces is a mnemonic for communicating designated meanings to daily users (Low, 239) through the emotional engagement it conjures. At the same time it provides present meanings through the struggles to define what the place *should* represent. In Philadelphia, the conflict over the siting of the new Liberty Bell Pavilion at Independence Mall where an African American burial ground once stood is yet another example. The controversy around the issue is still unresolved. Likewise, there have been numerous conflicts between Native Americans and the federal government over the sites designated as public spaces and the social meanings they hold as native burial sites. Low's main point is that the social production and social construction of public space relies on both a continuing dialogic conversation *and* a dialectical relationship that is oppositional, often disruptive and contested.

According to Sorkin, today's American cities are very different from those of the early twentieth century (205-232). What is reflected on the landscape of today's cities is our concern with concentration and high- density living; our obsession with security and surveillance; our desire for new modes of segregation (gated communities); and a "happy face familiarity" (Disneyfication) that actually distances us from interacting with others who are different. Sorkin refers to these as "theme park" cities where there are no public spaces, only private spaces masquerading as public space. The appropriation of public space form by the private sector sends conflicting messages and destabilizes what we think and know about public space, about the nature of "public" and the significance of "space." He describes this new kind of urbanism as manipulative, dispersed, hostile, and, because of the mixed signals it sends, undermining of our ability to recognize and participate in the democratic process. Today's new public spaces take the form of the mega mall, the corporate enclave, pseudo historical marketplaces, and the theme park where nothing is as it seems.

A salient feature of contemporary American urbanism is the fact that basic forms of public space—pedestrian streets, squares, plazas, and promenades are increasingly becoming either obsolete or unrecognizable. Indeed, according to Herzog, public life is reconfiguring itself in a very different form so much so that we may be witnessing the end of public space as we know it. The idea that a common identity can be forged out of the American experience is a test to the strength of America's ability and will. It remains crucial to the national imagination, but has not been achieved. It will be interesting to see what happens to public spaces and civic society in Europe as European societies respond to the increasing influx of immigrants from Africa and the Middle East who likely have very different ideas about national/cultural identity, democracy, and the democratic process. Will France, the Netherlands, Italy, and England, for example, manage to construct public spaces that will allow groups to see each other in a culturally favorable light or will they begin to exhibit the anxiety and apprehension that characterize American public life and retreat into the privatized public spaces that characterize the United States?

Another reason for the decline in public space in America is that historically public spaces were the places where people stimulated each other to engage in conversation and share ideas. The powerful unwritten, unspoken codes governing behavior percolated up from the body politic and governed people's behavior and interactions. Modern society seems to call into question whether it is possible for this to happen today. Not only are the powerful codes lacking but the power of the state (which somehow seems different from the body politic) has increased. This shift away from traditional forms of civic engagement is most pronounced among the middle class, particularly those who have retreated to gated communities and the poor who come to the table with little preparation for public participation. What remains of the public sphere is mediated by technologies and politics that encourage free-for-all consumerism, privilege the privacy afforded by the automobile, and foster new technologies of communication that discourage face-to-face interactions. (Sheller and Urry; Graham and Marvin, 209-213). This raises the question, why are public spaccs important, what purpose do they serve? What do we lose if we lose spaces where people can debate questions that are important to them? For these questions, there is an easy answer. First, in places where there are vibrant, organic public spaces, what is important comes from the polity, not professional politicians. Second, in true public spaces where ordinary citizens have an opportunity to express and argue ideas, the cooperative nature of community unfolds, and a forum for rational discussion is established. Everyone becomes responsible for the decisions that affect everyone. But ensuring that this happens can get *very, very* messy.

Public Spaces as Sites of Theater

In addition to being sites of tension and contests, on a smaller scale, urban public spaces are the places people go to people watch. These more organic spaces are theaters full of sights and sounds where people perform in front of others and tell their stories. Architects are fond of describing cities as "theatres" of memory and of prophecy. If they are correct, small urban public spaces are the places where individuals—using only the material environment as a prop—tell their stories in front of an audience. Whether it is the domestic drama in which the focus is on the everyday or it is the spectacle, the stories of winning and losing, being happy or sad are told in dress, mannerisms, and localisms.

Jane Jacobs in *The Death and Life of Great American Cities* referred to the public space of Hudson Street where she lived as a ballet. All of the various inhabitants have their place in the ballet through their comings and goings: residents leave for work, store owners open their shops, longshoremen come in for a bite to eat, and mothers walk their children (Jacobs, 50-54). The performances of the people tell the story of a neighborhood working to keep itself alive—and safe from outsiders and interlopers.

Jacob's ballet is performed in the small and fairly homogeneous space of Hudson Street but it is not remarkably different from what Elijah Anderson observed along Philadelphia's Germantown Avenue, a much longer, less homogeneous street but nonetheless a theater where a play or (maybe several plays) are being staged. Anderson notes that in Chestnut Hill where a "code of civility" applies, the narrative is one of amicableness, economic success, and racial integration. People of all races mingle, are polite with one another, and seem comfortable in their roles. They know how to act in this space, which itself gives countless clues—outdoor coffee shops, banks, bookstores, high-end retailers, flower shops, and ice cream parlors, not to mention cobblestone streets, quaint architecture, and the appearance of "old money" aristocracy.

Further down the Avenue, however the street itself—with its riot gates, check cashing establishments, pawn shops, dollar stores, and thrift shops—provides other clues about how one must act. Young men adopt the clothing styles, body language, and speech of the streets. According to Anderson, "[p]eople 'profile' here, 'representing' the image of themselves by which they would like to be known: who they are and how they stand in relation to whom" (22):

> A young man in his twenties crosses the street after taking care of some business with another young man, gets into his new black BMW and sidles up next to his girlfriend [. . .] he is dressed in a crisp white t-shirt with "Hilfiger" emblazoned across the back, nice shorts and expensive sneakers. He makes a striking figure as he slides into his vehicle and others take note. He moves with aplomb, well aware that he is where he wants to be and for the moment at least, where others want to be as well. His presentation announces that he can take care of himself should someone choose to tangle with him. (Anderson, 23)

Anderson's work highlights the theatrical dimension of public space but it also foregrounds the powerful unwritten, unspoken codes that govern how we act in public spaces. But what happens in public spaces when the codes governing people's behavior change from block to block or when there is no consensus regarding roles or how to act?

Both Jacobs and Anderson treat the street and the sidewalk as quintessential urban public space—theaters and places to "people watch." But at the same time, these spaces carry other meanings and house our fears—of crime, difference, and strangers. It is this perception that has led to increasing privatization, consumer malls, and gated communities. The new American street (and the new American public space) is now nicely ensconced in the shopping mall.

Public Spaces as Sites of Exclusion

Understanding the various practices of exclusion and how they work is critical if we are to understand the social geography of urban public spaces in advanced capitalist societies. Not only is exclusion a practice that shapes public spaces, it is also the principle threat to the future of public space. The sidewalks, streets, and small neighborhood parks are a valuable lens to observe some of the more opaque instances of social exclusion. Some practices are seemingly innocuous (such as glances, gazes, and glares) and rely on mutual understandings; however, others are more problematic and troubling. The "no snitching" phenomena in Philadelphia and other major cities can be seen as an exclusionary practice that is more insidious than the typical glances, gazes, and the like. The no snitching message has been delivered in rap lyrics and on t-shirts and caps. Proponents claim that "vigilante policing" keeps outsiders at bay and prevents interlopers (the police) from interfering in the affairs of the community. In truth, its intent is to discourage crime witnesses from providing law enforcement with information that can lead to arrests. Community members are cautioned not to inform on criminals. The trend has frustrated police and prosecutors, who point out that the real intent is to intimidate witnesses. The practice remained out of sight until recently and even now remains poorly understood by social scientists who don't fully know how it is enforced much less why people consent to it and how it can be countered.

Exclusion is fed by inequality, difference, and fear. These are the issues that have come to occupy the American imagination. The public spaces of cities are the places where these problems are most visible and, as the Atlanta city ordinance suggests, they are the spaces where policing and enforcement is most vigorous.

Sibley, in his work on geographies of exclusion, describes an incident that took place in the 1960s in Philadelphia's Rittenhouse Square that illustrates the practice of exclusion. He notes:

> One warm Sunday afternoon, there was a group of hippies sitting on the wall surrounding the park, some playing acoustic guitars, but mostly just chatting and enjoying the sunshine. At some point, a park guard started to order people off the wall on the grounds that it was not a place to sit. The wall, he asserted, was there to separate the path from the grass. It was definitely not to sit on. Almost everyone acquiesced. This might have been because the park guard [. . .] was equipped with a revolver [. . .] and appeared intimidating. It could also be the case that this group of middle class youth, having been brought up in conformist communities was accustomed to accepting authority despite their trappings of non-conformity. (Sibley, 3)

One wonders what would happen today. Would the youths not have moved? Would the officer not have ordered them to move? Would the young people have verbally and/or physically challenged the officer? To what extent does the exclusion or inclusion of the youth threaten the integrity of public space?

Inasmuch as public spaces are highly dependent on codes, institutions, norms, and customs rather than laws and rules, the occupant of a public space is at the mercy of culture. The regulation of social behavior is conducted as much by the individual's knowledge that he or she is being watched as it is by armed guards or by more obvious forms of discipline. But what happens when the cultural cues are no longer read and disciplining practices (e.g., gazes) mean nothing? When being looked at disapprovingly is not interpreted as a sign to act right? Urban public space represents one of those arenas where not only is it possible to study the disciplining effects of society, it offers an opportunity to concentrate attention on the way these disciplining strategies may fail, succeed, or be disrupted.

Most traditional urban public spaces cater to groups who are homogeneous in some way and who understand the norms and codes and discipline themselves accordingly. Many of the spaces themselves have been designed in such a way that they attract certain users and exclude others. Penn's Landing in Philadelphia offers a good example of the delicate balance that must be achieved between exclusion and keeping public space public. Each year during the annual Independence Day festivities, a diverse crowd of different ages, races, income groups, ethnicities, and sexual orientations assemble to watch the fireworks and celebrate the day. Police presence is increased significantly and the crowd is rigorously managed and surveilled to make certain the events come off without a hitch—and they usually do. What's interesting about Penn's Landing is that at other times, when police presence is less, it is often empty. While there might be many reasons for this, what I propose here is that when public spaces are truly open and accessible to all, people feel uncomfortable and afraid of what might happen. Exclusion on the one hand is a necessary (but perhaps not sufficient) condition for public space to perform its function and on the other hand is one of the primary tools contributing to its demise.

Public spaces that cater to middle and upper income groups seem to be increasing (or rather private spaces masquerading as public spaces are increasing). At the same time, in spaces that welcome the poor, graffiti, gang takeover, and vandalism are increasing. Where do the poor and marginalized go to make sense and give meaning to what is happening around them? How do their understandings get incorporated into politics and the national ethos? How do they participate in democratic life? Where are their public spaces and places?

These questions are not only related to equity of access but as Sorkin and others point out, they also pertain to private versus public control of spaces. Diminished revenue streams and fiscal constraints in many cities make it necessary to place the management and control of public space in the hands of corporations whose goal it is to generate profits. Thus, many of the new so-called public spaces (such as sidewalk cafes) restrict access to those who can afford to pay. Restricting access, à la Garrett Hardin, is increasingly becoming the preferred strategy for reappropriating space to attract more welcome users and excluding those who *look* like they cannot afford to pay. Thus, the important question is not how to make *more* public space but how to make public space *more public*.

Conclusions

I began by examining the nature of public space and suggesting that public space is not just an architectural form or material construction but is part of a dynamic and fluid socio-spatial dialectic which ultimately reflects our tastes, values, and visions. Its relationship with society is both dialogic and dialectical. Its very presence as a material thing and a conceptual framing of the world and the ideas we hold makes it constitutive of the processes that created it in the first place—either through the materiality of the tangible, visible scene or through the symbolic qualities embedded in it that make it undeniably and unmistakably normative.

My goal for this essay has been to disturb the surface of how we think about public space and hint at alternative readings. Public spaces are not only the big and the beautiful. They include all of the sites where meaning is made—streets, sidewalks, and neighborhood parks. I suggest here that we can interrogate them as objects and also interpret them for what they might tell us about our social world. Obsolete housing, inadequate schools, racial and cultural diversity, conflict, discomfort, high taxes, and feelings of helplessness have pushed many Americans away from civic engagement and into safe, homogeneous, gated communities. As public spaces become more policed and disciplined, they reflect certain values and preferences but at the same time they present us with a way to think about the nature of our society.

There is nothing permanent about urban public space. Rather, it is rapidly becoming a relic of a by-gone era and is being abandoned in favor of the safe, non-theatrical, public spaces of the internet, chat rooms, and blogs. There is no guarantee that public spaces as we know them will continue to survive and the question is, if we lose them, are we prepared to accept what we have lost?

Bibliography

Anderson, Elijah. *Code of the Streets: Decency, Violence and the Moral Life of the Inner City*. New York: W.W. Norton Publishers, 1999.

Carr, Stephen, Mark Francis, LeAnne Rivlin, Andrew Stone. *Public Space*. Cambridge: Cambridge University Press, 1992.

City of Atlanta. The City of Atlanta's City Council Proposed Legislation to establish Free Speech Zones. Ordinance # 07-0-1100. City Utilities Committee. Legislation introduced by Jim Maddox. http://apps.atlantaga.gov/citycouncil/2007/images/proposed/07O1100.pdf.

Foucault, Michel. "Space, Knowledge and Power." *The Foucault Reader*. Edited by Paul Rabinow. New York: Pantheon Books, 1984.

French, J. S. *Urban Space: A Brief History of the City Square*. Dubuque: Kendall/Hunt Publishers, 1978

Graham, Stephen and Stephen Marvin. *Telecommunications and the City: Electronic Spaces, Urban Places*. London: Routledge, 1996.

Hardin, Garrett. "The Tragedy of the Commons." *Science* 162 (1968): 1243-1248.

Hayden, Dolores. *The Power of Place: Urban Landscapes as Public History*. Cambridge, MA: The MIT Press, 1995.

Henaff, Marcell and Tracey Strong, eds. *Public Space and Democracy*. Minneapolis: University of Minnesota Press, 2001.

Herzog, Lawrence. *Return to the Center: Culture, Public Space and City Building in a Global Era*. Austin: University of Texas Press, 2006.

Jacobs, Jane. *The Death and Life of Great American Cities*. New York: Vintage Publishers, 1992.

Low, Setha. *On the Plaza: The Politics of Public Space and Culture*. Austin: University of Texas Press, 2000.

Roussopoulos, Dimitrious. *Public Place: Citizen Participation in the Neighborhood and the City*. London: Black Rose Books, 2000.

Schein, Richard. *Landscape and Race in the United States*. London: Routledge, 2006.

Sheller, Mimi and John Urry. "The City and the Car." *International Journal of Urban and Regional Research* 24 (2000): 737-757.
Sibley, David. *Geographies of Exclusion*. London: Routledge, 1995.
Sorkin, Michael. "See You in Disneyland." In *Variations on a Theme Park: The New American City and the End of Public Space*, edited by Michael Sorkin, 205-232. New York: Hill and Wang Publishers, 1992.

Fig. 1. Hiking and Skiing permitted, Biking NOT. Photograph by author.

Fig. 2. No Skateboarding Allowed ("NO" removed). Photograph by author.

Fig.3. Sanctioned improvements to public playground. Photograph by author.

Fig. 4. Untagged graffiti. Photograph by author.

Fig.5. Untagged graffiti. Photograph by author.

Fig. 6. Control of public space. Photograph by author.

Fig. 7. Along Germantown Avenue in Chestnut Hill.

Fig. 8. Along Germantown Avenue in Mt. Airy.

Fig. 9. Along Germantown Avenue in North Philadelphia. Photograph by author.

Fig. 10. Along Germantown Avenue in North Philadelphia. Photograph by author.

Fig. 11. New American Street, enclosed in mall. Photograph by author.

Fig. 12. "Outdoor cafe," enclosed in mall. Photograph by author.

Figure 11-design barrier to prevent sleeping

Fig. 13. Design barrier to prevent sleeping. Photograph by author.

Fig. 14. Sleeping benches. Photograph by author.

Part Three: The Mutability of Public Space

Walking the High Line

Eric J. Sandeen

The High Line Canal, a marvel of engineering constructed in the early 1880s to bring water to the plains east of Denver, persists as an 84-mile long artifact of settlement history along the Front Range of the Rocky Mountains in Colorado. Now accompanied by a 68-mile long trail, this public space offers a platform from which to view the development of metropolitan Denver. The Canal and Trail constitute a linear landscape that holds its own narratives: the continual negotiation of the boundary between public space and private property and the sequence of experiencing this landscape as an environment of its own. The High Line could be classified as a public amenity, but, as the naturalist Robert Michael Pyle has pointed out, it is essential to maintaining a day-to-day connection between a burgeoning metropolitan population and nature, however "second hand" or vestigial that connection might be. This essay walks the High Line with particular care through the city of Aurora and examines its function as a coherent landscape amid barely restrained metropolitan growth.

The relationship between nature and the city can be seen most clearly in the interior West, where the demands of rapidly expanding settlement collide with the limitations of natural systems in an open terrain that both reveals and obscures. In plain view, Front Range development seemingly sweeps over the eastern horizon, creating what the naturalist Robert Michael Pyle has termed "a hell of a mess" (130). Tucked into this terrain of development and overlooked by the placeless rhetoric of boosterism and ambition are compensatory environments that Pyle embraces in his account of personal growth and suburban development, *The Thunder Tree*. Following a trail marked by Pyle's narrative, this essay, "Walking the High Line," enters the contradictory territory of metropolitan Denver, Colorado by way of an anomalous landscape of enduring temporal and spatial significance, the High Line Canal and Trail (fig. 1). Looking at the High Line confirms the importance of water in the semi-arid West and presents a version of the settlement history of the region.[1] The history of the Canal and Trail

[1] The most accessible source for High Line Canal history is the 1994 Historic American Engineering Record documentation, available online through the American Memory database at the Library of Congress: http://memory.loc.gov/.

leads to the central purpose of this essay: examining the High Line as a particular landscape form, a linear landscape that allows us to order exuberant, even chaotic, metropolitan growth and see the interplay between competing demands for public amenities and private development. As an example of the sometimes surprising views that a linear landscape can provide of this contested metropolitan cartography, I will follow the High Line through Aurora, a suburb that has grown large enough to count itself as a midsized city, ranking (in 2003) between Corpus Christi, Texas and Raleigh, North Carolina. The view from the High Line may provide an opportunity to assess Aurora's claim to represent a new urban form. More broadly, establishing the High Line as a distinct landscape, with its own relationships to natural history, human settlement, and public management, provides solid footing from which to survey urban sprawl in this fragile environment.

The High Line has a venerable position in the history of irrigation in the eastern Colorado piedmont. Following the successful development of the Union Colony in Greeley, in 1879 the Northern Irrigation Company proposed a long-distance irrigation canal that would draw from the South Platte River to water the eastern plains. The result of the Company's three-year effort was a well-engineered anomaly: a band of water, dropping two feet per mile by following the "high line" that inscribed the original topography of the land on an area already gridded for settlement.

The importance of the High Line as a supplier of water to agricultural and urban users has diminished as the area east of the Front Range has developed into metropolitan Denver. During the late nineteenth century the Canal was an important conduit for water drawn from increasing distances. As the municipalities along its banks developed their own wells and water delivery systems, the Canal became an inefficient delivery system, surrendering as much as 60% of its contents through leakage into the surrounding soil and by evaporation. Other canals disappeared underground to escape the sun's rays and the infrastructure of city services took care of the rest, leaving the High Line as a reminder of an antiquated technology tied to a landscape long since platted and paved. Over its 84-mile length, derelict headgates that once opened to deliver High Line water to individual users and disused laterals—subsidiary ditches connecting irrigation projects to the main canal—signify a history of use by agriculture, municipalities, and federal facilities.

Two systems of power flowed through the Canal during the initial decades of its operation. The first, the legal right to put public water to private use, was energized by the evolving Colorado law of prior appropriation, under which water was allocated to potential users in the order in which it had been claimed—first in right, first in law. This was a significant departure from procedures in eastern states, which drew from riparian water rights that had their origins in English common law. Prior appropriation assumed scarcity and required "beneficial use" of this resource, thereby creating an economy of water founded on development and the application of complex technologies of water distribution and curtailment (Wilkinson, 231-235). This was a complex web, indeed, in both its legalities and its landscapes. More claims to a stream could be registered than there was water to be allocated, even in wet years. Canals or other diversions intended for a distant user could cross private property without surrendering any of its contents before reaching its destination. In dry years, those with the most recent claims, holding "junior" rights, might not receive any water at all while older, "senior" rights holders drank their fill. The High Line held a junior right to the South Platte and two drought years in the 1880s proved the tenuousness of its claim to water. Water police manned locked headgates as the meager flow was distributed according to priority within its system; at night, parched farmers broke the locks (Kienast, 47).

Second, and even more problematically, the High Line served as a conduit connecting elements of a dynamic landscape of settlement. Agricultural and urban users drew from the same scarce sources. Urban areas of competing ambitions, such as Denver and its smaller eastern neighbor, Fletcher, viewed access to water as a guarantee of future prosperity. The growth potential of the smaller community, rechristened Aurora in 1891, depended on the number of taps and diversions allocated by the Canal owners. By 1911 that ownership had passed to the newly established Water Board of the city of Denver from the Northern Irrigation Company. The latter was a subsidiary of the Colorado Mortgage and Investment Company, known colloquially as "The English Company" because of its foreign investors. For Aurora, the Canal was then made doubly unreliable, subject to the vagaries of the weather and the volatility of political relationships.

The entry of the federal government through the Bureau of Reclamation intensified competition within these systems of power. The early twentieth-century agricultural historian William Smythe portrayed the 1902 founding of the Bureau as providential. The news coming from the Roosevelt administration was received by people in the semi-arid region "like the stroke of high noon on the clock of destiny" (Smythe, 284). Expressed less grandiloquently, the founding of a bureau that could reshape Western landscapes provided added impetus for water projects west of the Front Range that would benefit settlements to the east. The High Line was just the final manifestation of a complex system of government programs, engineering marvels, and ecological marauding that extended out of view, hidden by the 14,000-foot peaks of mountains to the west.

Within twenty years of the digging of the High Line the two methods of controlling the fundamental resource of growth were established. Water could be managed either at the moment of need through appropriation of existing sources or for the predictable future through construction of dams and reservoirs. The control of the tap and the headgate have been durable powers in the West, guiding economies, determining development, and shaping large-scale modifications to the landscape. The appropriation of water and the ability to construct dams and reservoirs are among the few transformative forces that Charles Wilkinson has called "the lords of yesterday" (17), although these two still exert their power today. Having been founded along the unreliable High Line Canal, Aurora has been searching for water ever since its founding, tapping groundwater accessible on the plains, sharing resources with Denver, and commanding its own resources west of the horizon.

The digging of the High Line tied Denver and Aurora together in a dynamic, competitive relationship that pointed both cities upstream in search of water. Both sought to build reservoirs that would assure water flow during the dry months by impounding water that could not be put to immediate use—the normal runoff from spring melting in the high country and rainy periods that occurred during inconvenient times of the year when water was not needed downstream. These torrents rolled uselessly down drainages, into rivers like the South Platte, and, eventually, out of the area in the direction of thirsty Great Plains states. Denver and Aurora, among others in the developing urban area on the Colorado plains, pioneered dam sites and capture areas in the narrow canyon formed by the South Platte as it broke through the Front Range. These water projects joined the rail lines of the previous generation in a technological transformation of these landscapes, through tunnels, flumes, and other massive structures, the engineering marvels of the age (fig. 2).

Soon the water available on the east side of the Front Range was not enough; first Denver and then Aurora explored the western slope. In part, this impact on landscapes more than 100 miles west of the metropolis is a familiar Western story most dramatically portrayed in the search for water in 1920s Los Angeles. In fact, in 1914 Denver employed J. R. Lippencott, later complicit in the exploitation of water in California's Owens Valley, to explore water sources in the moister, sparsely settled valleys west of the divide (Skari, 25-26). His recommendations extended Denver's reach more than 100 miles to the west. The effect in the wetter valleys of central Colorado was not as dramatic as water diversions would later be for the eastern slope of the Sierras, but, by establishing the profitable tactic of inter-drainage appropriation supported by massive engineering projects of water tunnels and pumping facilities, the Denver Water Board, in the words of Marc Reisner, became "a kind of understudy of the Metropolitan Water district of Los Angeles" (430).

A second impact is harder to trace, although, because of its greater extent, more profound. Let Crowley County, far to the south and east of Denver and Aurora, stand as a representative for many areas that lay, like the developing metropolis, east of the Front Range. By the completion of the massive Homestake project in 1967, Denver, Aurora, and a few other developing cities, drew large amounts of water from the Arkansas River basin west of the Front Range. There was a heavy investment in this water: Denver constructed a system of tunnels and pumping stations; Aurora transferred water between drainages in order to reach the South Platte, where it could easily be reached (fig. 3). In order to guarantee this supply, both cities bought up senior water rights, in some cases offering ten times what agricultural users were willing to pay. As a consequence, when the Arkansas broke through the mountains and finally reached Crowley County, it had no water to give

(MacDonnell, 54-60). The desertification occurring miles to the south and east of Aurora and Denver was caused by water purchases made more than 100 miles west of both cities.

By containing the power of water, the High Line embodies a particular kind of control over the Western landscape, so trenchantly documented by environmental historian Donald Worster. In this "hydraulic society" human ambitions are drawn to this prime commodity, encouraging the observer to read the development of the West as a capitalistic exercise in appropriation and exploitation (Worster, 7). This reflection from a ditch, to modify Worster's expression, reveals systems and structures of control. Viewing a particular ditch, such as the High Line, reveals more clearly specific landscapes and the built environment of the evolving metropolis.

Equally significantly, the High Line is itself a linear landscape that interjects its own narratives as it lays claim to space within the metropolitan area. A linear landscape is much longer than it is wide and, in its length, represents an element on the map large enough to have to be dealt with, or at least recognized. Thus, an airport runway would not qualify, but a jogging path would. Such landscapes possess clear boundaries, defining their linearity; a person sauntering down a linear landscape follows a designated path and is not a trailblazer. Fully articulated landscapes of this type twist and bend, enticing the eye of the observer onward. Because of their unprofitable dimensions—an unruly length and an inconvenient width—linear landscapes tend to be publicly owned. Again because of their sometimes meandering or purposeful nature (a mountain path, a pipeline right of way), linear landscapes tend to cross one another, creating in the crosshatched intersection a composite landscape of often contrary intentions, a confounding entity unto itself. Finally, because a linear landscape must be experienced sequentially, rather than ascertained with one glance, narrative is an essential element, in the case of the High Line, articulating an unfolding tale of how public space inserts itself into private development and negotiates the economies of cultural values in the metropolitan area. Linear landscapes encourage epiphanies, one after the other.

The High Line Canal and Trail represents a striking example of such a landscape form that allows the pedestrian or rider to traverse metropolitan terrain (fig. 4). Over its 68-mile course the Canal Trail corridor presents a variety of surfaces for the pedestrian, jogger, bicyclist, and rider on horseback: shaded country lane, improved dirt track, and macadamized or concrete path in various states of repair, the most recent sections of which conform to the design requirements for bike paths and green belts. The Trail runs alongside the Canal for most of this journey, except where private property does not permit access or the Canal is isolated in a piece of water-moving technology, such as a flume carrying the Canal water above the landscape or a siphon allowing it to flow under an obstruction. On such occasions, the Trail ventures onto urban streets, marked by a line in the pavement or a directional sign. The experience of transporting oneself down the Trail is also syncopated by the many at-grade crossings with urban streets and avenues.

The Trail follows a canal that is strikingly intermittent. According to the season and the geography of water allocated to the Canal, one can come upon a lush riparian environment, green with water-loving plants and endearing fauna, such as ducks and other waterfowl. More often, particularly in the lower Canal, or Eastern Canal, one views a dry ditch, shaded frequently by cottonwood trees that have enough stamina to persist through the many dry weeks of high summer.

East of Denver, mostly in Aurora, the Trail and Canal is a managed corridor, consistent and relatively uncomplicated. The Trail is generally concrete and is sometimes paralleled by a narrow dirt jogging track. The Canal, a ditch 12 to 20 feet wide from lip to lip, is dry for most of the year. Weeks for which water is allocated to support the cottonwoods are prized returns to the relatively lush environment of irrigation days. Encounters with a well-watered Canal after the chance inundations of summer storms are experiences all the more savored because of their serendipity. At times, the High Line is a channelized trench (fig. 5), its concrete embankments offering a canvas for the occasional suburban graffiti artist. The Canal and Trail lose their canopy of trees as they meander east through newer subdivisions, offering increasingly unobstructed views of the land patterns—public, private, protected by covenant, or opened by easement—that abut the corridor. The linear landscape of the High Line is both an historical and visual vantage point from which to survey the area and a functional element of the metropolis.

An anomalous landscape like the High Line can help us understand the growth of Aurora. Now, Aurora is a major component of the metropolitan area but, in 1891, it was the small, four square mile settlement of Fletcher. At the end of World War II, Aurora had a population of 11,000. By 1970 this number had increased to 74,000 and doubled during the next decade. As of 2007 Aurora is poised to pass the 300,000 mark. The territory governed by the city of Aurora now consists of 164 square miles, some of its municipal area separated from the core, spread out over three counties.[2] Pointing to these remarkable statistics, Aurora's comprehensive plans dating back to the early 1980s have proclaimed the city a new urban form, not signaled by the skyscrapers of the traditional city to its west, but a de-centered, twenty-first century metropolis.[3] The High Line gives a narrative path to the history of a city that has defined itself in terms of development—in terms of the future rather than the past. The insistent geography of the Canal insinuates its way among the development schemes and general buying and selling of post-World War II suburban expansion, helping us to understand the highly differentiated banality of a form which, if not new, is at least oxymoronic: the suburban city.

Looking at Aurora from the High Line brings into higher relief the shaping hand of the federal government during the first century of the city's growth. During World War I, the Gutheil nursery, which had been supplied with water from the High Line, was uprooted in favor of the Fitzsimons Army Hospital, a major recuperation stop for American soldiers during both world wars and, famously, for President Dwight D. Eisenhower, who made this spot his Western White House as he recovered from a heart attack suffered while visiting Colorado in 1955. Nearby Lowry Air base, connected to the Canal by laterals, served the Army Air Corps during World War II and became the first home of the Air Force Academy, also in 1955. A major client for the Canal through the 1970s was the Rocky Mountain Arsenal, site of munitions assembly, repository for chemical and biological agents and, for a time in World War II, the site of asparagus fields tended by prisoners of war. Countless soldiers passed through these facilities and their families stayed at the burgeoning motel areas along Colfax Avenue, particularly

---

[2] All statistics are drawn from the 2003 City of Aurora "Comprehensive Plan."
[3] First articulated in the 1983 "Comprehensive Plan" and often repeated thereafter.

those clustered near Fitzsimons and its convalescing patients. After World War II, many of these armed forces families decided to stay in Aurora, thereby contributing to the boom during the 1950s, embodying in tract housing the benefits of the GI Bill, and embedding a strong military culture in Aurora.[4]

The federal government's decision to subsidize the automobile, of which highway construction was only the most visible manifestation, had an effect on Aurora measured by the connecting tissue of the Canal. Avenues like Alameda and Mississippi, the day-to-day conduits of increased automobile traffic, provided the easiest way out of the older urban core of Denver, a movement abetted by the climate of opinion among planners at both the municipal and federal levels that cities needed to be relieved of congestion (Gutfreund, 87-95). A Colorado law passed in 1947 allowed for the creation of "improvement districts," enabling developers to create quasi-civic entities on technically unincorporated land, premised on the easy availability of automobiles to reach jobs and services not provided within the subdivision.[5] Such was the case, for example, with Hoffman Heights, developed in 1950 between Denver and Aurora along the banks of the High Line Canal. In 1954, amid great controversy, Aurora annexed Hoffman Heights, initiating the first era of its post-war expansion (Mehls, Drake, and Fell, 135-140).

The initial, eastward expansion of Aurora is more straightforwardly marked by Colfax Avenue (US 40), which during the first post-war decade served as both Main Street and means of escape. The High Line weaves around Colfax Avenue just as it breaks free of the line of pre-interstate-era motels that lie to the east of the only traditionally urbanized portion of Aurora. Widened to four lanes and given a median divider in 1951, Colfax was the main entry to town from the east. On the eastern outskirts, not far from the High Line, Aurora erected a grand arch proclaiming itself "Gateway to the Rockies" to legions of tourists who ventured forth on family vacations. During the 1950s, Colfax Avenue was Aurora's Miracle Mile, a

[4] Facts about post-World War II Aurora are drawn from Mehls, Drake, and Fell, Chapter 5.

[5] City of Aurora, "Aurora, Colorado Performance Report," 2.11. Through the 1980s Aurora offered few jobs in relation to its size.

designation that began to lose its luster in the late 1950s, when Havana Street, a section-line or "mile road" (Pyle, 93) intersecting Colfax, competed for commercial locations, reorienting the axis of traffic and pushing residential networks to the south. By the 1970s, Aurora was the fastest growing city in the United States but this activity did not take place near Colfax, which began to suffer the fate of other urban core areas. A sure sign of blight was its eligibility for urban renewal funds.[6]

The High Line landscape establishes a necessary axis for mapping the tensions of Aurora development over the past fifty years. By following natural contours, the High Line shows that Aurora development was at first pointed east, following the possibilities that water brought to both farmland and human settlement. As Owen Gutfreund has so aptly documented in *20th-Century Sprawl*, later expansion in metropolitan Denver was directed by the development of the interstate highway system. In the case of Aurora, this is evinced by the construction of an inner ring bypass around Denver, I-225, that reoriented the expanding suburb along another axis pointing south, toward the Denver Tech Center (a massive complex of technical and financial services facilities that by 1994 housed more than 900 firms in 165 office buildings) with competing developments clustered nearby (Gutfreund, 98). The 1983 "Comprehensive Plan" presented a Cartesian mapping of Aurora's future: planners anticipated that 65% of Aurora's growth would occur in the city's southeast planning district during the remaining years of the twentieth century and the proximity of lucrative employment drove the population of this area beyond the predicted 160,000 by century's end. Meanwhile, the older and relatively poorer planning districts and census tracts of Aurora, located in the north and west of the city close to the High Line, were predicted to experience only moderate growth. These were already the areas with the largest proportion of multi-family housing, the highest vacancy rates, and the greatest ethnic diversity.

Twenty years later, the 2003 "Comprehensive Plan" graphed the challenge clearly according to these axes of development. Aurora had now to balance the needs of the heterogeneous, older city, rich in amenities, arranged from west to east along Colfax and the High Line,

---

[6] Today all three of Aurora's urban renewal areas are located along Colfax Avenue.

with the overwhelming demands of increasingly homogeneous developments of larger homes and private facilities aligned to the interstate flowing toward the high-tech employment centers in the south. Meanwhile, the construction of a second ring highway, E-470, pushed Aurora beyond its circumference, farther to the east, into rural Douglas County, where over 80 square miles of annexed land awaited development. The High Line and I-225 have operated as directional markers for development in a city that exploded in population after World War II.

Within a grid of development challenged by the vectors of growth, a center becomes hard to identify. The idea that Colfax was the heart of Aurora was challenged by the construction during the early 1980s of a municipal center for the city, around which the High Line runs in a concrete channel. Here is the civic core, at least the city's governmental offices, the main branch of the library, the historical society, and the detention center. Strong, centripetal forces in Aurora flow to the several covered shopping malls, among which the Aurora mall, near the civic center off Alameda Avenue, holds pride of both place and name. This indeterminacy, the unmanageability of a gridded city plan with a moving target instead of a bull's eye, gives the High Line the status of more than a leitmotif. It is a reminder of the original geography, a narrative line that connects elements of Aurora's exuberant, often unplanned growth, and a platform from which the metropolis can be continually resurveyed. Within the last generation, this linear landscape has also become a functional element of the metropolis, through its requirements for management and negotiation interjecting public, and sometimes inter-governmental, conversations into terrain commanded by the ambitions of private enterprise.

In 1970 the High Line was opened to the public as a "strip park," to use the terminology articulated by a contemporary *Denver Post* article (September 11, 1968), a walking and bicycle path supplanting the trail pioneered by ditch riders in former times. The making of this 68-mile-long public space required the cooperation of a variety of governmental entities—the South Suburban Parks and Recreation District, the Denver Water Board, the City of Aurora, and, later, Colorado State Parks and Trails. Changes in appearance and texture signal a complex negotiation of responsibilities (and liabilities) among the jurisdictions. The fashioning of the Trail through Denver connected major parks—DeKoevend, Eisenhower, and Bible. Farther downstream, the city of Aurora assembled open spaces and natural areas that would build from the High Line.

The High Line helps make Aurora a legible civic entity. If Aurora is, in fact, a new urban form and not just a suburb with pretensions, then spaces of sociability that carry symbols important to community identity and allow room for inhabitants to negotiate the future shape of shared space are important. We can view the significance of the High Line Canal and Trail as a generative and representational Aurora landscape in three ways.

The first is a kind of cartographic precondition, represented by the sketchy map on the back of the 2005 tourist brochure for Aurora (fig. 6). Reduced to its essentials, Aurora is nothing more than the laconic grid, with nodes of interest designated by the numbers. In order to create the semblance of a center, a stylized logo is superimposed, roughly where the civic center of town would be. Underneath this branding of the map—the sun, Aurora—is the ubiquitous grid, familiar enough not to be inviting: major avenues on the section lines, residential development within the frame of these squares, clusters of shopping and entertainment at major intersections or along interstate interchanges. Any non-Cartesian part of this picture would draw attention to itself, more so, even, than the number that might be assigned to it.

The High Line acts as a centripetal force in a metropolitan area that seems to be all periphery. This linear landscape announces its importance to the city at the beginning of the Aurora section, as the Trail changes from worn blacktop to freshly poured concrete. At major intersections, the Trail competes with street and sidewalk by attaining the markers of another avenue, or, more mundanely, a driveway. Along the path, small environments in public space offer up vignettes to the eyes of passing strangers. Some of these work in the vocabulary of public landscape design, such as irrigated lawns with trees and other plantings that sweep to the very edge of the path (fig. 7). Others appropriate valuable public space through self-evidently private areas, such as a garden or a horseshoe pit (fig. 8). City landscaping requirements encourage public use but attempt to hold permanent, private incursions at bay, creating a boundary between public land and private property along the High Line that resembles similar borders in larger environments, such as municipal parks, designated open spaces, or even National Parks. The development of these landscape design requirements can be read along the High Line as can the reaction to this wide appeal: signage enumerates prohibitions and rules of the road and some property owners aggressively mark their territory, either through signage or privacy fencing.

Secondly, the High Line reminds us of one of the enduring confusions of Western development—that even though the rhetoric of expansion is driven by a growth industry including banks, the chamber of commerce, and the city government itself, and is trussed up by the latticework of faith in technology and an infrastructure of contractors and computer services companies—the main issue remains access to water. Just as Denver controlled Aurora development in the late 1940s by limiting the number of new taps on the water system, later on access to more water was the precondition for growth. The Two Forks Project, a massive scheme for water development that would have inundated 22 miles of Front Range canyon, most of it belonging to the South Platte, including the surviving flumes of the High Line Canal, offered one way of providing more water. Even before the plan was rejected by the Environmental Protection Agency in 1990, Aurora scrambled to assemble a water plan from other sources, a nervous search that produced reassurance only at the end of the decade. The importance of water to the future can be seen graphically through these two illustrations, part of a triptych composed by the city to advertise Aurora's bright future at the end of the 1980s. Here "Opportunity" is associated with the newly constructed detention facility. "Security" means water for the future (City of Aurora, "Aurora, Colorado Performance Report"). (See fig. 9.)

Along the High Line Trail, water means history as well as prospect. The most visible ties to the history of settlement promoted by the Canal's water are the cottonwood groves that were planted to stabilize the banks of the ditch and then propagated along the narrow riparian zone. A more typical urban canal would have been landscaped with residential use in mind. A canal in the nearby city of Boulder, for example, offered late nineteenth-century inhabitants the possibility of ornate gardens, incorporating bridges and picturesque tableaux along the canal's banks (Halloren and Sanders, 3). The High Line was built to deliver water for agricultural use and the cottonwoods offer a rougher, more overpowering, and yet fragile presence of nature (fig. 10). The trees are roughly as old as the Canal and are therefore reaching the end of their lives. The loss of each one is mourned and the replacement of this canopy becomes an exercise in historic landscape preservation.

Third, the High Line provides the backbone for an expanding system of parks and trails that both break up the grid and give Aurora an identity. During the booming 1970s, inhabitants of this area famously identified themselves with their subdivision and not with the city of Aurora. Named after one of the most aggressively marketed and luxurious of these Aurora developments, this was sometime called "the Mission Viejo Syndrome."[7] During the same period, citizen satisfaction surveys indicated that city parks and open spaces were among the most favorably received elements of the city (City of Aurora, "1991 Citizen Survey Results"). By the time of the production of the tourist brochure cited above, green spaces had come to represent quality of life. Clashing with the anonymity of the map of Aurora, the rest of the brochure is a swathe of green. Bicycles, many of them to be seen along the High Line Canal Trail, replace cars and greenswards and barns stand in the place of the more mundane four-lane avenues with protected left-turn lanes (fig. 11). Such highly desired spaces have been transferred from planning maps to solid ground. Present-day Aurora boasts an extensive array of trails, parks, natural areas, and open spaces, armed with the easements and zoning buffers to make them permanent (fig. 12).

A good example of this influence lies just downstream from the City Center of Aurora, in a large natural area called the Delaney Farm. This property, acquired by the city as a natural area in the 1980s, is defined by the long loop of the High Line Canal Trail and its intersection with the Toll Creek Trail, a more recent addition to the Aurora recreation system. The Farm consists of historic buildings—one of them the last surviving round barn in the area—and an organic farm that serves as a nature education area. The John Gully Homestead, marked by the small house occupied by that family, attests to the

[7] Mehls, Drake, and Fell, 169. Pyle, who grew up in Hoffman Estates, christened Front Range boosterism "the Aurora syndrome." During his ten formative years, the population exploded and expansion bulldozed most of his childhood landscapes. Such unreflective destruction, he believed, could only be described clinically, as a neurosis.

hardships of settlement in this area: never successful at farming, the Gullys tended a toll gate and stage stop.[8] Overlooked by many who observe the thriving prairie dog colony and plump raptors in this area are the technological ganglion of siphons and headgates that announce the confluence of the High Line and Tollgate Creek.

One could read an account of Aurora told from the position of the High Line as a conceit. In fact, much of Aurora's development had its origins in policy rather than topography. For example, the suburban boom of the 1970s was caused in no small measure by the passage by the Colorado legislature of the Poundstone Amendment, which prohibited any city from annexing land without the approval of all the residents of the county in which the proposed annexation was located. Since Denver was uniquely a city and a county, the law, in effect, circumscribed only Denver's expansion. The effect of the Amendment was exacerbated by the ongoing battle over the desegregation of Denver schools, hastening the flow of population and capital to the periphery (Gutfreund, 93-94). Also during this decade, the Aurora Planning Department inscribed the city map with a blue line, resolving not to annex land beyond its boundaries but to encourage infill development that would lay claim to unincorporated areas and improvement districts instead. Over the course of the next decade, expansion out onto large parcels of unregulated, less expensive land on the eastern plains overwhelmed control and changed the political map, overpowering Aurora's intentions and pushing the city to embrace annexation once again in 1984.

The High Line ties together larger narratives of resource use and urban development but it also functions as a distinct landscape with its own history of construction and preservation and a distinct public value as utility and amenity. Environments like the High Line Canal and Trail are more commonly labeled linear features or as a contributing element of a system of landscape components devoted to the same purpose—a park system, for example, or a network of deeded open spaces. The High Line is more than this. As a linear landscape it unfolds through narrative—manifested both as written histories and as the sequential experience of encountering it.

---

[8] "Delany Farm." Trails.com. http://doi.contentdirections.com/mr/trails.jsp/10.2126/HGR244-014.

Robert Pyle articulates what could be called the naturalist's tale, drawing together these strands of narrative. In *The Thunder Tree*, Pyle appeals for an acknowledgment of the persistence of nature amid development and shows how fruitful close attention to the terrain commanded by the High Line Canal and Trail can be. Amidst the wrenching dislocations and wholesale appropriations of suburban sprawl, this narrative sees compensatory spaces in patches of green and even the vacancy of abandoned lands. Pyle, dubbed by the Denver *Post* "the Thoreau of the High Line" for his eloquent autobiography, weaves into this work the natural history of the Canal and the development of present-day Aurora. Pyle reminisces about his own interest in butterflies and presents an inventory of species that can be found along the Canal. Equally important, he makes a case for "secondhand" lands, the vacant lot, the dirt pile, the field left fallow in anticipation of development (Pyle, xvii). He rails against "the extinction of experience" that comes with the pavement and close-cropped lawns of the suburbs (145). Everyone connects with nature somewhere, he points out, and for most suburbanites that communion occurs *here*. "Everybody," he asserts, "has a ditch, or ought to" (xix). Thus, Pyle represents the green tradition of American literature, in which Thoreau, indeed, is a luminary figure. The parallels are sometimes striking. In *Walden* Thoreau observes nature in the entrail-like rivulets of mud on a railroad embankment and draws philosophical inspiration from a battle between ants or the act of hoeing beans. Pyle nets butterflies and, on a summer's day in the late 1950s, hides under a cottonwood tree on the banks of the High Line during a storm—the thunder tree. Most significantly, both writers see the relatedness of narrative, motion, and discovery. Thoreau walks, canoes, and climbs; Pyle attempts to follow the entire course of the High Line.

Robert Pyle moved to a four-bedroom house in Hoffman Heights as a child, before this development was annexed to the city of Aurora. Racine Street lay at the eastern edge of a suburb of 20,000 then; by the time he moved away, Hoffman Heights was well east of the center of a city of 200,000. In the 1950s Pyle played in the waters of the High Line, venturing as far as Toll Gate Creek, near the Delaney farm, even then a vestige of the agricultural past that would later be designated an historic site. Later, he visited a high school teacher's brick ranch house, "God's Little Acre," an outpost of settlement soon to be engulfed by subdivisions. The center point of Pyle's narrative is the thunder tree, an ancient cottonwood which offered his brother and him refuge from hail of increasing caliber during a squall in July 1954.

The High Line offers more than a narrative path through this naturalist's autobiography. It embodies both stability and movement: "The moisture it carries gives a stream its character, but it is movement of water over territory that makes it a place" (Pyle, 63). This essay has presented the High Line as containing a metaphor, channeling a commodity, and constructing a settlement history. Through Pyle's work it is also clear that exploring the landscape of the Canal alters space, time, and experience. His own experience of being pummeled by hailstones in the hollow of a cottonwood tree shaped his youth and ordered his experience of the Canal. It turns out that the tree had its own, unexpected history to contribute. It did, in fact, date to the period of the Canal's construction. He had assumed that it had been hollowed by a long-ago lightning strike. Only as an adult did he learn from George Swan, the retired Canal ditch rider, that in 1938 boys had lost control of a fire they had set to smoke out the bees in this honey tree. Even youthful play, it turns out, has a history along the Canal. Much of this is intangible, or in the case of the thunder tree, mute. Occasionally, though, small memorials reveal the continuing importance of this public landscape in the lives of its visitors (fig. 13). Travel down the mile roads was predictable, expeditious, even laconic, Pyle points out. Venturing down the High Line lengthened distance—the Canal would meander three miles or more for every metered mile of pavement—and extended experience. Thus, an intersection between the Canal and a mile road, such as Havana Street, signified the crossing of two environments, two different realms of time, and two sequences of narrative possibility (fig. 14). Many of these sequences are played out by simply experiencing the landscape—a process that transcends the title "amenity."

In 2010 the water in the High Line Canal will be reallocated according to protocols that are already being negotiated in 2007. This is a new conversation that may reveal a lot about Aurora's claim to have elevated itself to equal status with Denver. More than 500,000 visitors each year have put the High Line under public scrutiny. The cottonwoods have been recognized as historic features that must receive enough water to sustain life. Canal headgates have been documented for the Historic American Engineering Register and the High Line advertises itself as a National Landmark Trail. Public facilities—schools, parks, golf courses—crowd the Canal's boundaries. The traditional argument about water allocation sounds more antiquarian than ever, as most of the largest users switch to other sources. Moving to the center of the debate is the meaning of this landscape and how experiencing it ties together the broken strands of contemporary metropolitan life.

Bibliography

City of Aurora. "Aurora, Colorado Performance Report." 1989.
—. "Comprehensive Plan." 1983.
—. "Comprehensive Plan." 2003.
—. "1991 Citizen Survey Results." 1992.
Gutfreund, Owen. *20th-Century Sprawl: Highways and the Reshaping of the American Landscape*. New York: Oxford University Press, 2004.
Halloren, Michael and Judith Sanders. "Urban Canals in the American West." In *Landscapes of Water, Proceedings of the First International Conference,* Bari, Italy, 2002.
Kienast, Kate. "Historic Irrigation Ditches and Canals in the Denver Metro Area: New Concepts for Old Waterways." Master's thesis, University of Colorado at Denver, 1996.
MacDonnell, Lawrence. *Reclamation to Sustainability: Water, Agriculture, and the Environment in the American West*. Niwot: University Press of Colorado, 1999.
Mehls, Stephen, Carol Drake, and James Fell, Jr. *Aurora, Gateway to the Rockies*. Evergreen, CO: Cordillera Press, 1985.
Pyle, Robert Michael. *The Thunder Tree: Lessons from an Urban Wildland*. Boston: Houghton Mifflin, 1993.
Reisner, Marc. *Cadillac Desert: The American West and Its Disappearing Water*. New York: Viking Press, 1986.
Skari, David. *High Line Canal: Meandering through Time*. Denver: C & M Press, 2003.
Smythe, William. *The Conquest of Arid America*. Seattle: University of Washington Press, 1964 [1899].
Wilkinson, Charles. *Crossing the Meridian: Land, Water, and the Future of the West*. Washington, D.C.: Island Press, 1992.
Worster, Donald. *Rivers of Empire: Water, Aridity and the Growth of the American West*. New York: Pantheon Books, 1985.

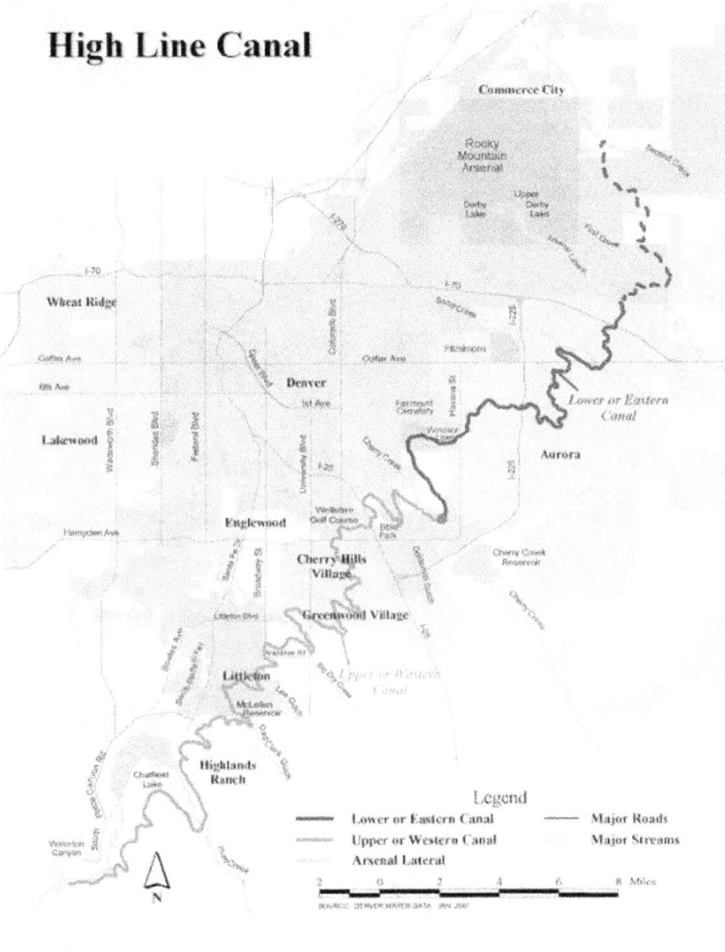

Fig.1. The High Line Canal. Map. 2006. Denver Water Board.

Fig. 2. *High Line Canal in Platte Canyon, Colorado*, 1882-1890. Photograph. William Henry Jackson. Western History Collection, Denver Public Library.

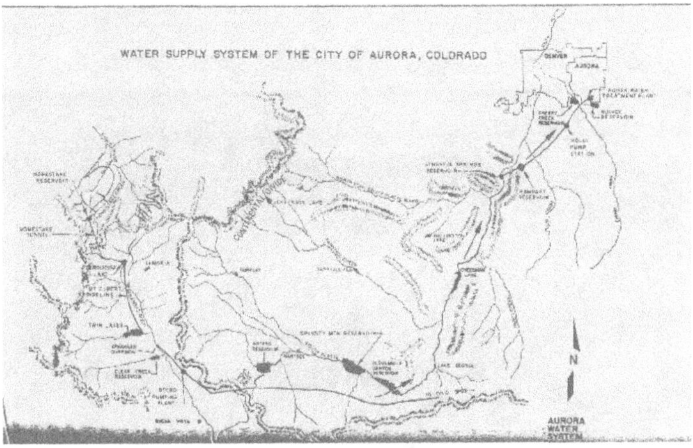

Fig. 3. The water supply system of Aurora as of 1967. Map. Stephen Mehls, Carol Drake, and James Fell, Jr., *Aurora, Gateway to the Rockies*, Evergreen, CO: Cordillera Press, 1985.

Fig. 4. The High Line Canal Trail south of Denver: old trees, crushed gravel path. 2007. Photograph by author.

Fig. 5. The channelized High Line Canal near the municipal center of Aurora. 2005. Photograph by author.

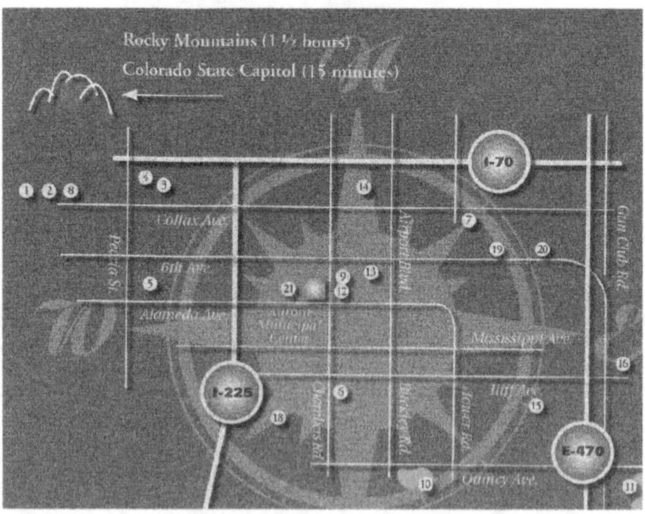

Fig. 6. Aurora's map of the new urban order. Stylized map. 2005. City of Aurora, tourist brochure.

Fig. 7. Along the High Line in Aurora. 2005. Photograph by author.

Fig. 8. Along the High Line in Aurora: public space privately claimed or generously tended? 2005. Photograph by author.

Fig. 9. Aurora – Rising to new challenges: security. 1989. City of Aurora.

Fig. 10. The disappearance of aging cottonwoods along the Canal bank (left) is anticipated by newer plantings at a discrete distance. 2005. Photograph by author.

Fig. 11. Aurora promotional brochure. 2005.

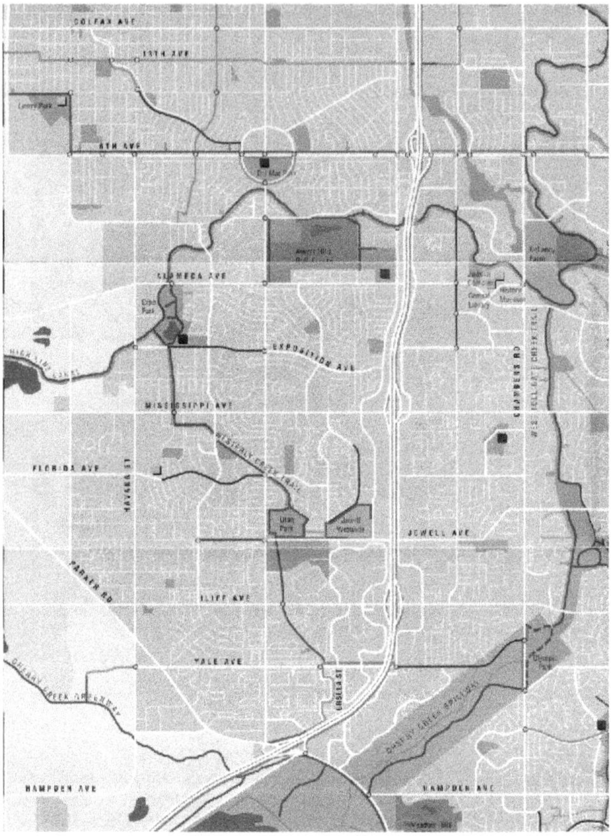

Fig. 12. Open lands and major parks in Aurora. Map. 2004.
Department of Parks and Recreation, City of Aurora.

Fig. 13. A Trail-side shrine. 2005. Photograph by author.

Fig. 14. The High Line at Havana Street, Aurora, marked by cottonwoods. 2006. Photograph by author.

The Search for a Democratic Architecture: A New Sense of Space and the Reconfiguration of American Architecture

Kerstin Schmidt

Chicago architects Louis Sullivan and Frank Lloyd Wright developed a new style of building American public and private space that they identified as "democratic architecture." Their way of building also reflects and expresses the more encompassing cultural and scientific changes regarding the concept of space that characterize the second half of the nineteenth century and the beginning of the twentieth century. In this essay, I propose that the concept of negative space captures this reevaluation of space and argue that it is precisely through negative space that Sullivan and Wright realize their idea of a democratic American architecture.

The notion of public space in America underwent significant changes around the turn of the last century. In the 1890s, for instance, the commercialization of culture gained momentum and was at the root of a skirmish about the use of public space for purposes of advertising. As Laura E. Baker shows in her recent essay on the responses to the commercialization of public space, the rapid proliferation of billboards in public space triggered a controversy between those who criticized the permeation of urban landscape by outdoor advertising and those who welcomed it as a sign of a new mass culture. Chicago alone had about 50 miles of billboards to frame its streets in 1905 (Baker, 1191). The billboard and the concomitant commercialization of culture changed the urban landscape thoroughly and raised a more encompassing aesthetic debate on how to deal with public space, and, quite literally from an architectural point of view, on how to build it. Among the most pointed confrontations on the architecture of public space that Baker presents is the Dewey Arch-Heinz Pickle incident of 1900. Commemorating Admiral George Dewey's victories, the city of New York had built a heavily adorned, triumphant Roman-style arch. As a piece of civic art, the arch stood for both the building of an empire and of a city and clashed heavily with a quite different public sight that was erected just across the street: a gigantic 45-feet long illuminated pickle spelling out the names of a number of Heinz products. The exigencies of pure marketing in a mass cultural society thus met with a diametrically opposed concept of building urban space, that is, adding classical-style buildings that represented a more conservative taste in architecture. In 1869, John Wanamaker built his famous Philadelphia department store, which by itself changed urban space, but Wanamaker also felt the need to use public space for advertising in new ways and to a new degree; the universal "Wanamaker & Brown" was chiseled on street crossings, painted on rocks, and put on house tops. The space of the store had been expanded and literally taken out into the pedestrian zone of the pavement, namely, into public space. But the time that saw the increasing use and abuse of public space for capitalist purposes also witnessed an oppositional movement as more and more urban parks and recreational areas were created in the United States. Believing that green space must be accessible to all classes, Frederick Law Olmsted, among others, had designed numerous urban parks, most famously among them Central Park and Prospect Park in New York City and several parks in Chicago for the World's Columbian Exposition. Following Olmsted and others, urban space was thus deliberately left untainted by commercial interests and was seen as catering to a growing urban population's need for nature and natural surrounds in the rapidly expanding metropolises.

These few examples indicate that public space was a highly contested issue for the public at large in early American modernity. It was avidly and widely discussed, and questions were raised as to how public space was conceived and how it should be built. In this essay, I would like to draw attention to the architectural side of this debate and look at new concepts of dealing with public space as well as its important counterpart, private space; if public space had been subjected to change, it is unlikely that private space was left entirely untouched. My paper focuses on Louis H. Sullivan and his more famous disciple Frank Lloyd Wright, two architects that had reconfigured the concept of space in American architecture. Sullivan and Wright are often subsumed under the label "Chicago School of Architecture," a term that loosely bundles together a quite distinct group of innovative architects based in Chicago around the turn of the century. Despite all the differences in their works, architects such as William le Baron Jenney, Henry Hobson Richardson, John Root and, earlier in his career, his partner Daniel Burnham have, together with Sullivan and Wright, renewed the nature of American building and contributed significantly to the development of modern American architecture.

It is a well-known fact that both Sullivan and Wright sought to develop what they referred to as "democratic architecture." Throughout their numerous writings, they expounded their ideas regarding democracy and architecture and saw democratic architecture as intrinsically connected to ideas of organicism and metaphors taken from biology.[1] Thus it is hardly surprising that, so far, interpretations of their work have indeed been controlled by these ideas, mainly because the architects themselves were reiterating them time and again and critics were eager to believe. While I don't think this approach is entirely wrong, it has blinded us in many ways to other aspects that powerfully influenced their work. This essay tries to shift attention away from organicism as the dominant model of explaining the new architecture and argues instead that buildings by both architects show a changed concept of space that is simultaneous with, if not related to, major changes in contemporary culture, science, and society at large. I will briefly sketch this wider cultural as well as scientific context on which their work is predicated and which is largely responsible for the changes that the concepts of space underwent in the second half of the nineteenth and early twentieth centuries. Then I will show how a new concept of space informs the work of both architects and discuss selected buildings in light of it.

In what ways have both architects reconceptualized current notions of public and private space and what does this tell us about American culture and society on the threshold of modernism? In my discussion of these issues, I am indebted to the work of historian Stephen Kern to

---

[1] Sullivan talks at length about his idea of democratic architecture in his book *The Autobiography of an Idea* as well as in *Kindergarten Chats*, apart from numerous addresses and papers delivered at meetings and conferences and subsequently published in architectural journals and records. For an extensive bibliography, see Samuelson's recent compilation in Morrison's *Louis Sullivan*. Wright also frequently deals with architecture and democracy in his many writings; see, for instance, *An Autobiography* and the multi-volume *Collected Writings*. See also: Wright's "An Organic Architecture," "An Autobiography," and "Democracy in Overalls." With regard to the function of organicism in architecture, see Eck's more recent comprehensive study as well as Fields's earlier discussion of selected concepts of organicism.

grasp theoretically the new concepts at work in the buildings of both architects. Kern's book *The Culture of Time and Space* is a magisterial *tour de force* study of the development that the concepts of time and space underwent in early modernity, more precisely, during the time span between 1880 and 1918. Kern puts the concepts of time and space at center stage and traces significant changes in these concepts, relating them to a variety of social, scientific, and cultural developments that characterize the time under consideration. Needless to say, Kern is not the first to notice the crucial role of these years for promoting modernism. He thus corroborates the belief that Western culture changed significantly during the second half of the nineteenth century and around the turn of the century. In very general terms, the development from an agrarian republic to a full-fledged industrial nation characterized the nineteenth century and brought along political, sociological, and economical as well as technological changes with long-lasting effects on a rapidly growing number of Americans. Major innovations characterize the century, from the steamboat and the locomotive to photography, telegraphy, and the telephone—to name but a few.[2] Some of these inventions, such as reinforced concrete with increased load-bearing strength; the electric light bulb; and, in the early twentieth century, air conditioning, obviously had a direct impact on architecture and changed the way of building almost immediately.[3]

What is more, the concept of space changed dramatically. For one, the new means of transportation (e.g., railroad) and communication (e.g., telegraph, telephone) triggered the feeling that spatial distances were shrinking and eventually catered to the effect that space and the

---

[2] For more detailed descriptions of major innovations, see also: Kern, Introduction and Chapter 1; Trachtenberg, Preface and Chapter 2; and Giedion, *Mechanization Takes Command*.

[3] That technological innovations can hardly be without socio-cultural repercussions was cogently argued by, for example, Lewis Mumford. In *Technics and Civilization*, he pointed out the interrelationship between technology and technological innovations on the one hand and culture and society at large on the other hand. He developed the concept of "technics" to refer precisely to the interplay between both, whereby technology would only be one part of technics. However, he also argued that changes of this kind had always taken place over a much longer march than most scholars tended to believe (*Technics and Civilization*, 4). Mumford's 1931 study *The Brown Decades* can be considered as one of the most important works in the reception history of art and architecture during the time considered here.

experience of spatial distances collapsed altogether. Perhaps more significantly, the concept of space also underwent drastic changes in other fields such as physics. For centuries, space was seen as an empty container that could be filled by diverse concrete material objects. Einstein, perhaps most prominently, cast doubts on this view when he denounced the belief that space was empty and instead suggested that it should rather be seen as a field full of energy. When he stated in the "Special Theory of Relativity" of 1905 that "there is an infinite number of spaces, which are in motion with respect to each other" (*Relativity*, 159), he had crushed the notion of space as neutral, and as the mother and mere "receptacle" of things that Plato envisioned in the *Timaeus*. According to the new physics, the universe was full of fields of energy in various states, and space thus had become just as "substantial as a billiard ball or as active as a bolt of lightning."[4]

Einstein's discoveries were not disseminated widely in America, and it is quite unlikely that Sullivan and Wright were familiar with them. There was, however, an awareness in the larger cultural climate of the time—call it *zeitgeist*—of the large-scale scientific and cultural changes, above all new discoveries in the so-called new physics, that could not be without societal repercussions for long. This is evident, for instance, in the work of Henry Adams who famously wrote about the changes he perceived in science and culture in the last chapters of *The Education of Henry Adams*. Though published as late as 1918, the manuscript circulated privately as early as 1907. Though it is doubtful to what degree Adams and his contemporaries were conscious of Einstein's publications per se, it is all the more noticeable that Adams registered signs of what he called a "new order" of the time. In "The Grammar of Science," he talked about this "new order" that, even if it could perhaps not be captured fully at the time, was nevertheless visible to him and could, as he maintained, no longer be ignored. For Adams,

---

[4] Kern, 154. In his Foreword to Max Jammer's history of theories of space in physics, *Concepts of Space*, Einstein explained the new concept of space: "Under the influence of the ideas of Faraday and Maxwell the notion developed that the whole of physical reality could perhaps be represented as a field whose components depend on four space-time parameters. [. . .] That which constitutes the spatial character of reality is then simply the four-dimensionality of the field. There is then no 'empty' space, that is, there is no space without a field" (xvii). On the history of space in physics, see Jammer, especially Chapter 5 and the newly added Chapter 6.

it encompassed new meanings of what he called "Force itself" (Adams, 456). The new dictum of science was that "Motion seems to be matter and Matter seems to be Motion" (456). According to Adams, this new principle led to the "infinite chaos of motion," which he considered "unknowable and unthinkable" (460). The world around 1900 had ceased to be a safe place and had been changed radically by the "new avalanche of unknown forces" (457). As Adams pointed out, it was no longer a "unity but a multiple" (457), and as a multiplicity, it had become indeterminable. Even if Sullivan and Wright were not consciously integrating concepts of the new scientific innovations into their work, their buildings reflect the influence of these developments. If architecture can be seen as the art of building space and building with space, a fundamentally changed concept of space can hardly be without repercussions for contemporary architecture.

One of the major accomplishments of Kern's study is the impressive collection of material that frames his argument. But the book's strength is perhaps also its major weakness. He collects a wide array of material and shows an acute awareness of cultural traditions, but at times merely glosses over most of the examples and discusses them almost only in passing. Most of his sources and examples are taken from a European context, so that too little attention is given to the American context; he mentions both Sullivan and Wright, but deals with them in only a few lines. This weakness notwithstanding, his theoretical conceptualization of a new sense of space proves quite fruitful: Kern presents a concept which he calls "negative space" that is predicated on the changes described above. By negative space, he means a reevaluation of space that sums up what happens to space in early modernism. Whereas formerly certain parts of a given space were regarded less important, such as the supposedly empty space around the object, Kern argues that this power relationship had now been turned upside down. Concrete, solid objects that had earlier been the center of attention were losing importance as the very issue of solidity and objectivity was questioned by ideas from the new physics and other scientific developments. What used to be thought of as merely negative, actually as not even existing, that is, empty space, was now important and taken as a carrier of meaning. Negative space had, as Kern argues, been turned into "positive negative space" (Kern, 153).

The term itself is not entirely Kern's invention; he borrowed it from the visual arts, where the subject of a painting—its center of attention—is described as positive space and the background as negative space, as empty and more or less meaningless. Seen in light of the changes that the concept of space had undergone, "negative space" had now acquired a constitutive function in the process of constituting meaning, a function that goes beyond the role of merely supporting an object. Let me give two examples to illustrate the point: in Edvard Munch's well-known painting of existential anguish, *The Cry* (1893), it seemed as if the space surrounding the screaming face reverberated with the sound waves of horror. The background, formerly seen as empty and mere negative space, was energized and, as a matter of fact, produced a substantial part of the powerful effect that this painting has exerted on viewers for decades. In a sense, the sound waves transferred the scream into the world, made it seem virtually ubiquitous; the scream could be felt, the painting seems to say, anywhere in the world and thus had the possibility of affecting the whole world. In other words, the effect that the painting has had stems to a large degree from a reevaluation of negative space, redefined and carrying meaning to unprecedented degrees. Another example is the sculptural work of Alexander Archipenko. He redefined the notion that space was a mere framing of the mass of sculptural material and frequently used voids, or better, alleged voids, to explore issues of solidity and emptiness. Maintaining that "sculpture may begin where space is encircled by the material," Archipenko sculpts the torso in his *Woman Walking* (1912) as a void, suggesting that space and spatial form had become the "guts" of the sculptured figure (qtd. in Kern, 159). What used to be considered essential in art (e.g., a torso) and consequently rendered with great care had thus been subjected to a major shift in emphasis, from so-called positive space to a reinterpreted negative space.[5]

The reevaluation of negative space was not limited to painting and the arts. Seen not only in aesthetic terms but in a larger political and social context, this new approach to negative space encompassed a leveling of former differences between what was considered primary and secondary in culture and society, or, put differently, this reevaluation had the potential of smashing long-established cultural and

[5] See also: Kern, 163, 159-160.

social hierarchies. As Kern argued, it brought about a "breakdown of absolute distinctions between the plenum of matter and the void of space in physics, between subject and background in painting, between figure and ground in perception," and so on (153). The resulting "breakdown of absolute distinctions," or, as I would rather call it, this leveling of hierarchies in both art and society, can also be read as a democratic movement—though different from the organicist models of architectural democracy that architects such as Sullivan and Wright envisioned in their writing. This democratic movement was achieved precisely by resorting to a new concept of space, negative space, which expressed the cultural and social transformation characteristic of emergent American modernity.

Sullivan and Wright began to change the way space was conceived of in architecture and their new approach reflected larger socio-cultural developments at the turn of the century. They moved away from the traditional notion of space as a negative, non-existing element between concrete objects such as ceilings, walls, and floors and developed an architectural design that comprised space itself as a constitutive element in the process of designing public and private spaces. Both architects work with supposedly empty spaces. As we will see, Sullivan reveals their substance, while Wright sculpts and redesigns them.

*

Louis Sullivan was the designer of many buildings and spaces of public life in urban Chicago and smaller cities throughout the Midwest. Before dealing with Sullivan's department store *Carson, Pirie, Scott* in Chicago, let me briefly set the stage for the discussion of the new skyscrapers by resorting to a quite revealing literary account. Henry James's description of the changed New York skyline with the dominating skyscrapers is wonderfully evocative with respect to the immateriality of the new tall buildings and the new sense of space manifest in their structural design. He described the skyscrapers as "extravagant pins in a cushion already overplanted" (76). Additionally, they were "stuck in as in the dark, anywhere and anyhow" (76). The architectural "giants" consisted of "reflecting surfaces" with a "thousand glassy eyes" (77) and are "suspended in the New York atmosphere" (83). Being devoid of history and historical depth, these "towers of glass" could "only afford lights" (95); they were but "piercing notes in that concert of the expensively provisional" (77). Henry James's observations emphasize once more that this new way of building goes way beyond the somewhat outmoded organicist and form-follows-function ideas. Rather, a new approach is in demand that would read these extravagant skyscrapers—the "towers of glass," "stuck in as in the dark, anywhere and anyhow"—against the foil of new interpretations of space and new theories in science.

The skyscraper that James described so beautifully indeed epitomized the changed dynamics of architectural form and space in the early twentieth century. While the skyscraper was made possible by a number of features and inventions such as the elevator, artificial illumination, the ventilation system, and reinforced concrete, it thrived, above all else, on the idea of the skeleton structure. If traditional art, and architecture just the same, relied on the sharp distinction between figure on the one hand and ground on the other—in other words, between the object at the center of a given work of art and the background surrounding it—this particular division was turned upside down by the skeleton which foiled the traditional aesthetic expectation. The inside of the building was turned outside, making the building's underlying, previously hidden structure its outside shape.

Louis Sullivan's famous *Carson, Pirie, Scott Store* (1899–1906), the former Schlesinger & Mayer warehouse, is a case in point.[6] It was situated on what was known as the "World's Busiest Corner," one of the most public places in Chicago, the intersection of Madison and State streets. The front of this department store fulfills primarily one function, and that is the admission of light. The building's basic elements are the characteristic horizontally elongated "Chicago windows," that contribute to the homogeneous and sober surface

---

[6] Even though *Carson, Pirie, Scott* is a department store and not a skyscraper proper, it was made possible only by skyscraper technology, that is, the strong, slender steel skeleton that could support itself as well as the weight of many floors. The first building, completed in 1899, consisted of nine stories and the department store was extended in 1904 by a twelve-story corner building.

appearance and that are arranged according to the framework of the steel cage.[7] The skeleton shows as huge squares into which the outer wall is divided and provides the dominating aspect of the building. The wall as a substantial solid given is nearly invisible and only alluded to in this new type of construction; what used to represent positive space here (i.e., a solid wall) is questioned in its materiality and replaced by transparent windows as a representation of negative space. The purity of form that *Carson, Pirie, Scott* shows is so important precisely because it draws one's attention to the construction of the building—instead of concealing that very construction. By exposing its innards, the architect makes us aware of the building's structural and mechanical makeup. He thus bridged the gap between architects and engineers, between art and mechanics. Unlike the predominant assumption that saw Art as epitomizing high culture and thus as strictly separate from and inherently different in quality from mechanics and engineering, the skyscraper changed this hierarchical relationship, turning it upside down by laying bare its own building plan and redefining it in new aesthetic terms.

This new building style did not meet the taste of most contemporaries. In 1896, the noted architectural critic Montgomery Schuyler, for example, wrote that the skeleton when it is "undraped" is something that "nobody can succeed in admiring." He said this with regard to Burnham and Root's *Reliance Building* (1890–1895). For him, an influential contemporary voice, aesthetic consideration should be devoted to a building's envelope and not to its construction; he downplayed the skeleton structure, dismissing it as merely a "mechanical problem" ("D.H. Burnham & Co.," 412). Given his scorn for the skeleton structure, it is surprising that Schuyler offered the highest praise for Louis Sullivan who epitomized the "steel cage," sometimes also called the "Chicago construction." He considered Sullivan the architect who had succeeded in finding an adequate artistic expression for the "steel cage," which Schuyler, in 1896, deemed so

---

[7] *Carson, Pirie, Scott* is described in great detail in the book-length study by Siry who discusses the building with regard to the development of the department store in Chicago (see especially Chapter 3). Surprisingly, Morrison devotes only little attention to the building (see Chapter 6) in his otherwise encompassing study of Sullivan's life and work. See also: Elia's reading of the building in the chapter "Uncertainty and Inspiration at the Turn of the Century."

radical that it had "abolished the wall" ("Architecture in Chicago," 387). It may be tempting to agree simply with Schuyler's claim of the wall's destruction, yet a more cautious evaluation may be in order. The steel cage structure had not abolished the wall altogether; what it accomplished was a thorough questioning of the relation between inside and outside, between visibility and invisibility. In other words, it put the relationship between solidity and nothingness, that is, empty space, at the center of a new architectural notion of space.

The significance that this new style of building had for a democratic aesthetics of architecture was also understood by the French traveler Paul Bourget. He made an interesting observation with regard to the new architecture during a journey across the United States. In *Outre-Mer* (1895), he wrote that, in these buildings, utilitarianism, science, and beauty coincided in such a way that a new kind of art, an art of democracy had emerged, and he described it thus:

> The sketch of a new kind of art becomes apparent, an art of democracy, triggered by the masses, an art of science where the certainty of the laws of nature gives to the seemingly most unrestrained boldness the quietness of geometrical figures.[8]

By comparison, the dominant aesthetics of the nineteenth century preferred an architecture of surfaces that did not build from the inside out and sacrificed, in many ways, reality to appearance. The envelope, as Schuyler noted, was considered of prime importance and should, if considered successful, conceal all structural or mechanic aspects.

The relationship between utilitarianism and beauty in aesthetics that Bourget observed points toward a more general "American" concept of aesthetics that has triggered substantial controversy in the field of American studies over the past decades. The frequent provocation that a concept of the aesthetic is necessarily underdeveloped in a society which is as heavily influenced by utilitarianism and individualism as America was recently addressed in

[8] *Outre-Mer*, 162. "L'ébauche d'une espèce nouvelle d'art s'y dessine, d'un art de démocratie, fait par la foule, d'un art de science où la certitude des lois naturelles donne à l'audace en apparence la plus effrénee des tranquillités de figures de géométrie." (My translation)

an essay by the German literary critic Winfried Fluck. He reassessed the concept of an American democratic aesthetic as an anti-elitist concept of art, recovering a struggle toward an "artless art" as its main objective. By showing how ideological work in American democracy is done precisely through the aesthetic dimension of works of art, he concludes that the search for a democratic aesthetics is a "venerable American tradition."[9] One way of producing "artless art" toward the end of the nineteenth century would be to include new sciences and physics, purportedly antithetical to a more traditional concept of Art, and, for architecture, to strip buildings of their classical surface decorum. Sullivan's and, as we will see, Wright's approach to architectural space epitomized the search for an "artless art" and thus confirmed an American democratic aesthetics of art and architecture.

*

In a recent paper on art and aesthetics, Philip Fisher concluded that "We are a civilization of the domestic" and one could think that an analysis of Frank Lloyd Wright's prairie houses may well have triggered such an observation. In the history of building private homes in America, Wright had found an architectural design in the so-called prairie-style houses that impressively catered to the basic needs of a "civilization of the domestic." The house, for Wright, was primarily a shelter, and its preeminent function was to make its inhabitants feel secure and safe from a cold, inimical world outside. The well-known prairie style and other buildings by Wright would redefine the nature of both private and public space not only in America. As an architect of a truly international reputation, he is known for the Taliesin houses in Arizona, Oak Park in Chicago, and the Guggenheim Museum in New

[9] Fluck, 37. Fluck also points out that it was F.O. Matthiessen who first linked democracy with formalism in his seminal study *American Renaissance* (31). Later, in "The Problem of Ideology," Bercovitch repudiated the long held belief that art was distinguished by the fact that it transcends ideology (Bercovitch, 639).

York City as well as for the Imperial Hotel in Tokyo. His architectural vision would influence, perhaps also define, much of twentieth-century architecture, in America and elsewhere.

But Wright's concept of housing aimed at more than simply building what are often called cave-like structures and are seen as catering, for the most part, to merely psychological needs. His architectural design goes beyond this idea and works in mainly two ways: on the one hand, it indeed rigidly closed the house to outside inspection, but it also went into the oppositional direction when, on the other hand, it opened up interior space and focused on the space that is created by walls and ceiling. Wright was aware of his major reconception. This new focus on space itself was predicated on his conviction that "the reality of the building consisted not in the four walls and the roof but inhered in the space within, the space to be lived in" ("An Organic Architecture," 58). One did not start a building by thinking of single rooms, and entrances, and walls, and so forth, rather the focus in this new theory of building was on space, on composing with space and not so much with particular elements. Wright explained that his initial conception was to "keep a noble room for worship in mind, and let that sense of the great room shape the whole edifice," or, in other words, "[l]et the room inside be the architecture outside" (*An Autobiography*, 178). Wright shifted the focus to a kind of space that would formerly have been identified as merely supportive and possibly empty, in other words, on negative space, that now had become meaningful and, if you will, positive.

Wright's *Unity Temple* epitomizes this aspect of his theory of architecture; as a semi-public space, the temple was built in Chicago in 1906.[10] The building consists of two single cubicles connected by an entrance hall. One cubicle is the actual temple for community worship, whereas the other is meant for other community purposes and houses offices. The actual temple, the room for worship, on which I will focus here, consists of a simple cubical interior that is visible on the outside of the building. The building is constructed with simple blocks of cement and an unadorned concrete slab roof. The single space/room shapes the edifice. Even if Wright stayed within the traditional

[10] For more detailed descriptions of *Unity Temple*, see also: Wright, "Designing Unity Temple"; Blake, 281-285; Lipman, 270-272; and Levine.

architectural lingo when he talked about "rooms," he also explained why he thought that this was something new, something modern:

> The first conscious expression of which I know in modern architecture of the new reality—the space within to be lived in—was Unity Temple in Oak Park. True harmony and economic elements of beauty were consciously planned and belong to this new sense of space-within. In every part of the building freedom is active. Space is the basic element in architectural design. ("An Autobiography," 313)

Architecture thus had become a true "art of space," and the subjects of architecture were not so much material objects, such as walls and ceilings, as the spatial enclosures created by them. The wall as a solid dividing unit had been questioned, and Wright replaced it by glass as a transparent material, with the admission of daylight to unprecedented degrees as a consequence. But what is more, Wright's particular handling of architectural space also evoked a certain sense of freedom, as Wright claimed in the above quotation, a liberating movement away from the overstuffed, heavily ornamented Victorian households, cluttered with knickknacks, both inside and outside, and rigidly separated into cell-like structures. Whereas the Victorian response to empty or negative space was to fill it up, Wright, on the contrary, did not consider space as merely empty, as a vacuum to be abhorred, but as an active and most important architectural element. The resulting purity and sobriety of this new style of building aimed at a reevaluation of the hierarchically structured building style that had dominated contemporary architecture. It also shows, as Wright implied, a sense of aesthetics identified as particularly American, an alliance of ideology and democracy with science and the economical.

Even if Wright believed that interior, private space could not be separated from outside space, from nature, and he saw both as part and parcel of what he liked to call the "organic whole," it was much more than the organic, democratic sense of open spaces that fed into this work.[11] Wright's prairie houses exemplified the endeavor of building

---

[11] See Wright's use of "organic whole" in, for example, *An Autobiography* (168, 170). Natural space was paramount for him and, as such, he echoed the larger role attributed to space in America with regard to the myth-ridden "virgin" land that has shaped

new spaces for private use. *Falling Water* (i.e., *House Kaufmann*), built in Pennsylvania in 1935–36, is probably his most widely known design of this kind, but the earlier *Robie House*, built on Woodlawn Avenue in Chicago in 1908, is another example. Building *Robie House*, Wright used several plane surfaces, receding and advancing at different depths.[12] One of these planes was the porch which he adopted as a traditional American element and used widely in his designs, but his reverence for tradition did not make most contemporaries like his building any more. The porch in Wright's building was changed quite radically; it became an extension of the wings, or it thrust out into space as a pure cantilever hovering above the earth. To these and similar hovering horizontal elements, Wright added planes of vertical surfaces. He organized these variously, multiplying them, intersecting them, or incorporating different advancing or receding planes, so that eventually the solid volume of the house almost disappeared. Wright consciously destroyed the established order of building and of building space. These destruction processes were contingent on modern concepts of space and had very little to do with organic metaphors as architects were quick to point out and critics were too eager to take at face value. Wright readily admitted his dedication to architectural revolt in a letter to Charles Ashbee in 1910 when he wrote,

> But you see I am cast by nature for the part of the iconoclast. I must strike—tear down, before I can build—my very act of building destroys an order. ("Ten Letters from Frank Lloyd Wright," 67)

Seen from an aesthetic perspective, Wright's houses displayed a constant endeavor to find and foster interrelationships between the various spaces and elements such as walls, ceilings, openings, and enclosures. When we look at Wright's buildings and the plans on which

---

American character so thoroughly and has also been linked to the progress of democracy. Wright himself always maintained he would build programmatically according to and in alliance with the given natural surroundings and claimed that only natural, organic architecture had the capacity to be truly democratic ("An Organic Architecture," 58-59).

[12] For a more detailed description of *Robie House*, see Treiber, 18-21; Giedion, *Space, Time and Architecture*, 410-411. For a historical contextualization of Wright's prairie houses, see Seligmann.

they are based, one feels instantly reminded of contemporary avant-garde developments in the European art scene. Much more than ideas of organicism and biological interpretations that his work has always been linked to, Cubism and the abstract art movement in Russia, fostered by Kasimir Malewitsch, come to mind when one looks at Wright's *Robie House*. Along with several others, Malewitsch was protesting against the object in his art and was trying to eliminate it altogether by reducing painting to signs and shapes and the spaces they create. In Malewitsch's architectural studies, cubes and rectangles and strips float, they penetrate, or hover among each other, thereby creating spatial interrelationships that can be seen as mere research into space itself. Perhaps even more radically, the Dutch painter Mondrian and the group of artists associated with Theo van Doesburg tried to sacrifice entirely the representation of external objects and sought to achieve a sense of purity in art. As their periodical *Stijl* (for which they were named) shows, their art was focused on planes of different colors and shapes and the distributions and juxtapositions of those planes in space. Most of Wright's contemporaries, though, were seeking to deride *Robie House*—in terms similar to those that were later brought against Le Corbusier—by comparing it to a steamship. By pointing out this analogy, they had, perhaps unintentionally, captured the spirit of a time increasingly defined by the aesthetics of the machine, an aesthetic later identified with modern America.

*

Upon reading a first draft of Henry James's essay *The American Scene*, the eminent American philosopher and psychologist William James made a very perceptive observation regarding his brother's essay that is also quite revealing for a discussion of a changed concept of space in America—the current "American scene." The observation pointedly, if involuntarily, captures my preceding discussion of space in early American modernism. On May 4, 1907, William commented, in a letter to his brother, on Henry's notably circuitous approach to his subject and wrote that it seemed to him as if Henry wanted to avoid naming something explicitly in the essay. He continued that:

> by dint of breathing and sighing all round and round it to arouse in the reader who may have had a similar perception already [. . .] the illusion of a solid object, made [. . .] wholly out of impalpable materials, air, and the prismatic interferences of light, ingeniously focused by mirrors upon empty space [. . .] your account of America is largely one of its omissions, silences, vacancies. You work them up like solids. (*Letters of William James*, 277-278)

William James's reference to empty or silent space, which, for him, is made out of light and air rather than solid materials, corresponds to the concept of negative space as introduced above. On the threshold of modernity, William James noticed that "omissions" had taken over solidity, that America seemed to have lost solid objects and objectivity in general. In other words, these conspicuous "omissions" had taken over the place that was formerly occupied by concrete objects. By the same token, James beautifully perceives the revaluation between figure and so-called ground, between positive and negative space that looms large in the architectural works considered here. Rather than to grasp an object firmly, there is a "breathing and sighing all around" the space where there used to be an object. William James's remarks powerfully underscore the role and importance of negative spaces for the making of modern America.

If Horace Kallen called William James "the first democrat of metaphysics" (30), Sullivan and Wright may be called the first democrats of American architecture. Both Sullivan, who built skyscrapers and tall commercial buildings in urban centers for a rapidly industrializing nation, and Wright, who reconfigured semi-public and most private spaces of the home, conceived of a democratic architecture by resorting to a changed concept of space, negative space, in their buildings. Their new building style destroyed the established order of building by working with this new concept that was itself predicated on major innovations and changes in science and culture at the time. Even if the architects were probably unaware of the scientific theories per se and, in the case of Sullivan, almost predate them, the congruities and assonances that arise in this cultural moment, as a part of the *zeitgeist*, are striking. Even if the development I have sketched had only little immediate political or social impact for American democracy, I would argue that the revaluation and affirmation of negative space had one defining feature in common with the general idea of democracy: the leveling of hierarchies in terms of a revaluation of neglected "empty" space that before had only supporting roles, and was now brought into the center of attention. Starting from space—negative space—revaluating it and overturning the long established hierarchy between figure and ground, between subject and background, Sullivan and Wright developed a democratic aesthetic of American architecture that paved the way for architectural and artistic styles identified with a typically American form of modernism that would soon flourish at home and overseas.

Bibliography

Adams, Henry. *The Education of Henry Adams*. 1918. Boston: Houghton Mifflin, 2000.

Baker, Laura E. "Public Sites Versus Public Sights: The Progressive Response to Outdoor Advertising and the Commercialization of Public Space." *American Quarterly* 50, no. 4 (Dec. 2007): 1187-1213.

Bercovitch, Sacvan. "The Problem of Ideology in American Literary History." *Critical Inquiry* 12, no. 4 (Summer 1986): 631-653.

Blake, Peter. *Le Corbusier, Mies van der Rohe, Frank Lloyd Wright*. München: Piper, 1963.

Bourget, Paul. *Outre-Mer*. Paris: Alphonse Lemerre,1895.

Eck, Caroline van. *Organicism in Nineteenth-Century Architecture: An Inquiry into Its Theoretical and Philosophical Background*. Amsterdam: Architectura & Natura Press, 1994.

Einstein, Albert. Foreword. In *Concepts of Space*, by Max Jammer, xiii-xvii. New York: Dover, 1993.

—. *Relativity: The Special and the General Theory*. Authorized translation by Robert W. Lawson. New York: Three Rivers Press, 1961.

Elia, Mario Manieri. *Louis Henry Sullivan*. New York: Princeton Architectural Press, 1996.

Fields, Ronald F. *Four Concepts of an Organic Principle: Horatio Greenough, Henry David Thoreau, Walt Whitman, and Louis Sullivan*. PhD diss. (authorized facsimile), Ohio University, 1968.

Fisher, Philip. "Wonder and Attention: Aesthetics of Singular Objects." Paper delivered at the International Conference *The Power and Politics of the Aesthetic in American Culture*, Munich, June 9-11, 2005.

Fluck, Winfried. "The Search for an 'Artless Art': Aesthetics and American Culture." In *The Power and Politics of the Aesthetic in American Culture*, edited by Ulla Haselstein and Klaus Benesch, 29-44. Heidelberg: Winter Universitätsverlag, 2007.

Giedion, Siegfried. *Mechanization Takes Command*. New York: Oxford University Press, 1948.

—. *Space, Time and Architecture: The Growth of a New Tradition*. 1941. Cambridge, MA: Harvard University Press, 1997.

James, Henry. *The American Scene*. 1907. Bloomington: Indiana University Press, 1968.

James, William. Letter to Henry James, May 4, 1907. *The Letters of William James*. Edited by Henry James, 2:277-280. Boston: Atlantic Monthly Press, 1920.

Jammer, Max. *Concepts of Space: The History of Theories of Space in Physics*. 1954. New York: Dover, 1993.

Kallen, Horace M. *William James and Henri Bergson: A Study in Contrasting Theories of Life*. Chicago: University of Chicago Press, 1914.

Kern, Stephen. *The Culture of Time and Space, 1880–1918*. 1983. Cambridge: Harvard University Press, 2003.

Levine, Neil. *The Architecture of Frank Lloyd Wright*. Princeton: Princeton University Press, 1993.

Lipman, Jonathan. "Consecrated Space: Wright's Public Buildings." In *On and By Frank Lloyd Wright: A Primer of Architectural Principles*, edited by Robert McCarter, 264-285. New York: Phaidon, 2005.

Morrison, Hugh. *Louis Sullivan: Prophet of Modern Architecture*. 1935. Introduced and revised by Timothy J. Samuelson. New York: W.W. Norton and Co., 2001.

Mumford, Lewis. *The Brown Decades: A Study of the Arts in America 1865–1895*. 1931. New York: Dover, 1971.

—. *Technics and Civilization*. New York: Harcourt, Brace, and Company, 1934.

Schuyler, Montgomery. "Architecture in Chicago: Adler & Sullivan." *American Architecture and Other Writings*. Edited by William H. Jordy and Ralph Coe, 1:377-404. Cambridge, MA: Belknap Press of Harvard University Press, 1961.

—. "D. H. Burnham & Co." *American Architecture and Other Writings*. Edited by William H. Jordy and Ralph Coe, 1:405-418. Cambridge, MA: Belknap Press of Harvard University Press, 1961.

Seligmann, Werner. "Evolution of the Prairie House." In *On and By Frank Lloyd Wright: A Primer of Architectural Principles*, edited by Robert McCarter, 56-85. New York: Phaidon, 2005.

Siry, Joseph. *Carson Pirie Scott: Louis Sullivan and the Chicago Department Store*. Chicago: University of Chicago Press, 1988.

Sullivan, Louis H. *The Autobiography of an Idea*. 1924. New York: Dover, 1956.

—. *Kindergarten Chats and Other Writings*. 1918. New York: Dover, 1979.

—. "The Tall Office Building Artistically Considered." 1896. In *Architecture in America: A Battle of Styles*, edited by William A. Coles and Henry Hope Reed, Jr., 42-46. New York: Appleton – Century – Crofts, 1961.

Trachtenberg, Alan. *The Incorporation of America: Culture and Society in the Gilded Age*. New York: Hill & Wang, 1982.

Treiber, Daniel. *Frank Lloyd Wright*. Basel: Birkhäuser, 1988.

Wright, Frank Lloyd. *An Autobiography*. 1932. New York: Horizon Press, 1977.
—. "An Autobiography." *Frank Lloyd Wright: Writings and Buildings*. Edited by Edgar Kaufmann and Ben Raeburn. New York: Meridian Books, 1960.
—. *Collected Writings*. Edited by Bruce Brooks Pfeiffer. New York: Rizzoli, 1992.
—. "Democracy in Overalls." *Frank Lloyd Wright: Writings and Buildings*. Edited by Edgar Kaufmann and Ben Raeburn. New York: Meridian Books, 1960.
—. "Designing Unity Temple." *Frank Lloyd Wright: Writings and Buildings*. Edited by Edgar Kaufmann and Ben Raeburn. New York: Meridian Books, 1960.
—. Letter to Charles Ashbee, July 8, 1910. "Ten Letters from Frank Lloyd Wright to Charles Ashbee." *Architectural Record* 13 (1970): 67.
—. "An Organic Architecture: The Architecture of Democracy." In *Architecture in America: A Battle of Styles*, edited by William A. Coles and Henry Hope Reed, Jr., 58-59. New York: Appleton – Century – Crofts, 1961.

Designed Space vs. Social Space: Intention and Appropriation in an American Urban Park

Timothy Davis

Historians and other interpreters of urban space tend to focus on the production of the built environment. This approach emphasizes the perspectives and activities of architects, planners, and other prominent shapers of civic form. It privileges the act of building and the intentions of designers over the ways in which urban environments are used, altered, and given meaning by ordinary citizens through everyday routines. Sociologists such as Michel de Certeau have suggested that greater attention should be paid to the ways in which these patterns of everyday life enable ostensibly powerless subjects to participate in the construction of urban space. Close examination of social practices and subtle physical traces in Washington, D.C.'s Rock Creek and Potomac Parkway demonstrates how marginalized populations employ planned urban space for purposes that co-opt and subvert design intent. These groups include homeless people seeking refuge; gay men who have produced a widely recognized "queer space"; and suicides, who have adapted majestic bridges to their own ends. While these appropriations have made little permanent impact on the parkway's terrain, they exemplify the ways in which groups and individuals produce spatial significance through subtle and often surreptitious social behavior.

Invoking the theoretical constructs of French sociologists Henri Lefebvre and Michel de Certeau, this essay examines the ways in which marginalized populations have appropriated portions of one of Washington's D.C.'s preeminent planned urban spaces for purposes that co-opt and subvert the designers' original intentions. By articulating overlooked and in many cases almost invisible geographies, this approach provides a richer and more complex account of the history and current functions of one particular urban space while promoting an inhabitant-based paradigm for the interpretation of public spaces in general (fig.1). Just as reader-oriented criticism contests the traditional concept of inherent textual meaning and valorizes resistance to authorial hegemony, privileging the user's role in constructing spatial significance calls into question the designer-oriented approach to the interpretation of the built environment that dominates scholarly and critical discourse. By emphasizing the emergent, indeterminate, and ephemeral essence of spatial production, this perspective also challenges traditional assumptions that the built environment can be read as a stable text encapsulating the spirit, essence, or ideology of any particular age, especially if such readings are limited to the analysis of official pronouncements or the interrogation of popular culture in the usual print and visual media.

While traditional scholarly approaches provide considerable insight into prominent cultural concerns, the social implications of the built environment can be understand more fully by embracing broader perspectives that call attention to overlooked places, processes, and populations. Complementing conventional academic research with ethnographic fieldwork and paying close attention to quotidian details, ephemeral traces, and overlooked practices of everyday life can increase the depth and breadth of the historical record, illustrate subtle but significant social processes, illuminate strategies of dominance and resistance, and promote greater tolerance for the varieties of cultural experience. For those who care about such extra-academic matters, this expanded knowledge can also inform planning, design, and policy decisions, contributing to a more inclusive, equitable, and accommodating urban environment.

In his influential work, *The Practice of Everyday Life*, Michel de Certeau called for a new approach to the study of urban space that would focus on the activities of ordinary people and emphasize the ways in which fleeting actions and everyday routines gave form and meaning to the built environment (de Certeau, dedication, xi-xii). Acknowledging both the pervasive nature of society's disciplinary regimes and the limited ability of ordinary people to mold the modern urban landscape, de Certeau, Henri Lefebvre, and others argued that the uses people make of the built environment constitute significant forms of social and spatial production in their own right. [1] De Certeau characterized these activities as "ways of making do" or "tactics,"

---

[1] Important examples of this genre include Bourdieu, *Outline of a Theory of Practice*, and Lefebvre, *Writings on Cities*. Of course American scholars such as J.B. Jackson, William Whyte, Jane Jacobs, and Grady Clay had been "privileging" ordinary landscapes and everyday life for quite some time without generating the same acclaim in rarified academic realms as their more overtly "theorized" and politicized francophonic counterparts.

which, he observed, "insinuate" themselves into places defined by the dominant order in temporary, opportunistic, and often "devious" fashion. He frequently used the term "poaching" to underscore the furtive and illicit nature of these transactions. While such tactics are capable of fleetingly appropriating fragments of the spatial order, they rarely produce lasting physical changes or exert dominion over the whole. Nevertheless, de Certeau proclaimed, these "guileful ruses of *different* interests and desires" constitute significant "networks of antidiscipline" that function as notable—if generally overlooked and underappreciated—"victories of the 'weak' over the 'strong'"(xi-xix, 34-39). In order to broaden the discourse of urban studies—and provide a more richly textured, accurate, and equitable account of the nature of urban experience—de Certeau called for further investigation into what he called "these multiform, resistant, tricky and stubborn procedures that elude discipline without being outside the field in which it is exercised, and which should lead us to a theory of everyday practices, of lived space, of the disquieting familiarity of the city" (96).

This exploration represents an attempt to look below what de Certeau called the "thresholds where visibility begins" and articulate the ways in which the "weak" have insinuated themselves into the topography of the "strong" to inscribe subversive geographies in what many regard as a reassuringly familiar, thoroughly disciplined, and extensively examined urban space. Washington, D.C.'s Rock Creek and Potomac Parkway is a 2 ½ mile long ribbon of greenery stretching from Rock Creek Park to the National Mall and forming a geographic boundary between Georgetown and Washington proper (fig. 2). The parkway is a classic City Beautiful scheme that was intended to transform an "urban eyesore" into an attractive, efficient, and economically beneficial civic amenity. Beginning in the 1880s, various commissions emphasized the visual, social, and environmental degradation of lower Rock Creek Valley, which had long been used as an urban dumping ground. A succession of elaborate prospectuses demonstrated how the dangers and disorders of inchoate urban space would give way to the beauty, logic, and harmony of the planned city. Not only would the threatening residents of the primarily African-American district abutting the parkway be displaced, but the "dense thickets and somber groves" that gave shelter to criminals, tramps, and other undesirable elements would be replaced by open sunny landscapes, where middle-class families could promenade alongside graceful carriageways. This vision was largely realized by the 1930s. Dozens of houses and their low-income residents were removed and threatening groves gave way to sodded hillsides and artful plantings, whose primary audience was automobile commuters (fig. 3). By exorcising one of the most discomfiting sectors of Georgetown, the parkway contributed significantly to the district's gentrification.[2]

The basic contours of this process have been mapped in various official documents such as National Register nominations, park cultural resource studies, and, with more depth and complexity, in several art historical publications.[3] In contrast to the producer-oriented perspectives that dominate these narratives, this essay focuses on the ways in which consumers have appropriated the parkway's spaces for their own needs, some of which were not just unaccounted for in the original plans but antithetical to the practical goals and ideological imperatives that shaped the project's design and development. Since the needs and practices of commuters and conventional recreationalists have been addressed in various studies, this essay addresses the actions and environments of those who have traditionally been excluded from scholarly attention and participation in planning processes. These alternative populations represent the realms of sickness, deviance, and death that de Certeau claimed were marginalized or ignored by modern city planners: they are the homeless people who find refuge in the parkway's woodlands; the gay population that transformed a section of the parkway known as "P Street Beach" into a prominent component of Washington's queer geography; and suicides, who have long gravitated toward the parkway's towering bridges. While these appropriations have made little permanent impact on the parkway's terrain, they

---

[2] For reference to the "open and sunny" versus "dense thickets and somber groves," see U.S. Congress, Senate, *The Improvement of the Park System of the District of Columbia*, 141. Additional key documents related to the development of Rock Creek and Potomac Parkway include: *Rock Creek Park: Information for the Public in Relation Thereto,* 7-14; Cox, 107-122; "Photographs: Rock Creek Parkway, Present Condition"; U.S. Congress, Senate, *Report on the Improvement of Rock Creek from Massachusetts Avenue to the Mouth of Rock Creek*; U.S. Congress, House, *Report of the Rock Creek and Potomac Parkway Commission: 1916*, 10, 17.

[3] See Mackintosh; Bushong; Krakow; and Davis, "Rock Creek and Potomac Parkway: HAER Report No. DC-697" and "Rock Creek and Potomac Parkway, Washington, D.C."

exemplify the ways in which marginalized groups and individuals "poach" urban space and produce spatial significance through transitory and opportunistic social tactics that rarely receive official endorsement or scholarly recognition.

Reflecting nationwide trends, the presence and visibility of homelessness in Rock Creek and Potomac Parkway has fluctuated significantly over the past fifteen years.[4] After peaking in the early 1990s, Rock Creek Park officials maintain that the current homeless population of the parkway is relatively low, claiming that there are rarely more than one or two people residing in the area at any given time. The park police helicopter is used several times a year to locate homeless camps from the air, which are then targeted for removal.[5] While the current homeless presence in the parkway is clearly lower than it was in the early 1990s, officials may be expressing wishful thinking or depending too heavily on the panoptic powers of aerial surveillance, as close investigation on the ground reveals ample evidence of homeless presence. The resurgent woods and thickets along the west side of Rock Creek valley between P Street and Oak Hill Cemetery have long afforded shelter to homeless groups and individuals (fig. 4). This stretch of the parkway is well suited to the

---

[4] Information on homeless patterns is based on conversations with park staff and Washington residents, along with personal observation during fieldwork for HAER report on Rock Creek and Potomac Parkway and subsequent research, 1991 to the present. In the early 1990s, a non-profit organization operated a homeless shelter in a series of trailers adjacent to the south end of the parkway. During the day, the occupants of this facility maintained a highly visible presence, loitering with their belongings under the nearby freeway ramps and panhandling parkway drivers. The freeway ramps provided excellent shelter, and a number of individuals erected temporary habitations of cardboard, plywood, and other readily gathered materials. When the shelter was closed amid much protest in November 1992, there was a temporary upsurge in informal campsites until residents moved on to other accommodations, as reported in newspaper accounts such as "District Postpones Closing Homeless Shelter in Foggy Bottom for Second Time," *Washington Post*, 30 September 1992; "Shelter Residents Say They Damaged Trailer," *Washington Post*, 21 November 1992; "34 Arrested in Shelter Dismantling," *Washington Post*, 25 November 1992; "Homeless People Sleeping at Campsite in D.C. Park Get a Rude Awakening," *Washington Post*, 23 June 1993; "Former Shelter Residents Hold a Reunion," *Washington Post*, 22 November 1993.

[5] Telephone conversations with Rock Creek Park Superintendent Adrienne Coleman, 16 May 2006 and U.S. Park Police spokesman Sgt. Scott Fear, 19 May 2006.

role, as the topography and vegetation afford ample opportunities for concealment, while the proximity to Georgetown provides easy access to panhandling and foraging activities. Contributing greatly to the area's isolation, the main parkway path separates from the parkway drive at this point, crossing to the west side of Rock Creek and running along the base of the hillside. By early summer, a wall of brush obscures the path from the protective gaze of passing motorists and the hillsides above are sheathed in vegetation. While this isolation can be unnerving for regular parkway users, it provides an ideal opportunity for the homeless to create clandestine camps.[6] The ways in which they do this exhibit the furtive and transitory tactics identified by de Certeau. In fact, many of these "surreptitious creativities" and "ways of operating" are so subtle and well disguised as to be essentially invisible to most parkway users, and it appears, to the parkway's administering forces.

A closer look at the seemingly featureless, intractable, "undesigned" and by implication meaningless "thickets" bordering the official multi-use trail reveals that they are interlaced with a profusion of unofficial paths of varying degrees of permanence and legibility. Many are only faint traces leading to what appear to be individual campsites, some of which are barely more than flattish spots of clear ground. There were few active camps when this fieldwork was conducted in early spring before temperatures warmed and the vegetation leafed out, but food and liquor debris, discarded clothes, and castoff bedding attested to former occupancy. An area under the Q Street Bridge ideally configured for homeless occupation by virtue of its isolated location and sheltered configuration—and frequently used for this purpose in the past—was so eerily immaculate that it appeared likely to have been swept clean by a park service patrol. The remains of a disrupted fire ring afford a subtle reminder of past and future uses. The transient nature and inherent limits of both administrative strategies and resistant tactics was further underscored by the presence of a homeless cache hidden in the woods about 20 feet away. Judging

[6] Recurrent incidents prompted parkway forces to take a more aggressive approach to brush-clearing in this area in the early 1990s. The most recent of these attacks occurred in May 2006, as reported in "4 Robbed on Mall; Jogger Attacked at Rock Creek," *Washington Post*, 27 May 2006.

by the absence of pollen that coated nearby objects, these belongings had been left very recently; imperfectly concealed but carefully protected by a rubber mat weighted in place by meticulously arranged sticks and stones. Several hundred yards further south, a can collection stashed at the end of a faint trail provided further evidence of the clandestine activities and stubborn resilience of this largely invisible population.

Not all of Rock Creek and Potomac Parkway's homeless are invisible, of course. Some maintain a relatively discreet and deferential presence, inhabiting the margins of spaces used for the intended purposes of commuting and recreation (fig. 5). An elderly white male is often seen in the wooded area north of Q Street Bridge. When spotted on a weekday afternoon in mid-May, he was resting in the sun with his belongings gathered nearby. Taking sophisticated advantage of local topography and usage patterns, he was positioned on the opposite side of the creek from the main trail, but screened from passing motorists by the riverbank and associated vegetation. An African-American male of indeterminate age similarly exploits the defensive potential of the creek and parkway drive, locating his campsite in a nearly inaccessible brushy area between the two just below P Street Bridge. According to several sources, this individual has resided in the same location for ten to fifteen years. The park police tacitly condone his presence, having warned him that he would be evicted if he failed to curtail his habit of throwing liquor bottles into the stream. Though the man wears the excessive clothing characteristic of mentally ill homeless people, and can often be seen passed out by the roadside, he has apparently been able to sustain the terms of this informal agreement.

Other homeless people have similarly achieved the status of de facto parkway denizens. Recreational users of the amphitheater-like bowl on the east side of the parkway known as P Street Beach spoke of two men who had developed a successful tactic for sleeping in the area. One of these was known alternately as "Umbrella Man," for a favored accouterment, or "Pee-bag Man," in reference to an innovative hygiene regime employing Ziploc bags to dispose of bodily waste in an area with no public toilets. Pointing to what appeared to be random piles of cardboard placed for disposal beside public trash cans, park regulars described how both men used these materials to construct bedding and shelters, which they dismantled every morning and concealed in plain sight as waste products assigned to the appropriate administrative locations. While this system clearly depends on the tacit collaboration of the volunteer maintainers responsible for trash collection, it exemplifies the "guileful ruses" and "ways of making do" through which marginal populations appropriate spaces designed for more conventional uses.

By far the most prominent marginal use of P Street Beach is its long-standing status as a gay cruising ground. While anecdotal evidence of this role dates back at least to the 1950s, and portions of the parkway were decried as "a habitat of unsavory characters, perverts, and delinquents" in a 1962 park police memorandum, P Street Beach's status as a landmark in Washington's queer geography dates primarily to the gay community's broader appropriation of the adjacent Dupont Circle neighborhood, which began in the 1970s and became dominant in the 1980s and 90s. P Street Beach had originally been conceived as a genteel public pleasure ground, with multiple carriage ways, ornate bridges, and broad open vistas. Changing recreational patterns and the transition to automobile traffic necessitated a dramatic simplification of these schemes, but the general concept of a broad open bowl affording views to and from the parkway drive remained. Due to a combination of natural growth and limited maintenance budget, however, these viewsheds were rapidly occluded, producing a pleasant grassy area that was visually isolated from the parkway drive and the streets above (fig. 6).

During the 1960s and 70s, this area gained a reputation as a counterculture hangout where men and women enjoyed the sun, played music and Frisbee, and partook in illegal drugs. Park officials acted to discipline these activities by constructing a stage where youth-oriented concerts were held, including a 1971 Independence Day production by the cast of *Hair* that attracted 12,000 people. As the broader counter-culture waned in the mid-1970s and the gay community became more prominent and empowered, P Street Beach became increasingly identified as a gay cruising area, particularly at night, when it gained a reputation as one of Washington's primary locations for promiscuous public sex.[7]

---

7 "Celebrating the American Revolution," photograph and extended caption from *Washington Post,* 6 July 1971. Members of the Nighthawks, a popular local band of

It is beyond the scope of this essay to provide an overview of Queer Theory [8] or summarize the literature dealing with the well-known function of public parks as gay-cruising areas.[9] Nor is this the place to provide a detailed account of the practices involved, other than to note that various websites amply document the variety, volume, and volatility of sexual activities that draw participants to the area. This essay is intended to identify the ways in which P Street Beach's long-standing status as "Queer Space" shapes the area's social geography, elucidating a few basic patterns, relating them to some of the more prominent intersections between Queer Theory and writings on the social production of space, and reporting recent developments that have little impact on the physical landscape yet represent a significant transformation and potential loss of one of Washington's most notable subversive geographies (fig. 7).

Gay tour guides and travel websites have long highlighted P Street Beach's "recreational" function. The National Park Service and District of Columbia police are well aware both of what goes on there and of their limited ability to curtail it. Recently, at least, both express a commitment to maintaining good relationships with the gay community and assert that they do not harass gay users or engage in indiscriminate entrapment activities. Important exceptions include sting operations

---

long-standing, fondly recall participating in a "love in" at P Street Beach in 1967. (The Nighthawks Official Band Website, http://www.thenighthawkscom; Gay D.C. Walking Tours: Dupont Circle Rainbow History Project, Washington, D.C., 2003.) Additional information on P Street Beach in the shifting demographics of P Street Beach and associated activities supplied by numerous Washingtonians, gay and straight, including several current park service officials.

[8] Of particular relevance to this study are writings that deal explicitly with the history, theory, and practice of Queer Space. This includes Bell and Valentine; Sanders; and; most pointedly and comprehensively, Ingram, Bouthillette, and Retter.

[9] John Grube's essay, "'No More Shit': The Struggle for Democratic Gay Space in Toronto," in *Queers in Space*, (127-145) affords an insightful overview on some analogous issues involving Toronto's public parks and ravines. Ingram's "'Open' Space as Strategic Queer Sites" (*Queers in Space*, 95-125) and Bell's "One-Handed Geographies: An Archeology of Public Sex" (*Queers in Space*, 81-87) provide more geographically varied and theoretical introductions to the topic, while Maurice van Leishout's "Leather Nights in the Woods: Locating Male Homosexuality and Sadomasochism in a Dutch Highway Rest Area" (*Queers in Space*, 339-355) gives a detailed account of the practices and social significances associated with one particular site.

aimed at curtailing prostitution and drug dealing, and routine nightly sweeps designed to disrupt nocturnal activity. Since the park is officially closed at night, anyone found inside can be arrested, though most are not cited until they've received multiple warnings. Officials maintain that the primary goal is to protect park users, as robberies and assaults are associated with both activities. Anyone caught engaging in sexual acts will also be prosecuted. While park and police officials maintain that this is a sexual-orientation-neutral policy, some queer theorists assert that proscriptions on public sex constitute a form of hetero-normative repression. Some activists have deployed conspicuous public sex for political purposes, but there is little evidence of conscious political intent in the ongoing sexual activities at P Street Beach.[10]

The gay community has, however, appropriated the broader space of P Street Beach for more conventional political purposes. As early as 1972, P Street Beach hosted organized Gay Pride celebrations, continuing in this role for a number of years until the event moved to the larger, more accessible, and better-serviced school grounds nearby. In 2000, P Street Beach hosted the first Youth Pride Festival and continues to be used for this event today. By formally appropriating official urban spaces for alternative purposes, these activities represent significant victories for gay activists, on the one hand, and of the weak over the strong, in de Certeau's formulations. At the same time, the temporary and conditional nature of these appropriations underscores the disequilibrium addressed by both perspectives. Furthermore, the official endorsement of Pride activities reflects the increased mainstreaming of gay culture, a development that incites mixed feeling within this highly diverse community.

For many gay men, P Street Beach and associated activities represents a side of the culture and/or period in time that they would prefer to distance themselves from. Nocturnal cruising and public sex in marginal spaces continues to appeal to some elements of the gay community, but increased social acceptance has produced a wide array

---

[10] Telephone conversations with Rock Creek Park Superintendent Adrienne Coleman, 16 May 2006, and U.S. Park Police officers Sgt. Scott Fear and Lieut. Phil Beck, 19 May 2006; "Police Seek End to 'P Street Beach' Cruising," *Washington Blade*, 12 November 2004.

of overt meeting places, while the internet has created an alternative—and in some ways safer—medium for facilitating spontaneous or anonymous liaisons. The disdain with which many gay Washingtonians responded to questions about P Street Beach suggests that those who continue to rely on such practices are not just behind the times, but something of an embarrassment to efforts to destigmatize gay culture.[11] On the other hand, several longtime members of Washington's gay community expressed sadness and resentment about recent trends that appear to be leading to a de-gaying not just of P Street Beach, but of the Dupont Circle area in general. As is happening in many formerly gay enclaves around the country, heterosexual young professionals are replacing gay urban pioneers and transforming territories once celebrated for their queerness, first by other queers and then by the community at large. One of Dupont Circle's most prominent gay bars is now a Chipotles, Starbucks runs rampant, and Rock Creek Park, a signature local clothing store half a block from P Street Beach, has decamped to the predominantly gay beach resort of Rehoboth Beach. Even the Fireside, a popular gay watering hole located catty-corner across P Street from the beach itself, recently spruced up to keep up with the neighbors.

Crossing P Street at this location, one encounters a dense screen of vegetation that screens the beach from public view. Further south along 23rd Street, the park service has dramatically improved surveillance by removing bushes and clearing tree limbs to a height of 6 or 8 feet.[12] A graveled driveway was extended into the middle of the

---

[11] Other issues of class and cultural politics may also come into play. P Street Beach and similar places continue to appeal to those who seek to conceal their sexual identity, a practice that runs counter to contemporary Gay Pride agendas. As a public space ostensibly devoted to conventional recreation, P Street Beach affords a convenient cover for those whose individual or social circumstances mediate against conspicuous expressions of homosexual identity. In addition to attracting a significant proportion of seemingly lower-class men of color, P Street Beach draws a steady stream of middle-aged and older men whose dress and demeanor are more aligned with bourgeois heterosexuality than with contemporary gay expression. According to several informants, many of the businessman-types cruising the park's periphery are cultural conservatives seeking clandestine liaisons.

[12] The Park Police would prefer an even more aggressive vegetation-limitation campaign to improve surveillance in the P Street Beach area. Rock Creek Park Superintendent Adrianne Coleman observed that "law enforcement would like to cut

beach to facilitate supervision. No police and no obviously illicit activity were observed in the main portion of P Street Beach on numerous daytime visits in the spring of 2006. In fact, the area appeared both dramatically under-utilized and decidedly heterosexual. A few thonged and greased male sunbathers were scattered about, but the primary users appeared to be heterosexual dog walkers and Frisbee players, who identified themselves as being mainly recent arrivals residing in new apartments and condominiums being constructed and converted nearby. Washington natives or those who had lived in the area for more than three years were generally aware of P Street Beach's status as "queerspace," but many newcomers had no idea of its reputation. Only when the scantily clad nature of the few remaining sunbathers was pointed out and the procession of men into the nearby woods alluded to did it dawn on them that their quiet and peaceful park was not as subservient to conventional mores and official discipline as it seemed. While Queer Theorists have written about the ways in which the presence of explicitly gay bodies can effectively "queer" space without requiring actual physical interventions, this process apparently requires a greater density of queer bodies than is present in P Street Beach on most occasions today.[13] Contemporary beach users who are aware of the sexual proclivities of their neighbors generally maintain that the two groups coexist harmoniously, and asserted that as long as the sex occurred out of public view they were largely unconcerned. They also observed that sexual activity increased markedly once the trees and bushes leafed out in the spring, performing a concealment function that was welcomed by both populations. Despite this daytime

---

down all the trees in Rock Creek Park, replace everything with cement, and paint it green." Police officials expressed somewhat similar sentiments, while acknowledging that broader park management goals had priority (see note 10: Coleman; Fear; Beck). Just days before the Stonewall Riots that many herald as the beginning of the gay revolution, a vigilante group cut down dozens of trees in a Queens park in an attempt to displace homosexual activity; news accounts suggest that this was done with the tacit approval of the police ("Trees in Park Cut Down as Vigilantes Harass Homosexual," *New York Times*, 1 July 1969; "Queens Resident Says the Police Stood By as Park Trees Were Cut," *New York Times*, 2 July 1969).

[13] The "queering" of space by explicitly gay bodies, activities, and imagery is discussed in Ingram, "Marginality and the Landscapes of Erotic Alien(n)nations" (*Queers in Space*, 27-52) and James Polchin, "Having Something to Wear: The Landscape of Identity on Christopher Street" (*Queers in Space*, 381-390).

harmony, complaints about nighttime activities in P Street Beach have increased as demographic changes bring in residents who are not as tolerant of traditional gay activities.[14]

One gay regular who had recently returned after several years' absence observed that he was struck by the changing population, as well. A practiced observer, he noted that daytime cruising was still occurring, though cruisers were being more selective and only approaching men who appeared to be checking out other men. He also pointed out the significance of subtle spatial gradations within the beach itself: the thong crowd favored the flat area toward the bottom of the bowl, dog walkers held sway mid-slope, and proximity to the woods bounding the south edge of the beach or the thickets near P Street Bridge implied one was available for an excursion into these visually impenetrable zones.

The wooded area stretching along the steep hillside above Rock Creek south of P Street Beach proper is where most daytime sexual activity occurs. It has long been known in the gay community as The Black Forest, based partly on its natural characteristics and, more pointedly, on the dominant ethnicity of the men who engage in sexual activity there, often, though not always, for pay. On every visit to the area a number of African American—and, increasingly, Latino—males, generally appearing to be in their twenties or thirties, could be seen sitting at the edge of the woods, lounging along the well-worn trail that ran next to the chain link separating the parkway from the adjacent school grounds, or lurking half-hidden on the many informal trails that interlaced the hillside. Other men, usually white, well-dressed, and ranging in age from twenty to well into their sixties, passed back and forth along the top trail, occasionally stopping to talk to one of the woods-dwellers before descending into the obscurity of the forest below. While the upper trail appears little different from a typical informal footpath, condom wrappers, used condoms, breath mint packs, ATM receipts, and prolific concentrations of cigarette butts serve as fleeting reminders of the way in which marginal populations circumvent designers' intentions and elude networks of discipline and

[14] "Police Seek End to 'P Street Beach' Cruising," *Washington Blade*, 12 November 2004.

surveillance to give fleeting priority to subversive needs and desires (fig. 8).

Suicide is arguably the most transgressive of acts, and certainly not one that is planned or intentionally designed for. Nevertheless, topography and man-made constructions combine to make the parkway extraordinarily well suited for this purpose. The towering bridges afford ideal launching spots while the isolation, quiet, and pleasant surroundings apparently facilitate contemplation and produce a sense of serenity that may serve as mental preparation for the final leap (fig. 9). Park officials are loath to release statistics for fear of advertising the bridges' efficacy, but allow that a significant number of suicides occur in Rock Creek and Potomac Parkway every year. Statistics from the District of Columbia medical examiner's office released in the 1980s suggest that approximately four or five people end their lives this way annually. The Duke Ellington Bridge at Calvert Street affords the most dramatic drop and was long favored for this purpose, gaining local renown as "The Suicide Bridge." Local authorities note that the most popular time for this activity is the period between Thanksgiving and Christmas, which they indelicately describe as "Jumper Season."[15]

While bridge-related suicides are the most fleeting and statistically anomalous of the subversive tactics addressed in this essay, the practice has produced the most conspicuous design response—or disciplinary strategy, in de Certeau's formulation—to date. After much political maneuvering and public debate, a controversial eight-feet-high suicide prevention fence was erected on both sides of Duke Ellington Bridge in the mid-1980s. The circumstances surrounding the railing's construction underscore both the prominent role that even relatively minor architectural details play in the everyday lives of urban citizens and the limited ability of designers to comprehensively control fundamental human behaviors. Barrier supporters presented the project

---

[15] Newspaper articles during the 1980s Calvert Street/Duke Ellington Bridge railing controversy cited figures in this range, and various spokespeople noted the rate had been comparable in previous years. The leader of the pro-railing forces, former Undersecretary of State Ben Reid (whose daughter committed suicide from the bridge in 1979), asserted that hundreds of people had jumped from the bridge since its completion in the 1935 ("Troubled Waters over Bridge," *City Paper* [Washington], 17-23 February 1989). The "Jumper Season" reference appears in "Bridge Fence Debate Renewed," *Washington Post*, 8 December 1985.

as a clear-cut case of humanitarian duty, painting adversaries as craven aesthetes prizing architectural purity over public safety. Opponents did invoke historic preservation regulations and emphasize the structure's architectural significance as a masterwork of designer Paul Cret, but they also underscored the unique and invaluable role that the unobstructed views of the surrounding scenery played in the everyday lives of thousands of neighborhood residents and commuters. Critics asserted that fencing off the Duke Ellington Bridge was more of a political gesture than a meaningful response, deploying a dramatic architectural intervention in what was likely to be an fruitless attempt to combat a fundamental social problem more appropriately addressed through less-visible but more effective investments in mental health counseling, suicide-prevention hotlines, and other social services in which the district was notoriously deficient. As predicted, both by local prognosticators and de Certeau's theories about the creativity and resilience of urban inhabitants, the primary effect of the district's disciplinary strategy was to redirect the offending behavior rather than eradicate it. While the death rate decreased dramatically at Duke Ellington Bridge, those seeking to end their lives with fatal leaps simply moved elsewhere, and the overall incidence of bridge-related suicides remained essentially unchanged. The Connecticut Avenue (or Taft) Bridge has emerged as the platform of choice, though even the lower bridges at the southern end of the parkway have been employed for suicidal purposes, albeit with mixed results.[16]

---

[16] Coverage of the prolonged controversy over the railings' erection appeared in the *Washington Post*, *Washington Times*, *Washington City Paper*, and even the *New York Times* between 1985-1990. Key articles include: "When Landmarks Lure the Suicidal," *New York Times*, 12 June 1985; "Suit Filed to Halt Rock Creek Bridge's Fencing," *Washington Times*, 28 August 1985; "Groups File Suit to Block Suicide-Prevention," *Washington Post*, 28 August 1985; "Bridge's Suicide Barrier in Place," *Washington Post*, 24 February 1986; "Follow-Up on the News; Suicide Problem at Twin Bridges," *New York Times*, 4 May 1986; "Suicide Barriers Don't Work," Letter to the Editor, *Washington Post*, 19 April 1987; "Span of Life: The Case Against the Fences on the Ellington Bridge," *Washington Post*, 4 July 1987; "Public Still Railing Over Ellington Bridge," *Washington Times*, 10 July 1987; "Suicide Fence Debated Again," *Washington Post*, 11 July 1987; "Controversial Ellington Railing Pits 'Design' Against Saving Lives," *Washington Times*, 13 July 1987; "Ruling Approves Antisuicide Fences," *Washington Post*, 26 August 1987; "Troubled Waters Over Bridge," *City Paper* [Washington], 17-23 February 1989; "A View from the Bridge – And Its Cost," Letter to the Editor, *Washington Post*, 18 March 1990; "Fences that Have Failed,"

This has been a necessarily brief overview of several complex phenomena, any of which could be expanded into a full-fledged study. More comprehensive ethnographic research would undoubtedly reveal additional information about each set of practices, but there is considerable value in broad-ranging explorations that present key observations without burdening readers with overwhelming detail. Since de Certeau himself rarely produced the fine-grained analyses of urban environments he advocated, surveys such as this represent a middle ground between his theoretical postulations and the microscopic investigation of urban spaces and quotidian practices he championed. Though they are neither exhaustive nor definitive, these vignettes demonstrate the ways in which ordinary—and even socially marginal and ostensibly disempowered—people take environments produced by designers and administrators and make them their own. Not only are the claims they stake and the interventions they make often fleeting, furtive, and fragmentary, but the processes through which they act can be so subtle and surreptitious that they are all-but-invisible to most observers, complicating efforts to identify, record, and interpret their contributions to the urban experience. Participatory planning and scholarly emphasis on "history from the bottom up" have made great strides in challenging the hegemony of bureaucratic policymakers and monolithic master builders, but we still have a long way to go in recognizing and empowering those who are affected by design and planning decisions but may be unwilling or unable to express their needs and desires. Combining the ethnographer's or urban sociologist's methodological frameworks with de Certeau's theoretical concerns with the deeper implications of ephemeral and seemingly innocuous spatial behaviors could inform design and planning decisions while adding important dimensions to the historical record. On a more fundamental level, calling attention to ephemeral geographies and overlooked practices can enhance our understanding of the social

---

Letter to the Editor, *Washington Post*, 24 March 1990; "New Debate on Suicide Barriers," *Washington Post*, 24 May 1990; "Congress Saves Suicide Barriers on Bridge," *Washington Times*, 25 June 1990. Telephone conversations with U.S. Park Police officials Sgt. Scott Fear and Lieut. Phil Beck, 19 May 2006 attested to the continued prevalence of the practice and its relocation to other bridges. An unsuccessful suicide attempt from Taft Bridge was described in "Would-Be Bridge Jumper Rescued," *Washington Post*, 28 May 2006.

functions and cultural significance of our human interactions with the built environment, and with each other.

Bibliography

Bell, David and Gill Valentine, eds. *Mapping Desire: The Geographies of Sexuality*. London: Routledge, 1995.

Bourdieu, Pierre. *Outline of a Theory of Practice*. Cambridge: Cambridge University Press, 1977.

Bushong, William. *Historic Resource Study: Rock Creek Park, District of Columbia*. Washington, D.C.: U.S. Department of the Interior, National Park Service, 1990.

Cox, William V. "Notes on the Establishment of a National Park in The District of Columbia and the Acquirement and Improvement of the Valley of Rock Creek for Park Purposes." In *Park Improvement Papers*, edited by Charles Moore. Washington, D.C.: Government Printing Office, 1902.

Davis, Timothy. "Rock Creek and Potomac Parkway: HAER Report No. DC-697." Washington, D.C.: U.S. Department of the Interior, National Park Service, Historic American Engineering Record, 1993.

—. "Rock Creek and Potomac Parkway, Washington, D.C.: The Evolution of a Contested Urban Landscape." *Studies in the History of Gardens and Designed Landscapes* 19 (April-June 1999): 123-237.

de Certeau, Michel. *The Practice of Everyday Life*. Berkeley: University of California Press, 1994.

Ingram, Gordon Brent, Anne-Marie Bouthillette, and Yolande Retter, eds. *Queers in Space: Communities/Public/Places/ Sites of Resistance*. Seattle, WA: Bay Press, 1997.

Jackson, John Brinckerhoff. *Landscapes: Selected Writings of J.B. Jackson*. Edited by Ervin Zube. Amherst: University of Massachusetts Press, 1970.

—. *The Necessity for Ruins and Other Topics*. Amherst: University of Massachusetts Press, 1980.

—. *Discovering the Vernacular Landscape*. New Haven: Yale University Press, 1984.

—. *A Sense of Time, A Sense of Place*. New Haven: Yale University Press, 1994.

—. *Landscape in Sight: Looking at America*. Edited by Helen Lefkowitz Horowitz. New Haven: Yale University Press, 1997.

Jacobs, Jane. *The Death and Life of Great American Cities*. New York: Random House, 1961.

Krakow, Jere. *Historic Resource Study: Rock Creek and Potomac Parkway, George Washington Memorial Parkway, Suitland Parkway, Baltimore Washington Parkway*. Washington, D.C.: U.S. Department of the Interior, National Park Service, 1990.

Lefebvre, Henri. *Writings on Cities*. Translated and edited by Eleonore Kofman and Elizabeth Lennas. London: Blackwell, 1996.

Mackintosh, Barry. *Rock Creek Park: An Administrative History*. Washington, D.C.: U.S. Department of the Interior, National Park Service, History Division, 1985.

"Photographs: Rock Creek Parkway, Present Condition." Looseleaf notebook. Commission of Fine Arts Library, Washington, D.C., n.d., ca. 1915.

Rock Creek Park: Information for the Public in Relation Thereto. Washington, D.C.: Judd and Detweiler, 1889.

Sanders, Joel ed. *Stud: The Architectures of Masculinity*. New York: Princeton Architectural Press, 1996.

U.S. Congress. House. 1916. *Report of the Rock Creek and Potomac Parkway Commission: 1916*. 64th Cong., 1st Sess., H. Doc. 114.

U.S. Congress. Senate. Committee on the District of Columbia. 1902. *The Improvement of the Park System of the District of Columbia*. 57th Cong., 1st Sess., S. Report 166.

U.S. Congress. Senate. 1908. *Report on the Improvement of Rock Creek from Massachusetts Avenue to the Mouth of Rock Creek*. 60th Cong., 1st Sess., S. Doc. 458.

Whyte, William Foote. *Street Corner Society: The Social Structure of an Italian Slum*. Chicago: University of Illinois Press, 1943.

—. *Learning from the Field: A Guide from Experience*. Beverly Hills, CA: Sage Publications, 1984.

Fig. 1. The Social Space of Rock Creek and Potomac Parkway. 2006.
Photograph by author.

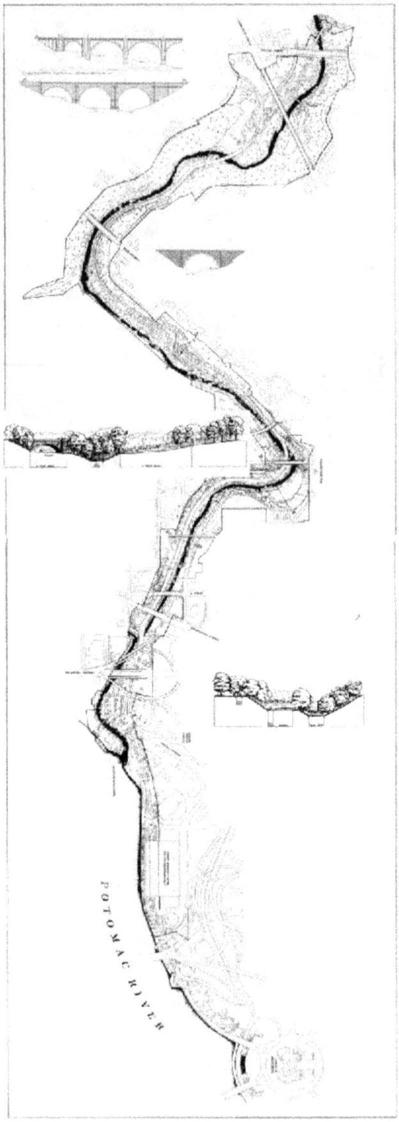

Fig. 2. Map of Rock Creek and Potomac Parkway. 1992.
Historic American Buildings Survey.

Fig. 3. Rock Creek and Potomac Parkway, before and after improvement. Bottom: 1901. Photograph. *The Improvement of the Park System of the District of Columbia*, 57th Cong. 1st Sess., Senate Report 166; top: 1937. Photograph. Washingtoniana Collection, District of Columbia Public Library.

Fig. 4. Official trail and vernacular paths and campsites. 2006. Photograph by author.

Fig. 5. Unofficial parkway residents. 2006. Photograph by Author.

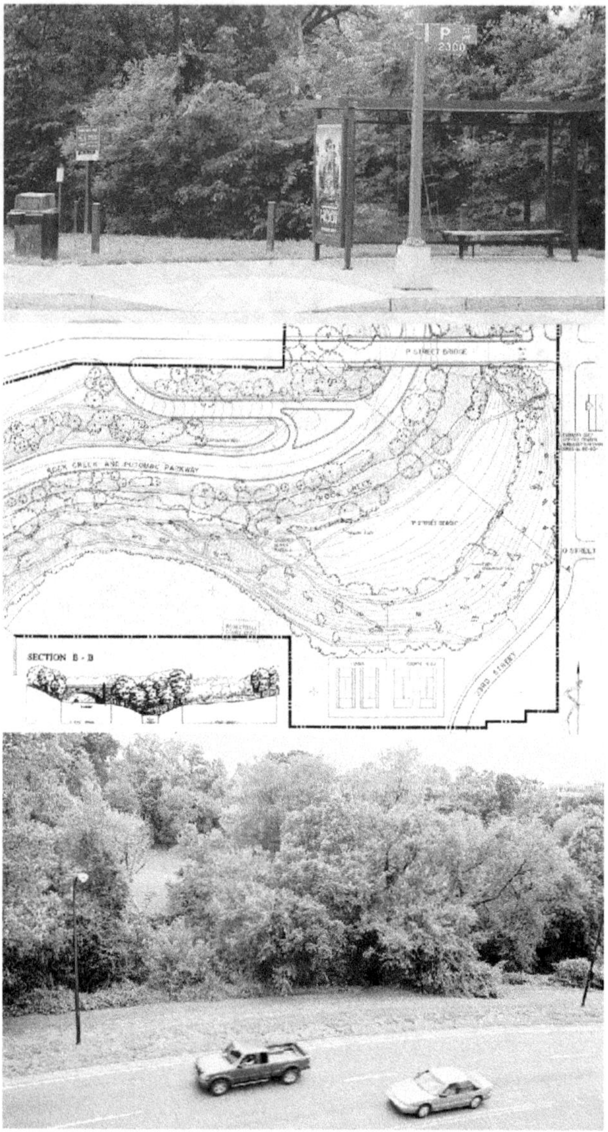

Fig. 6. P Street Beach: site plan (middle) and obstructed views. Plan, 1992. Historic American Buildings Survey. Photographs by author, 2006.

Fig. 7. Queer Space, P Street Beach, and Black Forest. 2006. Photograph by author.

Fig. 8. Trail in Black Forest strewn with condoms and ATM receipts. 2006. Photograph by author.

Fig. 9. Calvert Street and Connecticut Avenue Bridges. 2006. Photograph by author.

Public Space Transformed: New York's Blackouts

David E. Nye

A social history of power failures in New York City illustrates both the public's increasing dependence on electricity and its changing behavior during blackouts, as they shifted from being merely inconvenient (ca. 1910) to paralyzing events (1965 and after). Responses included aesthetic appreciation of the city as a new kind of landscape, a liminal experience of openness and sociability with strangers, criminal activity, especially looting and arson, and a heterotopian break with productivity, transforming the city into an anti-landscape that cannot sustain itself.

Introduction

Electrical blackouts are not only power failures. They disrupt the conventional use of space and time, creating opportunities for social improvisation in public space. How did the public behave when first confronted with power failures? How did this response change as blackouts became more familiar, but also more comprehensive in their effects? To answer these questions I will focus particularly on two blackouts that occurred in New York City in 1936 and 1965, and refer more briefly to later power failures.

Study of these events reinforces the conclusion of other historians of technology, who find little evidence of technological determinism as an explanation for public behavior.[1] Instead, blackouts are better understood as social constructions that vary over time, as well as over space. Not all cultures comprehend a blackout in the same way, although there assuredly are some commonalities to the experience. Even within the United States, rural people and those in small towns do not experience them in the same way as the residents of New York City. Here, I will explore the evolving meaning of blackouts in New York, stressing their symbolic and cultural dimensions, and

---

[1] For a discussion of this point, see Nye, *Technology Matters*, 17-48.

drawing on the concepts of liminality (Victor Turner) and heterotopia (Michel Foucault).

In most publications about blackouts, aside from journalistic accounts, technical, legal, political, and economic matters predominate, while the experiences of the majority of the public are scarcely mentioned. Official reports, whether by utilities or government commissions, focus not on those who are merely inconvenienced but unharmed, but rather on the concerns of power station managers, public officials, and those who suffered substantial losses. Such publications on electrical blackouts give the chronology of each event and assess the factors that led up to it. They explain technical failures, blame specific actors, and assess economic costs. For example, after the 2003 blackout that affected 50 million people in Canada and the Northeastern United States, a number of such reports appeared. A Canadian-American task force devoted twenty-seven pages to "how and why the blackout began in Ohio," and thirty pages to how failures there cascaded from one system to another. But their report contains almost nothing on the public's immediate response or subsequent behavior.[2] While it is useful to know that particular power failures were due to lightning, human error, poor maintenance, squirrels chewing on power lines, or inadequate generating capacity, these are all questions of supply and demand, or of cause and effect. The social and cultural history of blackouts remains outside such accounts, which treat each blackout as a unique event that is destined to become a legal case governed by the law of contracts and torts. After the 2003 power failure, for example, many lawsuits were pending, because billions of dollars had been lost. Elected officials also interpret blackouts, seeking the political advantage of putting the blame elsewhere. The reports and the public forums that follow a major power failure establish many facts, but they are too narrow to provide insight into the social meaning or the historical significance of blackouts.[3] They do not examine the public's often-intense experiences, and they do not consider non-technical interpretations of what it meant.

---

[2] U.S.-Canada Power System Outage Task Force, *Final Report on the August 14, 2003 Blackout in the United States and Canada*.

[3] See, for example, Federal Energy Regulatory Commission, *The Con Edison Power Failure of July 13 and 14, 1977*.

Yet for those caught in a blackout, these technical assessments do not describe the experience. More useful, if less comprehensive, are the newspapers and magazines, from which one can piece together the public experience of power failure. However, journalism necessarily focuses on immediate events and does not examine the shared characteristics of all blackouts or the cultural patterns that could emerge by looking at a *series* of these power failures. A blackout is a major cultural disruption, when the social construction of reality briefly breaks down. People must improvise temporary solutions to new problems, often solving them together with strangers. Furthermore, the intensity of this disruption has increased with each passing decade, as electrical networks have penetrated and controlled more aspects of everyday life, most obviously in recent years through broadband connections. In 2003 New Yorkers were dismayed to find that among the thousands of devices that did not work were mobile phones, gasoline pumps, escalators, ATM machines, and pumps in the water system. Because cultural dependence on electricity is intensifying, the response changes from one blackout to the next, revealing how society's dependence on electric power and communications is intensifying, The blackout unmasks the illusion that the city's form is "natural" or "normal," and confronts each individual both with specific difficulties and with the larger problem of making sense of the crisis as a whole.

Blackouts as Liminal States

Blackouts also reveal the degree of social cohesion at a particular cultural moment. Most immediately, every person caught in a blackout must redefine the potential uses of public space, and to some degree must improvise. Public spaces are ordinarily designed for certain forms of behavior that are implicitly understood by adults, though children often invent "inappropriate" uses for them. To take a simple example, sidewalks are intended for pedestrians. If children sit down to play on a sidewalk, their behavior is understood to be inappropriate because it blocks the flow of foot traffic. But during an electrical blackout people do sit on the sidewalks, and put their feet in the gutter. Instead of hurrying along, they have nowhere to go, and they have no immediate purpose. As the BBC reported on the 2003 blackout in New York City, there were "thousands of people lying on the sidewalks, propped up against shop fronts, lounging on the edge of fountains, huddled round small clusters of night lights, chatting."[4] To the considerable extent that public space in the United States is intended for commercial activity, it suddenly becomes useless. People caught in a blackout soon realize this, and improvise new activities for public space. Likewise, during a blackout privately owned commercial spaces are transformed. Inside stores people may begin to party, and in hotel lobbies and ballrooms they may sleep. Dislodged from their routines, the public soon realizes that a blackout opens up unexpected possibilities. Assigned social roles become more fluid. A businessman may take off his coat and go out into the street to direct traffic. Unable to leave a subway car or elevator, people break the usual code of anonymity and speak to one another. In a bar or restaurant, the patrons continue to drink and to eat, but the pace of these activities slows down, and the openness to complete strangers increases. As long as the power is off, there is literally no rush, social inhibitions weaken, and many hierarchies lose their importance.

Such moments suggestively recall Victor Turner's concept of liminal space, which he developed from Van Gennep's work on "rites of passage." In all such episodes, the first phase of detachment from a fixed position in the social structure is followed by liminal period, before the individual returns to a stable state at the end of the process. As Turner conceived it, liminal space emerges not through accident but through ritual. The moment is planned for, typically arising at a particular time each year. Turner noted in *The Ritual Process*, "liminality is frequently likened to death, to being in the womb, to invisibility, to darkness [. . .]" (95). The darkness of a blackout combined with the suspension of most normal activities suggests an experience that shares some of the characteristics of a "rite of passage." And yet, until it happens, the blackout is but a latent possibility, and therefore not like a ritual, which is an expected event. Each blackout arises as a sudden violation of the "normal." Its unframed liminality occurs in the ordinary spaces of everyday life rather than within a specially designated ritual space. Furthermore, the desired ending of a

4 "Blackouts Cause North America Chaos," BBC News, 15 August 2003, http://news.bbc.co.uk/go/pr/fr/-/2/hi/americas/3152451.stm.

blackout is not renewal but a return to precisely the condition one was in before it took place. Nevertheless, the uncertain transitory period of the blackout shares some characteristics with the liminal state. In both, distinctions of rank largely disappear and individuals become temporarily homogeneous parts of a *communitas*. By comparison with ordinary life, they often disregard social rank and personal appearance, and they become aware of the common human bond that is necessary for society to function (Turner, *Ritual Process*, 95-97). With these general characteristics in mind, consider blackouts in New York City.

New York Blackouts

New York experienced major blackouts in 1936, 1965, 1977, and 2003. Examining the first two establishes the nature of the blackout as a historical phenomenon, and will be the focus here. The 1977 and 2003 blackouts will only be referred to briefly, as they require separate treatment. In a basic sense, the same thing happened each time the power failed: lights went out and all electrical devices stopped functioning. Yet these blackouts were experienced far differently, in part because dependence on the system was increasing throughout these years, even as the knowledge of how to live without electricity was simultaneously declining. To take a simple example, someone writing at a desk in 1936 used either a manual typewriter or a pen. By 1965 they likely used an electric typewriter, which froze up when the power went off, but at least no work was lost. In contrast, in the office of 1977 and especially that of 2003, when computer systems went down they took millions of pages of information with them. In countless other ways, New Yorkers realized that their city was a technical system of ever increasing complexity, woven together through electrical connections. With each passing decade, more and more parts of the city, when without power, became parts of a vast, broken machine.

The city was not just becoming more vulnerable; in blackouts from 1965 onwards it ceased to be a sustainable landscape. J. B. Jackson defines landscape as, "A composition of man-made or man-modified spaces to serve as infrastructure or background for our collective existence" (8). Landscape is a shared creation that is not something outside of human beings that they merely look at. Rather, landscape is humanly modified space. It is not natural but cultural, not merely something seen, but rather the infrastructure of existence. When the power fails, however, the urban infrastructure becomes dysfunctional. For those trapped in subways and elevators it immediately becomes a prison, which they need to escape. A blackout that lasts more than a day or two places the entire population in similar jeopardy, for a city without electricity cannot long sustain human life. It becomes an anti-landscape. To take but a few examples: sewage backs up, food in freezers and refrigerators spoils, microwaves and stoves do not work, and normally air-conditioned skyscrapers heat up to over 40°C. As they lived through a succession of power failures, New Yorkers gradually understood the city's vulnerability during blackouts, and saw that in a week or even less, life would become unsustainable. This realization was not central to the experience of the first two blackouts, of 1936 and 1965, but emerged more clearly and became a central part of the meaning of the power losses in 1977 and 2003. Turner wrote that the "social dramas" of a crisis have "liminal characteristics, since it is a threshold between more or less stable phases of the social process" (*Dramas, Fields and Metaphors*, 39). The disruption of a blackout contains elements of liminality, but it teeters on the edge of becoming a disaster. It is a crack in the flow of time and in the structure of space, which at best can lead to a restoration. If a power failure breaches ordinary experience, creating an immediate crisis, it is followed by redressive action to sustain crucial social services, repair of the technical system, and return to "normal" life.

The Blackout of 1936

In the 1930s dependence on electricity was still recent and not yet entirely normalized. Local electrical systems often broke down, and people simply took outages in their stride. They did not yet seem to be extraordinary ruptures in experience. Indeed, the very word "blackout" was not yet in common use outside the contexts of the theater and military defense.[5] But as the electrical utilities merged into regional

---

[5] Comprehensive surveys of complete runs of the *London Times* and the *New York Times* show that "blackout" was scarcely in use at all before 1930, and that it referred

systems, the potential for widespread power failure increased. On January 15, 1936, after a short-circuit at the Hell Gate generating station of New York Edison, the power went off in the Bronx, in Manhattan above 59th Street, and in much of Westchester County. This happened at 4:16 PM, near the peak of demand. "Lights in offices, hospitals, department stores, homes, and industrial establishments throughout the area began to flicker and then failed completely."[6] Traffic lights went black, movies stopped, elevators stalled between floors, and the subways halted, stranding 60,000 people. When power began to come back on around 5:00 PM, the current was weak, and the subway switching system did not work properly. Cars could not climb grades on some parts of the line. As traffic lights began to come on, some stayed permanently on red, others on green. Life did not return to "normal" for three hours. The fact that the lower half of Manhattan, Brooklyn, Queens, and Staten Island, were not affected is a reminder that the electrical system was not yet tightly woven together into an interdependent system. In 1936, just one half of New York could lose power.

Temporary power failures had been quite common from Edison's first installation of an electrical system in New York's financial district in 1882. At the end of 1883 there were 500 customers, who did not yet expect the power to be available without any interruptions. When generation problems emerged, the nation's first power station had nowhere else to turn to buy electrical current, and the system went down. Expectations of continuous service developed only gradually, as the local grid was knitted together through mergers of many small power stations, creating Consolidated Gas (later Consolidated Edison) in 1901. Occasional outages still seemed "normal," however, and people kept candles, flashlights, and other backup technologies ready to hand. In the 1936 blackout, New Yorkers grumbled a bit, but expected that service would soon be restored. The same was true of the citizens of Newark, New Jersey, who experienced a five-hour blackout at the end of the same year, again in the late

---

to darkening the lights in theaters. The early history of the term and its relationship to the electrical system is traced in Nye, "Are Blackouts Landscapes?"

6 "Lights Out, City Paralyzed for Hours North of 59th Street," *New York Times,* 16 January 1936.

afternoon. There were no disturbances or problems due to the outage, and only one traffic accident. Bamberger's department store remained open, because it retained its own electrical generating system.[7] Otherwise, office workers climbed down ten, twenty, and thirty stories and joined factory workers, sent home early. "All hurried away from the gloom-swept center of the city, where the shadowy streets, lighted only by the dim beams of a bright moon" were interrupted by acetylene torches at intersections (ibid.).

As these two examples suggest, during the 1930s Americans regarded temporary power outages as being unavoidable. When a power system went out, most people still recalled everyday life before electricity, for the vast majority of Americans had grown up without it. Even in 1936 most American homes still had iceboxes, not refrigerators, and no one had a freezer. Many women had adopted electric irons, but they still knew how to heat up an iron on a stove, which was seldom electric at that time. Offices also could get along without the electric grid. Typewriters and adding machines were mechanical. Copies of documents were produced using carbon paper and saved in file cabinets. A modern office, even as late as 1940, primarily needed electricity for light and ventilation. A power failure was a nuisance but not a major problem.

The Blackout of 1965

By 1965, however, both offices and homes had many more electrical devices. Furthermore, the electrical system had been quite reliable for a generation, and power outages occurred less often and seldom lasted more than a few minutes. New York's Consolidated Edison had absorbed more than a dozen smaller companies between 1936 and 1960, notably Westchester Lighting Company and Yonkers Electric Light & Power in 1951. Furthermore, by the early 1960s a web of interconnecting lines linked the New York system with electrical companies 600 miles or more away, including Toronto and the hydroelectric plants at Niagara Falls. Lightning still destroyed some equipment, but the utilities had installed extra capacity that could be

---

[7] "Newark Dark for Hours as Fire Cuts Off Power," *New York Times*, 29 December 1936.

activated during an emergency, and they had could use regional interconnections to shunt power quickly to where it was needed. It therefore came as a shock on November 9, 1965, when a blackout hit most of New York at 5:27 PM and lasted as much as thirteen hours in some parts of the city.[8] Only Staten Island and parts of Brooklyn (served from New Jersey) escaped. All TV stations went off the air. Thousands of people were trapped in elevators, and while most were soon rescued, a few endured long hours there. Not everything shut down. The telephone system worked, some radio stations stayed on the air, and most hospitals carried on, using emergency diesel generators. But most of the city ground to a halt. "Initially, some people thought the blackout was to do with Soviet sabotage, or even UFOs, but there was a general sense that the government had things in hand."[9] Radio broadcasts, typically heard in automobiles, soon reassured the public that the blackout did not have an extraordinary cause, and that power would be restored the following day.

Tens of thousands of shoppers were caught inside stores; some ended up remaining for hours, or even all night. Others slept in hotel lobbies, ballrooms, and other public spaces. An estimated 80,000 people were trapped in subway trains. Evacuation took hours, and at midnight 10,000 were still stranded underground. One journalist recalled,

> a piece of television news which showed some people who'd been stuck all night in a subway car. A crowded mixed bag of young and old, well dressed and shabby, they seemed absolutely overjoyed at their predicament. And when the subway policeman who had been stuck with them congratulated them on the way they had been good comrades and otherwise passed the time courageously, they cheered him wildly. (Wainwright, 35)

The public generally behaved well. "Although persons in public vehicles (other than trains) and at work tended to regard the blackout as somewhat more unusual than others, there is no evidence that they

[8] A trove of documents on the 1965 blackout has been assembled under a grant from the Sloan Foundation. See The Blackout History Project, "Archive / The 1965 Blackout," http://blackout.gmu.edu/archive/a_1965.html.

[9] Kathryn Westcott, "New York's 'Good and Bad' Blackouts," BBC News, 15 August 2003, http://news.bbc.co.uk/1/hi/world/americas/3154757.stm.

perceived any real or potential threat in the situation."[10] Travelers at airports found themselves in unlighted buildings looking at dark runways, where no planes could take off. For those flying at the time of the blackout, the experience was surreal. Passengers invited to gaze at the lights of Manhattan saw them flicker out. It clearly was not fog or bad weather that obstructed the view—the lights of automobiles remained visible, tracing the lines of the streets, but the skyscrapers had become indistinct hulks, and the famous electric signs were gone. Some drivers found it hard to get home. Traffic signals stopped working, and those who made their way to the Queens Midtown and Brooklyn Battery Tunnels found them closed, for they lacked lights to guide traffic and fans to evacuate the exhaust. Those driving to New Jersey were more fortunate, for Lincoln Tunnel, Holland Tunnel, and the George Washington Bridge all got their power from New Jersey.

A few people could carry on regardless, notably Vladimir Horowitz, who continued a rehearsal, first playing from memory, later with the aid of flashlights. Many made extraordinary efforts, like the zookeepers who stayed up all night and found ways to keep their animals warm on a cold November night. In general, public order was good. At Walpole Prison in Massachusetts 300 inmates rioted under the cover of darkness for two hours, and there were scattered reports of petty crime or vandalism in some cities affected by the power outage. But the 12,000 policemen on duty in New York found the crime rate was lower than usual. However, there were four times more fires than normal, as people tried to improvise heat and light.

Although the blackout persisted all night, the public was almost uniformly peaceful, relaxed, and even exuberant. A party mood prevailed in many places, as the news media described in detail. *Life Magazine* devoted a whole issue to the experience, concluding that "the blackout quite literally transformed the people of New York," taking away their anonymity:

> Ordinarily smug and comfortable in the high hives of the city where they live and work, they are largely strangers to one another when the lights are on. In the darkness they emerged, not as shadows, but far warmer and more substantial than usual. Stripped of the anonymity that goes with full illumination, they became

[10] National Opinion Research Center, 11.

> humans conscious of and concerned about the other humans around them. In the crowded streets, businessmen, coats removed so that their light-colored shirts could be seen, directed traffic. Though the sidewalks were jammed, there was little of the rude jostling that is a part of normal, midday walking in New York. In the theatrically silver light of a perfect full moon (a must for all future power failures) people peered into the faces of passersby like children at a Halloween party trying to guess which friends hide behind which masks. In fact, the darkness made everyone more childlike. There was much laughter, and as they came down the stairs of the great office buildings in little night processions led by men with flashlights and candles, people held hands with those they could not see. (Wainwright, 35)

Networks of urban conveniences usually eliminate the need for much personal contact or fellowship. But the blackout forced New Yorkers back into contact, demonstrating their interdependence, temporarily rendering them childlike, as they held hands in the street. Numerous journalistic accounts of the 1965 blackout provide similar descriptions of the public reaction, which accords well with the idea of a liminal event, in which hierarchies break down and people from all levels of society feel at one, while stranded in a subway car or a hotel lobby. During subsequent blackouts *much* of the public responded in this way, and those who experienced a blackout generally regard it as a remarkable incident that was not actually unpleasant, despite some hardships. In short, the public has often experienced a blackout as an unexpected moment of social cohesion, which emerges spontaneously.

Contrast this response to blackouts with another striking moment of social solidarity in the history of New York City, which was a carefully constructed civic event, the rededication of the Statue of Liberty in 1986.[11] On that occasion, the organizers provided spectacular laser lighting and an immense fireworks display (40,000 projectiles, 100,000 bursts), which awed and pleased the immense crowds that lined the harbor on all sides, and filled much of it with small boats. While many security guards and extra police were on duty, prepared for unruliness or trouble, they had almost nothing to do. As the *Daily News* put it in a headline, "Nightmare Just Never Happened." Likewise, the *New York Times* found the crowds were "notably serene, almost

---

[11] For analysis of this event, see Nye, *American Technological Sublime*, 272-280.

reverent, as if paying homage to something righteous and inviolable." As one participant told a reporter, "There's a brotherhood and everybody's friendly." In this case, New Yorkers were seeking a rupture in ordinary experience, and public officials gave them a sublime event in the spectacular relighting of the torch of the Statue of Liberty, which induced a sense of unity and patriotism. At the climax of the display, "Strangers embraced one another. Some were so charged up that they jumped on top of benches" and began to sing "God Bless America."[12] In this moment of liminality, social divisions also broke down, but the moment was carefully structured in advance. In that sense, rededicating the Statue of Liberty also had affinities with the planned rituals that Turner analyzed. But a blackout was by definition unscheduled and unscripted. The crowds at the Statue of Liberty in 1986 were excited and amazed not by the accidental loss of illumination, but by a super-abundance of dazzling special effects. The sheer surplus of energy became both a visible sign of patriotism and a stimulus to revelry.

Hypothetically, the unexpected breakdown of the electrical system might have prompted the complete opposite: the public might have expressed depression, fear, or anger at public officials and the power company. Yet while one can find a few scattered examples of such responses, the general attitude toward the power failure was far more positive. To understand why, it is useful to look beyond Turner's work to a theory that posits not organized moments of liminality, but unscripted breaks in social time.

Heterotopia

New York blackout experiences have some affinities with what Michel Foucault calls a "heterotopia." He specifies that, "Heterotopias are most often linked to slices in time," and they begin "to function at full capacity when men arrive at a sort of absolute break with their traditional time."[13] Paradoxically, the traditional time of the modern

---

[12] Joseph Berger, "Rapt Crowds View Statue's Relighting" *New York Times*, 4 July 1986.

[13] Foucault, "Of Other Spaces," under "Third Principle." This text was the basis of a 1967 Foucault lecture.

city is based on acceleration, on the pressure to compress more experience, more work, and more reality, into every hour of the day. The blackout breaks this pressure, as it literally stops almost all public clocks and prevents most activity. It interrupts the flow of production, and confronts people with timelessness, with living and improvising in the present. The blackout thus breaks not with traditional time, but with modern, capitalist, productive time. If this accelerated time is money, then the timeless paralysis of a blackout uncovers other values and other uses of space.

Foucault also argues that every heterotopia "is capable of juxtaposing in a single real place several spaces, several sites that are in themselves incompatible" (ibid., under "Fourth Principle"). In a blackout, the electrified city is juxtaposed with a darkened twin of itself. The de-electrified city has the same contours and occupies the same location as the everyday city, but it is not the same place. Most obviously, it both looks and sounds different. The illuminated New York of skyscrapers and spectacular electric signs disappears, revealing an anti-landscape that inverts the values embedded in the electrified city. The most obvious of these values is the belief in progress, which electricity continually embodied at world's fairs from 1881 until at least 1939. The electrical building was one of the most popular at the Chicago Exposition in 1894, and electricity was the central theme of the Pan-American Exposition in Buffalo of 1901, which celebrated the opening of hydroelectric plants at Niagara Falls.[14] During the 1950s and 1960s, General Electric ran advertisements declaring that, "Progress is our most important product," and Ronald Reagan regularly delivered this slogan as the television host on the General Electric Theater.[15] For a century, the electrified city was presented as a dynamic utopian landscape. The de-electrified city was its static, regressive twin.

Second, the electrified city, and particularly New York, had always been intimately associated with the modern, in art, architecture, and literature, for example in the skyscrapers of Manhattan, the canvases of Joseph Stella, or the photography of Alfred Stieglitz.

---

[14] See Nye, *Electrifying America*, Chapter Two, and Nye, "Electrifying Expositions, 1880–1939," 113-128.

[15] The GE slogan is still widely referenced on the Internet, and even *The Nation* mentions it as an example of American progressive patriotism. See Green, "What Do We Stand For?"

During the first half of the twentieth century, Americans had learned to see New York's almost cubist skyline as the visualization of the modern. A blackout darkened that same skyline, transforming it into a hulking, static cliff.

A third value embedded in the electrified city was the possibility of incessant activity, based on ubiquitous energy. Darkness had been eliminated as an impediment for those who wanted to work, shop, stroll, or play sports at night. Electric current could be brought to any location, flexibly and easily, whether underground, atop a skyscraper, or in an airplane miles overhead. With electricity, it seemed, anything could be done almost everywhere. Americans had made a world where the radio and television were always broadcasting, where some shops were always open, and where many activities were available around the clock. If hyperactive, brightly lighted cities seemed novel in the early twentieth century, they seemed a necessity three generations later. Loss of power was no longer merely inconvenient; it paralyzed individuals accustomed to an enormous range of possibilities. Confronted with a blackout, people literally lost the freedom to select from a vast array of choices. A blackout was not merely a technical failure; it was a breakdown in the flow of time and in the visual continuity of the city. To borrow a term from literary criticism, it was a visible *aporia*.

The meaning of blackouts changed over time. In 1936, the electrical system itself had not yet been entirely assimilated into everyday life, and it was not taken to be "natural." By 1965, however, many New Yorkers regarded a blackout as a violation of the expected order of things. Yet it seemed an anomaly, with no long-term implications. The paralysis of that night could become a liminal moment. In later years, the perception of and response to blackouts shifted markedly, however. They became an annoying possibility every summer, as the demands of air conditioning pushed up the utility peak load. They also became less potential carnivals than opportunities for crime. New Yorkers endured not only many brown-outs and small system failures, but two massive breakdowns, in 1977, when primarily the greater metropolitan area alone was affected, and 2003, when more than 50 million people from Ohio to Long Island, were without power.[16]

A quarter century before the terrorist attacks of September 11, 2001 or the power failure of 2003, blackouts had acquired a sinister air. The vulnerability of the electrified world emerged when the lights flickered out, suggesting a dark side of urban life that had been ameliorated, though never entirely erased, by the introduction of street lighting in the nineteenth century. In 1977 parts of New York erupted in arson, looting, and riot. There was little delay. Within minutes of the power failure the streets of a few neighborhoods filled with looters, who stripped local stores of their goods.[17] More than 3,750 were not arrested because the police ran out of places to put prisoners. Certainly many more participated, and unlike the racial riots of the 1960s, which selected their targets, looting and burning was indiscriminate, destroying many Black-owned businesses. Clearly, the sense of community had broken down, and a liminal experience of unity was not possible in these areas. The 1977 blackout did not unleash a spontaneous, liminal moment of unity but revealed a deeply fractured society. The prosperity of 1965 had evaporated, and many New Yorkers believed the energy crisis was a permanent condition, the harbinger of widespread shortages. The City suffered from high unemployment and high-energy prices. It teetered on the edge of bankruptcy. The power outage was quite unlike the temporary disruption of 1965 that had scarcely upset a confidant and prosperous New York. Rather, the Blackout of 1977 appeared to be one more breakdown in civic order, as a divided society faced a grim and impoverished future.

The 1977 looting left its mark on popular culture. Take for example Kurt Russell's science fiction film from 1981, *Escape from New York*, in which New York City was a vast, walled ruin, inhabited entirely by criminals. When darkness falls, no lights come on, and it is fatally dangerous to go outside.[18] Likewise, during 1999 fears that the computer system would break down on December 31 at midnight,

---

[16] In 1977, a small but significant minority in a few neighborhoods chose to riot and loot, and revolt. The response to a blackout thus could measure the cohesion of society and the trust in authority.

[17] For a general account, see Mahler, *The Bronx is Burning*, 175-229.

[18] Carpenter and Castle.

evoked the dangers of massive computer system failures that would prevent the stock market from functioning, imperil the water supply, crash airline reservation systems, and lead to a long-term blackout. The de-electrified city was imagined as a terrifying, lawless place, a landscape of fear. The vision of New York in darkness and ruin also emerged in Billy Joel's popular song, "Miami 2017 (Seen The Lights Go Out on Broadway)" whose lyrics include the following lines: "I've seen the lights go out on Broadway, / I saw the ruins at my feet," But the singer immediately acknowledges that such disasters did not surprise. In fact, "we almost didn't notice it, / We'd see it all the time [in the movies shown] on Forty-Second Street."[19] There is thus considerable evidence in popular culture that Americans in general and New Yorkers in particular were quite aware of blackouts as grim potentialities, and that they were associated with apocalyptic cultural breakdowns.

By the beginning of the twenty-first century, blackouts were recognized as more than merely latent possibilities. They were unpredictable, but seemed certain to come. Breaks in the continuity of time and space, they opened up contradictory possibilities. From their shadows might emerge a unified communitas or a riot. The blackout shifted its meanings, and achieved new definitions with each repetition. For some, it remained a postmodern form of carnival, where they celebrated an enforced cessation of the city's vast machinery. Bakhtin once wrote that in carnival, "People who in life are separated by impenetrable hierarchical barriers enter into free and familiar contact" (123). Something similar occurred in New York's 1965 and 2003 blackouts, though to a far lesser degree in 1977, when a significant number of people in certain neighborhoods used the blackout as an opportunity for looting, that some perhaps justified as the social redistribution of goods, but which on the whole seemed remarkably devoid of support from any racial or ideological group.

Yet for many people, a blackout could still be an indeterminate moment of sociability and friendliness, as became clear during the 2003 power failure.[20] Once initial fears of a possible terrorist attack had

---

[19] See: http://www.sing365.com.

[20] "Blackouts Cause North America Chaos," BBC News, 15 August 2003, http://news.bbc.co.uk/go/pr/fr/-/2/hi/americas/3152451.stm.

been dispelled, New Yorkers flocked into the streets, sat on their front stoops, brought out battery driven ghetto blasters, and partied into the night. Unlike the ritual moments of renewal that Turner described, however, the blackout can also be interpreted as a shadowy harbinger of the end of high-energy civilization. The spontaneity of ritual aimed at the renewal of society. But after 1977, the public caught in an electrical blackout had an ambiguous, heterotopian experience. Caught in an anti-landscape that cannot long sustain life, they improvised liminal moments of social solidarity. But this solidarity was based on the implicit belief that the power would soon come on again. The public now knows that prolonged system failure will not merely paralyze a city, it will threaten citizens with food shortages, dehydration, and the failure of many essential services, while rendering many office buildings and apartments uninhabitable. After 1977, New Yorkers slowly realized that the blacked-out city might prefigure a metropolis that was breaking down, or even be the emerging shadow of ruin. After September 11, 2001, they realized that ruin was not only a possible outcome of bad energy planning or fuel shortages but also a terrorist goal. While the major utilities and the Department of Homeland Security prepared contingency plans to combat attacks on the electrical infrastructure, in Iraq insurgents were becoming adept at blowing up utility lines. In 2006 and 2007 the Baghdad electricity supply was erratic, with blackouts in some areas every day.[21] The history of American blackouts had entered a new phase.

Bibliography

Bakhtin, Mikhail. *Rabelais and His World*. Bloomington: University of Indiana, 1984.

Carpenter, John and Nick Castle. *Escape from New York*. Directed by John Carpenter. Kurt Russell Film, 1981.

Federal Energy Regulatory Commission. U.S. Department of Energy. *The Con Edison Power Failure of July 13 and 14, 1977*. Washington, D.C.: Government Printing Office, 1978.

Foucault, Michel. "Of Other Spaces." *Architecture/Mouvement/Continuité,* (Oct. 1984). Translated from the French by Jay Miskowiec, http://foucault.info/documents/heteroTopia/Foucault.heteroTopia.en.html.

[21] James Glanz, "Iraq Insurgents Starve Capital of Electricity," *New York Times*, 19 December 2006.

Green, Mark. "What Do We Stand For?: Progressive Patriotism." *The Nation*, 21 July 2004. http://www.thenation.com/doc/20040802/green.

Jackson, J. B. *Discovering the Vernacular Landscape*. New Haven: Yale University Press, 1984.

Mahler, Jonathan. *The Bronx is Burning: 1977, Baseball, Politics, and the Battle for the Soul of a City*. New York: Farrar, Straus and Giroux, 2005.

National Opinion Research Center. Commissioned by the U.S. Office of Civil Defense. "Public Response to the Northeastern Power Blackout." Typed report. Oct. 1966.

Nye, David E. *American Technological Sublime*. Cambridge, MA: The MIT Press, 1994.

—. "Are Blackouts Landscapes?" *American Studies in Scandinavia* 39, no. 2 (Autumn 2007): 72-84.

—. *Electrifying America: Social Meanings of a New Technology*. Cambridge, MA: The MIT Press, 1990.

—. "Electrifying Expositions, 1880-1939." In *Narratives and Spaces: Technology and the Construction of American Culture*, 113-128. New York: Columbia University Press, 1998.

—. *Technology Matters: Questions to Live With*. Cambridge, MA: The MIT Press, 2006.

Turner, Victor. *Dramas, Fields and Metaphors: Symbolic Action in Human Society*. Ithaca, NY: Cornell University Press, 1974.

—. *The Ritual Process: Structure and Anti-Structure*. Ithaca, NY: Cornell University Press, 1969.

U.S.-Canada Power System Outage Task Force. *Final Report on the August 14, 2003 Blackout in the United States and Canada: Causes and Recommendations*. April, 2004. https://reports.energy.gov/BlackoutFinal-Web.pdf.

Wainwright, Loudon. "A Dark Night to Remember." *Life Magazine*, 19 November 1965.

Air and Space

Sarah Luria

Twentieth century "urban renewal" was inspired in part by aerial views that let "master-planners" see how the city could be reconfigured on a large scale. In his still often quoted essay "Walking in the City" (1984), Michel de Certeau maintains that the planner's view from up high cannot see or understand the elusive ground experience of the city-users down below. Recent technologies suggest that it is time to update de Certeau's dichotomy. A "street-centered model" dominates city planning today and the ready access to aerial views through modes such as GPS and Google Earth let city-users integrate an aerial consciousness into life on the ground. This essay considers three examples that dramatize this change. The opening aerial tour of New York in the 1961 film *West Side Story* depicts the twentieth-century disconnect between city dwellers and master-planners. In contrast, two recent examples from the architecture of Zaha Hadid and the photography of Atta Kim reveal an integration of aerial and ground-level views that is shaping our planning and experience of public space today.*

Standing at the top of the Empire State Building recently, I experienced once again that "bliss of altitude," as Roland Barthes calls it, as I looked down on the streets below (Barthes, 10). High up on the eighty-sixth floor, one carries on a constant conversation with the ground. As I marveled at the top half of the skyscrapers I knew from the ground, a woman talked on her cell phone to a friend below, a couple mapped out how they would walk from their hotel ("there it is, over there") to Times Square and where they would stop for cheesecake along the way. The Empire State Building is a strange midpoint between air and ground, the peak of New York's skyline, at the top of—but still in—the city: an intimate aerial view.

My view of New York from the Empire State Building differs from that offered by Michel de Certeau in 1984 from the top of New York's World Trade Center, where he asserts how little one actually saw of the ground below from so high above. From the 110th floor, the streets and buildings were emphasized. One enjoyed the view of the

* I would like to thank Jesse Anderson of the Audio Visual Department at Holy Cross for his generous help in producing the images for this essay.

urban planner with his "totalizing eye"; one could believe one had grasped "the city" as a whole and could see how it worked and might work better. But in his essay "Walking in the City" de Certeau claims there was a whole level of city life that eluded the planner's "'all–seeing power'":

> The ordinary practitioners of the city live 'down below,' below the thresholds at which visibility begins. They walk [. . .] they are walkers, Wandersmänner, whose bodies follow the thicks and thins of an urban 'text' they write without being able to read it. These practitioners make use of spaces that cannot be seen.

Unlike my view from the Empire State Building, where a dialogue is staged by "ordinary practitioners" between aerial views and ground experience, de Certeau describes a radical disconnect between them. What the planners mistake for an overall understanding of the city is "an optical artifact" and a "misunderstanding of practices." As de Certeau concludes, "It's hard to be down when you're up" ("Walking in the City," 127-128).

I begin with these aerial views of Manhattan because they introduce a geographic plane of the city that is crucial to any discussion of public space. Aerial views have greatly influenced how we see, experience, and design our world. De Certeau's account of the urban planner's detached aerial view, in fact, voices a dominant critique dating from the mid-twentieth century by Jane Jacobs and others of the destruction of whole city blocks in the name of "urban renewal." Unable to see the full picture of life in the neighborhoods, "master-planners" such as Robert Moses attempted and often succeeded at bulldozing whole sections of a city in order to build their vision of a unified modern urban space. These ambitious plans were to an important degree inspired and enabled, as de Certeau suggests, by aerial views.

The airplane, and the increased access to bird's-eye views that it made available, had a huge impact on the built environment of the twentieth century. "*L'avion accuse*" Le Corbusier pronounced, "[the airplane] indicts the city." One of the first avid air passengers, Le Corbusier extolled the power of the airplane to see just what needed to be done to fix the mess that cities had become down below (Le Corbusier, *Aircraft*, title page and 11-13). According to Le Corbusier, he flew sketchbook in hand: "as everything became clear to me I sketched [. . .] the ideas of modern planning" (qtd. in Morshed, "The Cultural Politics of Aerial Vision," 201). Robert Moses's sweeping plans for New York City similarly relied heavily upon aerial views, as architectural historian Hilary Ballon explains,

> Moses grasped the city as a whole. Although he was oriented to the automobile, his preferred point of view for planning was from the sky, where people disappeared from sight and the city appeared as a physical tapestry of land masses, waterways, and structures. Moses used aerial photographs as a planning tool, a means that represented the strengths and weaknesses of his planning style. Aerial photos abetted his interest in the city as a whole. (Ballon and Jackson, 66)

Furthermore, aerial photographs played a key role in helping Moses to promote his admittedly impressive yet controversial projects. Moses's office produced "stunning brochures" for these projects with aerial drawings and photos to show before and after views of the "slum conditions" that existed and how they would be transformed (Ballon, "Robert Moses and Urban Renewal," 99-100).

In contrast to twentieth-century urban renewal, a "street centered model" and a "counter theory of urbanism" dominate city planning today (Ballon, 96). Just as today's planners are more focused on the ground view, today's "ordinary practitioners" have much greater access to aerial views. Both trends suggest that the conversation between air and ground views is arguably more noticeable today than ever before. Aerial views are everywhere and not just from airplane windows: the aerial photo in your morning newspaper; the peak you sneak at the roof of your house via Google Earth when you should be working. My son's computer game lets him scroll with dizzying speed between aerial and ground views of his battlefield. To be sure, we are reminded that a radical disconnect still exists between aerial views and what is now called "ground truth." A GPS working from satellite data will tell you to turn left where a left turn is no longer allowed. But such technology has accelerated the need to close the gap, to make sure this air to ground conversation goes both ways. Navteq, the company that supplies much of the GPS data, hires teams of researchers for "ground-truthing," but never enough it seems, to correct the faulty information that comes from satellites and note any changes, such as road construction, that alter were we can drive (Paumgarten, 88). Cell phones are used as makeshift GPS devices as well. Because the phones are mobile, speakers naturally begin by identifying their position: "Hi, I just arrived at Pennsylvania Station, where are you?" As a result, cell phones support a culture of self-surveillance, whereby we can construct our own aerial view of our situation on which to plot the progress of our friends and where we are in relation to them. Technology is empowering a split-level consciousness where we live between air and the ground, air and space. De Certeau's assessment of aerial views may still be correct: they are a "representation," a "facsimile," a "simulacrum." But it also seems clear that such images are no longer owned for the most part by the "spaceplanner urbanist, city planner or cartographer," nor is it clear that such images are "a misunderstanding of practices" (de Certeau, "Walking in the City," 128). Current technology has brought aerial experience closer to the ground so that de Certeau's World Trade Center view needs to be updated. The view from the Empire State Building provides perhaps a more representative image of the fluid relationship that currently exists between a map view and street-level experience. We drive through the city while we watch where we are on a GPS screen. We enjoy a "double gaze" that lets us "see" while we "go."[1] To revise de Certeau's claim, it's getting easy to be up when you're down.

Given that airplanes were instrumental to changes in twentieth-century urban planning, in this essay, I want to explore how today's split-level consciousness might be shaping our present experience and design of public space. I will analyze two examples that dramatize the change I am arguing has occurred in the relationship between aerial and ground views. The first example is the opening of the 1961 film *West Side Story*. The film opens high above Manhattan, with an aerial tour of modern day New York; it figures the perspective of the urban planner and highlights its lack of connection to life on the ground below, while it showcases the modern improvements to the public sphere wrought by urban renewal. The second example is the Car Park and Terminus

[1] The term "double gaze" is from Morshed's "The Cultural Politics of Aerial Vision," 202. De Certeau distinguishes between "seeing"—a knowledge of how a space is laid out, its map—and "going"—moving through a particular space ("Spatial Stories," 119).

Hoenheim North in Strasbourg, France, completed in 2001, by architect Zaha Hadid. Hadid is particularly pertinent here because she shares a deep affinity with the airplane-oriented urbanism of the twentieth century, but in this suburban tram station, air and ground views and experience become so identified with one another that they become interchangeable. The result is a site that combines air and space in innovative ways and that draws upon rather than overpowers the neighborhood that surrounds it. We have no doubt always imagined ourselves and our world from above as well as from below. Hadid's contemporary design underscores how present-day technology is enabling architects, planners, artists, and users alike to materialize that conversation in more concrete and subtle ways than ever before.

The twentieth-century disconnect between master urban planners, hovering high above the city, and city dwellers, threading their way through the city's tangle of streets, frames the opening shots of *West Side Story*. We begin high above the city, with an aerial eye, whose distance is marked by silence; we hear only the faint hum of traffic, but already this cool, detached view is brilliantly contrasted by a whistle, a call, which is coming from someplace in the streets far down below (figs. 1 and 2). "What will New York look like in the year 1999?" Robert Moses asked in 1949. "The air photographer in his blimp or helicopter will see the rolling ocean [. . .] Liberty guarding the magnificent harbor. The great arteries of travel will stand out" (qtd. in Caro, 909-910n). In *West Side Story*, we see New York's latest modern improvements master-minded by Moses himself: the Triborough Bridge, the expressways and their rational system of clover leafs, the pristine blocks of towers. We see a wealthy world-class city of tomorrow already realized today. We float over the internationally recognized landmarks of the Empire State Building and the United Nations Building and then onto the planned spaces of Columbia University's campus and a Robert Moses-type modern housing project with high residential towers clustered around a clean, attractive, green open space. During this aerial tour, the whistle persists, made more primitive still by the drum beat that joins in. At last, we drop down to where it is calling from, as we zoom in from public planned New York to private tenement New York, and the concrete, fenced in basketball court in the neighborhood known as Hell's Kitchen, on New York's West Side (figs. 3 and 4). Here we see, rather literally, what Jane

Jacobs would call the "ballet" of the street (fig. 5)—all that neighborhood-defining activity which de Certeau reminds us the panoptic power can't see or control (Jacobs, 50).

Rather than a city of the future, this is a city of regression, a threatening, primitive place, where gangs—not families—are the only means of protection. "The 'corridor street' must be destroyed," Le Corbusier proclaimed (Le Corbusier, *Precisions*, 169), and in this segment of *West Side Story* we can see why it should be: New York's grid is bad for visibility—your eye is limited to a single axis where you never know who is around the corner: friend or foe? Shark or Jet? The neighborhood's grid is fractured, full of dead ends and blind alleys where you can get cornered and attacked (fig. 6). We see the neighborhood, as the film's director Jerome Robbins described it, as a "pressure cooker" (Lehman, Introduction).

Filmed on location, the opening of *West Side Story* was shot on the very blocks of Hell's Kitchen that would soon be bulldozed by Moses to make way for Lincoln Center, the cultural plaza and new home of the Metropolitan Opera, and the New York Philharmonic, at 62nd through 66th Streets on Columbus Ave. Lincoln Center stands out among New York's "redevelopment" projects due to its ambitious scope and its size (it covered 45 acres) (Ballon, "Robert Moses and Urban Renewal," 107). The neighborhood had been the recent site of famous gang fights between Puerto Rican and Irish residents. The disconnect illustrated by the opening shots between the cool planner's rational aerial eye and the ground bound, hot-headed savage boys is clear, but the message of the film's opening is ambiguous. Is the Moses-style redevelopment of New York the problem or the solution? The opening suggests that the boys have been left to fight for a place they can call theirs in the cruel concrete jungle of the city. They long, as they later sing, to keep control over themselves, to "just play it cool," but they can't help but erupt into violence. In their primal working class world they lack the cool, technological distance of the aerial view that would give some perspective on their lives, the ability to see that it doesn't really matter if they are Puerto Rican or Anglo, they are all pretty much alike. The "star crossed lovers" of the movie, Maria (Puerto Rican) and Tony (identifiable only as Anglo, or perhaps Italian, in the film), are able to transcend tribal differences, to think outside the box, but they discover they cannot physically escape.

"There's a place for us," Maria sings as Tony, shot in a gang fight, dies in her arms, "a time and place for us."[2] But whether urban planners can provide that place—whether Tony and Maria's love might thrive in a better planned community (with some green space such as we never see in Hell's Kitchen)—or, at the very least, in the more civilized space of Lincoln Center, where their *Romeo and Juliet* based musical will certainly be appreciated—is left unclear.

That said, we can recognize that even the street scenes shot on the ground in *West Side Story* reveal an aerial, and urbanist, point of view. The neighborhood is portrayed negatively—its right-angled streets and alleys confine the gangs in a dead end. Behind this is the urbanist critique of nineteenth-century grid-plan cities such as New York. Le Corbusier's condemnation of 'the corridor street' cited above continues a long-standing prejudice amongst reformers, as Ballon notes:

> Urbanists and housing reformers as varied as Frederick Law Olmsted, Jacob Riis, Lewis Mumford, and Le Corbusier agreed on one thing: the traditional pattern of street-oriented, gridiron urbanism created unhealthy living conditions. It produced damaging population density and high land coverage that deprived people of basic human needs: open space, light, and air. ("Robert Moses and Urban Renewal," 96-97)

Lincoln Center offers a textbook example of how Moses and others sought to open up Manhattan's nineteenth-century grid by combining its blocks into "super blocks," where buildings could be set further back from the street (correcting the canyon effect of the street wall created by tenements and skyscrapers) and separated by open green space or plazas.[3] The fact that in *West Side Story* we see the streets of Hell's Kitchen through this urbanist ideology is important to note because it shows that even when we assume we are seeing a ground-level view, it can be filtered through an aerialist lens.

---

[2] "Just Play It Cool." Lyrics by Stephen Sondheim. 1956. Amberson Holdings LLC and Stephen Sondheim. http://www.westsidestory.com/site/level2/lyrics/cool.html.

[3] The original site plan for the Lincoln Square redevelopment project "reorganized 18 blocks into 5 superblocks, with one 19th century block left intact" (Ballon, "Lincoln Square Title 1," 280).

Although celebrated as one of today's most innovative architects, Zaha Hadid and her work continue the twentieth-century urbanist enthusiasm for flight and carry it, arguably, to extremes. One of Hadid's earliest designs featured a building that resembled an airplane lying on the ground.[4] Her drawings often depict her buildings as seen while flying by.[5] Hadid's airborne perspective lets her tilt and render space from unconventional angles and so free her buildings from the 90 degree, straight up and down angle that has ruled our design of life on the ground: "There are 360 degrees in the [world]," she notes. "So why stick to one?" (*Zaha Hadid*). At the same time, Hadid's work addresses its surroundings in a much more conscientious and respectful way than the earlier urbanists. She designs single buildings, not hubristic mega-city blocks, and intends her buildings to be non-invasive, and yet significant enough to exert a ripple effect on the area that surrounds them. She sees the city as an "urban carpet"—any change to it must attend to and affect the whole (Goldberger). Hadid's Phaeno Science Center in Wolfsburg, Germany (1995–2000) makes a physical point to tread softly on its surrounding area by being set on legs, so that the space of its neighborhood can continue to flow underneath.

In the Hoenheim-Nord Terminus and Car Park for Strasbourg's tram system, Hadid offers a particularly clear example of how this greater sense of movement inspired by air travel can blend in with and hopefully enhance the ground conditions of a surrounding neighborhood. Seen from the air, however, the site at first appears to present the aerial bias of a classic urbanist construct (fig. 7). The angled wing-like roof of the tram station takes off, lifts off, like a plane from the ground. The path of white that swooshes through the car park like a runway continues the roof's trajectory—this time on the ground. Most humorous are the parking spaces for the cars which seem to leave the cars in mid-flight, a fleet of planes, banking a turn together in perfect form, zooming around while you leave them at rest to hop the train (whose tracks are comically straight in contrast to Hadid's tippy world). This is a terminus that is all about movement—no dead-end

[4] The project, "The Peak," planned for Hong Kong, was never built.

[5] Philip Kennicott, "Zaha Hadid's Flying Ferocity; Visionary Architect's Designs Are Finally Being Unleashed," *Washington Post,* 23 July 2006.

corridor here. All its elements seem skewed, right down to the slanted columns that hold up the station's roof (fig. 8). The aerial self-consciousness the terminus exhibits points again to a motion that isn't restricted to the ground. It is as if the car park is performing for the air; it certainly looks most dramatic when seen from above, and it is from there that its pattern makes most sense.

And yet, seen from the ground, we can appreciate the many ways in which both the terminus and car park blur any boundary we might cling to between air and ground and show them to be interchangeable. Most dramatic is the fact that the roof and the floor of the station are made of the same material. The roof looks like the slab of concrete that forms the floor. The roof is not marked in the usual ways with cornices or gutters; nor does it follow the conventional practice of being made from separate materials, such as metal or tiles, to distinguish it from the rest of the building. There are no walls to the structure, instead the ground plane folds to form the roof. Without walls, the resemblance between the two planes is emphasized. We don't expect a ceiling to be concrete, a material which is so heavy, earthy, and grounded. Hadid expands our horizons and turns concrete into a neat, almost light, Bauhaus slab. Slits in the roof let in the sky, casting strips of light on the floor that echo the white strips marking the parking spaces in the car park beyond (fig. 9). Long sections of the roof are cut away to reveal the sky, making the roof's presence lighter and showing the terminal to be as much about shelter and ground-based solidity as about air and voids. Air and concrete are compatible; they fit together like interlocking pieces in a puzzle.

The car park continues this identity or middle ground between above and below as well as vertical and horizontal. The curve of the lot and its parking spaces that are so clear from the air are clear at ground-level too by the light poles, particularly the rows of black poles, which highlight the curve (fig. 10). The poles are also distinguished by their dark parking lines, which, like permanent shadows, project the vertical poles, horizontally on the ground. The poles themselves, though difficult to see here, are tilted, extending the tilted columns of the terminus out to the field of the car park. At night the white tips of the poles light up, again forming concentric curves, recalling the wands air traffic controllers use, and the lights used to mark runways.[6] Parked in their extended curves, the cars resemble a pack of race cars, as much as planes, banking a curve on a racecourse. The terminus and car park thus invoke the experiences of driving and flying and liken them to one another. The terminus itself resembles a car port and the parking spaces, at night, an airfield, so that we are between planes and cars, runways and racetracks. Hadid is as fond of cars as she is of planes, so that one critic styles her buildings "carchitecture" (Hoete, 136-149).

Both the aerial and ground views we've considered so far point to the many ways in which the car park and terminus work with the surrounding neighborhood, rather than overwhelm it, as a more traditional urbanist construct might have done. The suburb that surrounds the area is composed of both curves and straight lines (fig. 7). The housing development is largely planned on curving streets, but the allotment gardens that abut the car park are linear and based on a grid. Just in view in the aerial shot is the river that curves around this piece of land, which is also cut across the bottom by a set of straight railroad tracks. The angled planes that form both the terminus and the black field of the car park nicely bring together curves and straight lines. The styling of the terminus when seen in the larger context of the buildings that surround it seems suitable, and not incongruous. Perhaps one of the most significant achievements of the space as a whole is the way it manages to look innovative, but still familiar, retro and futuristic at the same time. The terminus has a minimalist, concrete, modern style that complements the concrete, modern-ish suburban houses beyond with their flat and angled roofs. Indeed, the entire effect, both from the ground and from the air, it would seem, is intended to make commuting less monotonous, and to give the surrounding neighborhoods an artsy style and an element of surprise that suburbs typically are seen to lack. The purpose of the whole site was to convince suburban dwellers to give up the convenience of their cars and to take mass transportation—no easy task (Ruby, 17). The car park and terminus can be viewed as one large advertisement for public transit; its dynamic, sleek space argues that commuting by tram will be faster and more efficient than commuting by car. The swoosh on the car park lot symbolically paves

---

[6] I am indebted to architect Tamar Warburg, who studied with Hadid, for helpful discussions about the terminus and car park, which I draw upon here.

the way to the tram and so forms an emblem for the way in which the site reaches out to and draws in the area that surrounds it.[7]

A recent photograph by artist Atta Kim offers one last image of today's intriguing interplay between aerial and ground views. Like *West Side Story*, Kim offers us a ground-level view of the Manhattan's grid, a view that is highly influenced by an aerial viewpoint. But Kim's resulting image is not so critical of the neighborhood he depicts. Like Hadid, in Kim's image the line between the air and the ground becomes very blurred. Kim's photograph "Park Avenue, 8 hours" (2005) makes even Manhattan's heavy corporate grid airy (fig. 11). With his camera posed at street level, it would seem that Kim's time lapse photograph of a Park Avenue intersection would be cut off from an aerial viewpoint, and yet the photograph is part of a series Kim entitles, very aptly, "On Air." Kim uses digital technology to create what is in effect an eight-hour time lapse image of this busy New York intersection. People are missing because they did not stay on stage long enough and were not luminous enough to register in the camera eye. The dominant color of the blur of cars is yellow, those New York taxis. Difficult to discern here is the faintest trace of yellow and headlights (less visible because they would have paused there less long) crossing the intersection before us.

"Park Avenue 8 Hours" substitutes a temporal overview for a spatial overview but it is nevertheless in several ways like a view from the air. As we saw in the aerial view at the start of *West Side Story*, here the buildings and streets with their traffic stand out and the people vanish. A New York intersection, an intense space at almost any hour, is rendered here, as it would be from the air, curiously quiet, peaceful, detached. Filmed from an urbanist's aerial view point, *West Side Story's* grid was claustrophobic, a maze full of dead ends. Here, in this

[7] Ruby offers this interpretation of the Strasbourg terminus: the space "works with" the surrounding area, rather than overpowers it, "the constantly intersecting flow of buses, trams, cars, trains and peoples" provides a dynamic focal point the area had lacked (19). Whereas Hadid's earlier projects have been criticized for being too theoretical—her Vitra Fire Station (1993) Hoete faults as a "sculptur[e]," a "caricature," even, of movement, one which itself remains "static"—Hadid's recent works are praised for their kinetic energy (Hoete, G2, 141-142). As Ruby concludes, the Strasbourg Car Park is "a landscape ready for take off" (20).

wealthy calm corporate canyon, an aerial view does not criticize but melds with a view on the ground.

Kim's photograph reveals the degree to which aerial experience can filter and transform our perception of life on the ground. Through the medium of time—and air—which produces a quiet bluish light like no light we will ever see, rigid distinctions, such as dark vs. light, are suggested but blurred. There is the contrast between the stillness of the buildings with the movement of the traffic. But they come to resemble one another, too. Like the cars that evaporate into an effervescent fizz, the buildings acquire a stillness and serenity that almost naturalizes them into a canyon seen at dawn, merging them with a natural topography that the city grid, by definition, denies, a fact underscored by that lone tree that seems so out of place here. Through Kim's use of technology and point of view, this landscape, which if photographed under a bright noon sun might look like a conventional view of a cold-hearted American corporate world, acquires a sense of mystery, of both eternity and dream-like fragility. Kim's image suggests the experience of artist Yvonne Jacquette who paints aerial views seen from the top of tall buildings or small planes. As art critic Lucy Lippard notes, this "has affected the paintings [Jacquette] does in the streets as well." Lippard explains, quoting Jacquette, "The way things look from a plane makes Jacquette 'want to break things down into small parts [. . .] Even the buildings aren't so solid anymore, they're more *relative*'" (qtd. in Lippard, 117, emphasis in the original).

But perhaps the most significant feature of Kim's aerial ground view is its omission of people. Through it, Kim's image advertises the power of the aerial view. With its fixation on the streets, the buildings, the traffic, Kim's view images the aggression of some twentieth-century urban planning. This clean, uncluttered view of Park Avenue recalls the solution found for the hot spots of Hell's Kitchen where the airy modern plaza of Lincoln Center now stands in the place of those feuding gangs. At the same time, Kim's view advertises its limits. It is a picture of the evanescent, elusive quality of city life and movement, the "sieve order," as de Certeau wonderfully describes it—the grid plan with holes through which move, unseen, human life and experience (de Certeau, "Walking in the City," 133). The photograph thus materializes de Certeau's "threshold of visibility," and it is surprisingly like the one he describes from atop the World Trade Center. Kim's "On Air" view of the ground records the absence of all that pedestrian activity and in this way calls our attention to it. It is an image of our invisibility, of all the camera eye cannot capture and therefore know. As such, it suggests an image, in de Certeau's terms, of the extensiveness of our power, our ability to live in the world in a way that can never be fully anticipated, controlled, seen, and hence monitored or repressed.

Gravity may dictate that we spend most of our life on the ground, and the limits of our vision may prevent us from seeing, when we're up above, much of what goes on down below, but today's satellite and computer technology accelerates the visual dialogue between air and space that our imaginations have no doubt always pursued. In an essay written in 1984, the same year de Certeau wrote of the split between ground and air, Adrienne Rich described how, as a child, she made a game of connecting them. She and her friend would address letters to each other "like this":

> Adrienne Rich
> 14 Edgevale Road
> Baltimore, Maryland
> The United States of America
> The Continent of North America
> The Western Hemisphere
> The Earth
> The Solar System
> The Universe

This verbal zoom would later be realized by Google Earth, and already existed visually in the wonderful "Powers of 10" project (which begins in the Milky Way and then zooms down to the Earth and then to Florida and then to a branch and then to a molecule of that branch), based upon the 1977 film of that name by Charles and Ray Eames. Rich explains that in her game she most enjoyed being able to alternate between zoom in and zoom out: "you could see your own house as a tiny fleck on an ever widening landscape, or as the center of it all from which the circles expanded into the infinite unknown."[8]

[8] Rich, 64. A website based upon the "Powers of 10" project can be viewed at http://micro.magnet.fsu.edu/primer/java/scienceopticsu/powersof10.

Rich points to the political implications of this accelerated shifting between aerial and ground view. Globalism, for instance, is the internalization of an aerial view—one that asks us to make a habit of zooming out and seeing ourselves as just one spot in the wide interconnected world. As Rich's description reminds us, an aerial view can make us feel immensely powerful and powerless—as if we are the center of the universe, or insignificant and small. The examples I've considered here underscore the politics behind point of view, which has everything to do with how we plan, build, and experience public space. Seen from the air, the nineteenth-century city looked terrible to certain planners whose design visions were influenced heavily by an aerial point of view. History and technology make it time to revise de Certeau's World Trade Center view with its polarized distinctions between planners and "ordinary practitioners," the watcher and watched, aerial views and ground experience. All these reasons, and more, compel us to consider any question about public space from both above and below.

Bibliography

Ballon, Hilary. "Lincoln Square Title 1." In *Robert Moses and the Modern City: The Transformation of New York,* edited by Hilary Ballon and Kenneth T. Jackson, 279-289. New York: Norton, 2007.

—. "Robert Moses and Urban Renewal." In *Robert Moses and the Modern City: The Transformation of New York,* edited by Hilary Ballon and Kenneth T. Jackson, 94-115. New York: Norton, 2007.

Ballon, Hilary and Kenneth T. Jackson. Introduction. *Robert Moses and the Modern City: The Transformation of New York,* edited by Hilary Ballon and Kenneth T. Jackson, 65-66. New York: Norton, 2007.

Barthes, Roland. "The Eiffel Tower." In *The Eiffel Tower and Other Mythologies*, translated by Richard Howard, 3-17. New York: Noonday Press, 1979.

Caro, Robert A. *The Power Broker: Robert Moses and the Fall of New York*. New York: Vintage Books, 1975.

de Certeau, Michel. "Spatial Stories." In *The Practice of Everyday Life,* translated by Steven F. Rendall, 115-130. Berkeley: University of California Press, 1984.

—. "Walking in the City." In *The Cultural Studies Reader,* 2nd ed., edited by Simon During, 126-133. London: Routledge, 1999.

Goldberger, Paul. "Artistic License." *New Yorker* (2 June 2003): 99.

Hoete, Anthony. "Zaha Hadid: Automobile and Architecture." In *Reader on the Aesthetics of Mobility,* G2.136-G2.149. London: Black Dog Publishing, 2003.

Jacobs, Jane. *The Death and Life of Great American Cities*. New York: Vintage Books, 1961.

Le Corbusier. *Aircraft*. London: Studio Ltd., 1935.
—. *Precisions: On the Present State of Architecture and City Planning*. Translated by Edith Schreiber Aujame. Cambridge, MA: The MIT Press, 1991. Originally published in French as *Précisions sur un état présent de l'architecture et de l'urbanisme*. Paris: G. Crès, 1930.
Lehman, Ernest. Introduction. *West Side Story Scrapbook. Collector's Set, Special Edition*. U.S. MGM Home Entertainment, 2003.
—. *West Side Story*. Robert Wise and Jerome Robbins. 1961. DVD [2003]. Los Angeles, CA. MGM Home Entertainment/Seven Arts Productions, Inc.
Lippard, Lucy R. *From the Center: Feminist Essays on Women's Art*. New York: E. P. Dutton, 1976.
Morshed, Adnan. "The Aesthetics of Ascension in Norman Bel Geddes's Futurama." *Journal of the Society for Architectural Historians* 63 (2004): 74-99.
—. "The Cultural Politics of Aerial Vision: Le Corbusier in Brazil (1929)." *Journal of Architectural Education* 55 (May 2002): 201-210.
Paumgarten, Nick. "Getting There." *New Yorker* (24 April 2006): 86-101.
Rich, Adrienne. "Notes Towards a Politics of Location." In *Arts of the Possible: Essays and Conversations,* 62-81. New York: Norton, 2001.
Ruby, Andreas. Essay (untitled). In *Zaha Hadid: Car Park and Terminus Strasbourg,* translated by Ishbel Flett, 17-20. Baden, Switzerland: Lars Müller Publishers, 2004.
Sennett, Richard. *The Conscience of the Eye: The Design and Social Life of Cities*. New York: Knopf, 1990.
Zaha Hadid. Exhibition. Audiotour. Solomon R. Guggenheim Museum, New York. June 3 - Oct. 25, 2006.

Fig. 1. *West Side Story*. Film. 1961. Robert Wise and Jerome Robbins, directors. DVD [2003]. MGM Home Entertainment/Seven Arts Productions, Inc.

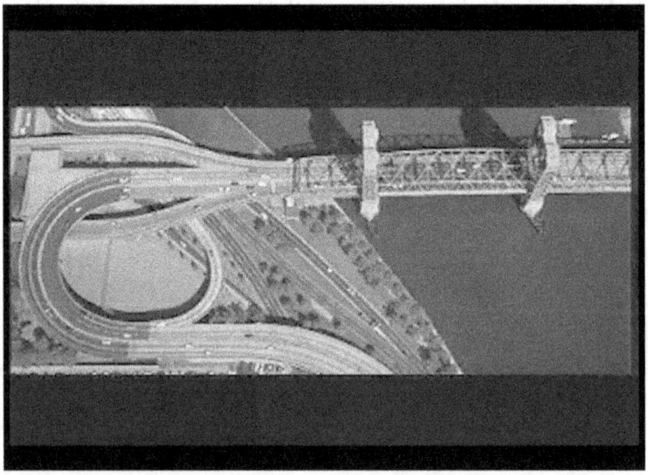

Fig. 2. *West Side Story*. Film. 1961. Robert Wise and Jerome Robbins, directors. DVD [2003]. MGM Home Entertainment/Seven Arts Productions, Inc.

Fig. 3. *West Side Story*. Film. 1961. Robert Wise and Jerome Robbins, directors. DVD [2003]. MGM Home Entertainment/Seven Arts Productions, Inc.

Fig. 4. *West Side Story*. Film. 1961. Robert Wise and Jerome Robbins, directors. DVD [2003]. MGM Home Entertainment/Seven Arts Productions, Inc.

Fig. 5. *West Side Story*. Film. 1961. Robert Wise and Jerome Robbins, directors. DVD [2003]. MGM Home Entertainment/Seven Arts Productions, Inc.

Fig. 6. *West Side Story*. Film. 1961. Robert Wise and Jerome Robbins, directors. DVD [2003]. MGM Home Entertainment/Seven Arts Productions, Inc.

Fig. 7. Car Park and Terminus, Strasbourg by Zaha Hadid. Photograph. Roger Rothan.

Fig. 8. Car Park and Terminus, Strasbourg by Zaha Hadid. Photograph. 2001. Hélène Binet.

Fig. 9. Car Park and Terminus, Strasbourg by Zaha Hadid. Photograph. 2001. Hélène Binet.

Fig. 10. Car Park and Terminus, Strasbourg by Zaha Hadid. Photograph. 2001. Hélène Binet.

Fig. 11. "On Air #110-8, from the New York Series." Chromogenic print. 2005. Atta Kim. Courtesy of the Yossi Milo Gallery.

Imagining the Interstate: Henry Miller, Post-Tourism, and the Disappearance of American *Place*

Andrew S. Gross

Automobiles, and the infrastructural developments necessary to produce and operate them, contributed greatly to regional standardization in the United States. However, automobile travel books, or road books, typically lament the disappearance of regional difference. This paradox is typical of the genre, and until the 1930s it could be more or less resolved by recuperating regional difference at a national as opposed to local level. Thus interstate travel became a symbol of national difference, and the patriotic domestic tourist was distinguished from the expatriate. This strategy was typical, for instance, of the American Guide Series of the 1930s and 1940s. Henry Miller's *Air-Conditioned Nightmare*, written upon his return to the United States in 1939 after a decade abroad, breaks decisively with this tradition by refusing to recuperate standardization in terms of national difference. His road book maps the disappearance of location as a site of tradition, identity, and experience; and in doing so it prefigures two of the major concerns of late modernism: existential alienation and the postmodern imaginary. Understanding the road book and its particular relation to the landscape as antecedent to both of these trends allows for a spatial definition of recent and contemporary cultural formations. If high modernism is the culture of literary exile, postmodernism is the culture of the interstate. In order to sketch the contours of this cultural and geographical transformation, my essay situates Miller's travelogue in a wider tradition of travel narratives, beginning with those of early automobile tourists, continuing with the returning expatriates, and concluding with the road novels of the 1950s and 1960s, when driving becomes the major trope of post-war protest and conformism.

Edmund Wilson and Henry Miller engaged in a telling exchange shortly before the latter returned to the United States in 1939. In one of the first American reviews of *The Tropic of Cancer*, a book then still unavailable in the United States, Wilson claims that Miller has written "the epitaph for the whole generation of American writers and artists that migrated to Paris after the war" (*The Shores of Light*, 706). The article, significantly entitled "Twilight of the Expatriates," mainly refers to the end of the bankrupt lifestyle pursued by the "expatriate bums," as Wilson calls them elsewhere (Review of "The Air

Conditioned Nightmare," 63). Miller replied in a letter to the *New Republic* that if Wilson meant to imply the narrator is a bum, "then it is me, because I have painstakingly indicated throughout the book that the hero is myself" (*The Shores of Light*, 709). The three salient themes in this exchange—the end of literary exile, the transient or mobile lifestyle of the bum, and the significance of personal experience—are central to the book Miller wrote after returning to the United States, *The Air-Conditioned Nightmare* (1945).

The travelogue, documenting Miller's return to the United States after a decade abroad, marks the moment when the previously fashionable term "expatriate" begins to lose its cultural significance. The geopolitical situation was extremely significant at this juncture, as the shifting alliances of World War II and then the Cold War produced a bipolar or global power structure that in many ways militated against more traditional definitions of national identity. My essay will focus on the "economic" changes affecting national identity, namely the tourism and consumer capitalism that Miller and others feared was standardizing the landscape, and hence leveling-out regional differences with homogenized products and architecture. Miller's travelogue charts the changing contours of a landscape being redefined as an *interstate* network. The technologies of transportation and communication responsible for this transformation simultaneously seemed to render *place* obsolete. Miller is the last expatriate because after him—not because of him—it becomes difficult to understand exile as a limited condition in relation to a particular place. His road book maps the disappearance of *location* as a site of tradition, identity, and experience; and in doing so it prefigures two of the major concerns of late modernism: existential alienation and the postmodern imaginary. Understanding the *road book* and its particular relation to the landscape as antecedent to both of these trends allows for a spatial definition of recent and contemporary cultural formations. If high modernism is the culture of literary exile, postmodernism is the culture of the interstate. In order to sketch the contours of this cultural and geographical transformation, my essay will situate Miller's text in a wider tradition of travel narratives, beginning with those of early automobile tourists, continuing with the returning expatriates, and concluding with the road novels of the 1950s and 1960s, when driving, as Morris Dickstein points out, becomes the major trope of post-war protest and conformism (175).

I. Interstate Travel and the Exile's Return

Road books have an ambivalent connection to consumer culture that manifests itself in their conflicted relation to place. On the one hand, they promote the use of one of the twentieth century's most important and obvious commodities, the automobile. On the other hand, they typically protest the impact of automobility on the landscape, and particularly the standardized roadside architecture which, as John Jakle and Keith Skulle have demonstrated, emerged concomitantly with standardized assembly lines.[1] Many of the earliest road narratives were advertisements for particular cars and automotive products, either published by automobile companies directly or printed alongside ads for tires and oil in magazines like *Sunset* and *Scribner's* (Gross, "Cars, Postcards, and Patriotism," 81). However, writers such as Emily Post, whose *By Motor to the Golden Gate* was serialized in *Collier's Weekly* in 1915, typically complained that the nationwide trends made possible by interstate travel were destroying regional difference (88). The paradox evident in Post is constitutive of road books as a genre; in their quest for authentic locations they are compelled to criticize the leveling effects of mobility.

Standardization finds its narrative equivalent in formal problem of parataxis: drivers with nothing new to report tend to slip into repetitive observations or litanies of complaint. This is a general problem of travel writing: the tourist must discover something unique or "authentic" to structure the itinerary and transform it into a narrative. The formal problem is often solved through a scenic view of the landscape which, as W.J.T. Mitchell has pointed out, produces correlative forms of subjectivity (1-2). Mitchell's analysis of the interdependence of sight and site in the construction of the scenic draws attention to how specific views of the American landscape have

[1] Jakle points out in *The Tourist*, "with improved highways and the rise of roadside commerce, regional differences were obscured beneath a veneer of roadside homogeneity" (199). See also: the trilogy of books Jakle has written with Skulle, *The Gas Station in America*, *Fast Food*, and *The Motel in America*.

always been central to constituting American identities and imagining the national whole. For Post, as for many of her contemporaries, the "authentic" American experience was pioneering and its location the American West. That her experience of authenticity is "staged," to borrow Dean MacCannell's term, is suggested by the fact that Post finally "discovers" the West when she imagines herself as a star in her own film (MacCannell, 98-100):

> We might have been taking an unconscious part in some vast moving picture production, or, more easily still, if we overlooked the fact of our own motor car, we could have supposed ourselves crossing the plains in the days of the caravans and stage coaches, when roads were trails, and bridges were not! (135)

The cinematic view is central to Post's project because it simultaneously constitutes the landscape as a scene for her enjoyment and erases the disturbing presence of modern technology ("the days of caravans and stage coaches"). Jeffrey Schnapp has pointed out that a landscape moving across a film screen mimics the way it slides past the windshield (22). In Post this metonymic link is key to reproducing and mastering automobility as a locus of authentic regional difference. Her use of the cinematic trope is related to what Leo Marx terms the "the rhetoric of the technological sublime," but with one important difference (Marx, 195). Marx is interested in authors for whom the machine is a visible symbol of progress (206); for Post machines are most interesting when they reproduce the conditions of their absence, presenting a technology-free vision of the world that is itself the height of artifice. Post deploys one technology of motion (film) to contain or naturalize the impacts of another (automobility), authorizing herself as a mobile subject vis-à-vis a nostalgically imagined region, which stands in the relation of synecdoche to the national whole.[2] This is the central mechanism of road narratives up until Henry Miller: *place*, threatened by the homogenizing effects of modern transportation, is recuperated as the sublime locus of the nation.

[2] The heady combination of modern technology and nostalgia for the West quickly established itself as one of the dominant tropes of early automotive advertising (Gross, "Cars, Postcards, and Patriotism," 80).

One threat to the recuperative value of national place vis-à-vis *interstate* travel was the popularity of *international* travel in the early twentieth century. Marguerite Shaffer has documented the advertising and cultural war waged to convince tourists to "see America first" (26-39). The advertising campaign can be understood as a rearguard action against the master trope of American high modernism: expatriation. However, it also reveals the predominant orientation of literary exile, which was nostalgic and domestic. By 1934 Malcolm Cowley had firmly established the circular trajectory of homecoming (related to nostalgia through the Greek *nostos*, "return home") in *Exile's Return*, still one of the prevailing accounts of "the lost generation."[3] According to Cowley, the expatriates who fled to Paris to pursue "the religion of art" reintegrated themselves in the American scene when the Great Depression made resources scarce: "Things had changed everywhere. The lost generation ceased to deserve its name; the members of it had either gone under [. . .] or else had found their places in the world" (284, 285-288). Finding one's place in the world, it should be noted, is not the same as finding one's *place* in strictly geographical terms. Cowley's exiles do not go back to the small towns where they were born; rather, they settle in New York to undertake projects that are national in scope, "selling out," for instance, to advertising firms or devoting their energies to social causes and/or proletarian literature. Homecoming, accord to Cowley, does not mean coming home; it is the continuation of an *international* journey in a *national* space by professional, political, or artistic means. Just as Post deploys one technology of motion (film) to recuperate the effects of another (automobility), Cowley deploys the circular structure of the return to reinscribe international travel in national space.

These early representations of twentieth-century travel, both interstate and international, provide complementary and interdependent images of a mobile nation in which trajectories are constitutive of *spaces* but not reducible to *places*. The emphasis on mobility is evident in the writing of Nathan Asch, an expatriate only briefly mentioned by Cowley, who after returning from Paris decided "to try to see America"

[3] See Georges Van Den Abbeele for an account of how the circular structure of the journey plays a role in the history of travel literature and political and philosophical thought.

by crossing the continent by car and bus (Cowley, 7). The book emerging from his travels, *The Road: In Search of America* (1937), finds America not in any particular location but in political solidarity with the underprivileged. Asch's relatively unknown domestic travel book is in many ways typical of what Michael Denning calls "the masculine romance of the road" in its preoccupation with hitchhikers, tramps, vagabonds, and the migratory poor (Denning, 187). There is, however, an important difference. Denning argues that the romance of the road tries but fails to represent the *place* of production, i.e. the factory assembly line (ibid.). The road book, however, is able to transform its internal resistance to place (which is to say its structural commitment to mobility) into a prescription for political *movement*. In his final chapter Asch reveals that his aim has been to "reach inside the miners and the field workers and the unskilled laborers and the poor farmers and the machinists and the construction men and the loggers, and their wives and girls and sisters and their children" to show them that their hardships unite them in a common cause (269). While this might seem a lot to expect of a short road trip, Asch's narrative is typical of the domestic recuperation of the international journey: the momentum acquired abroad "returns" by converting transportation into a political movement with national implications. Already evident in the trope of the "exile's return," and the progressive travel narratives emerging from it, is a depiction of modernization as what Anthony Giddens calls "disembedding": the decreasing significance of *place* in relation to symbolic systems spanning wide areas of time and space. Such systems include money, media such as radio and newspapers, and, I would argue, the interstate.[4] In the early part of the twentieth century, disembedding is still understood in primarily national rather than international terms.

Disembedding becomes a political and artistic factor in the project Asch was to join after finishing his own travelogue: the American Guide Series, published under the auspices of the WPA in

---

[4] Giddens, 2, 18. See also: David Nye's description of the effects of electricity and electronic media on the relation of subject to place: "what was once called the settlement of the United States begins to appear as a continual unsettling and uprooting. The journey into American space leads not to Wright Morris's *Home Place* but to more restless movement. The electrification of the countryside does not stabilize its population, but unsettles it" (187).

the late 1930s and early 1940s. Asch served as national editor on the series, as did many returning expatriates, including the series general editor, Henry Alsberg, who traveled extensively as a journalist in Europe and the Ukraine in the 1920s and 1930s. The enormous project, which produced a guide for every state and some regions and cities, set out to describe and preserve local traditions and communities already seen as endangered by modern developments in transportation and communication. The Guide Series also *was* one of these developments, organizing the country in a series of "tours," or road trips, which literally mapped national space from north to south, east to west, symbolically aligning small communities according to their relation to the starting point, Washington, D.C. (Bold, 36). This was the literary version of a process that began as early as 1919, when the Zero Mile Marker was placed on the Ellipse south of the White House as the starting-point for mile markers radiating out across the United States (Weingroff, "Zero Milestone, Washington, D.C."), and expanded in 1925 through the national standardization of road signs (Boorstin, 112). Like mile markers and road signs, the Guide Series organizes the American landscape from the perspective of the nation's capital; and like other travel books, it does so by emphasizing scenic aspects of the landscape. The editorial staff understood this scenic mapping as an innovation on par with the invention of a new genre. As Alsberg put it, "The tour form is a difficult form; it is like a sonnet; but, if you can learn it, you can be more interesting in the description of a tour than in any novel" (McDonald, 694). The tours are "interesting" due to their particular relation of text to geography: they inscribe progressive political narratives on the landscape by directing drivers to sites of social relevance, from historical battlegrounds to new bridges and damns. By converting the political into the scenic the American Guides make Asch's experience available to everyone: automobile tourism becomes an act of national solidarity. As in the other travel narratives of the era, the modernizing forces potentially disruptive to local tradition are reinscribed in a national whole.

W.H. Auden's long poem "New Year Letter (January 1, 1940)" offers a more cosmopolitan view of American mobility and thus represents one of the first steps away from the recuperative gesture of the sublime national space. Auden immigrated to the United States in 1939 shortly before the outbreak of World War II. His British contemporaries saw his expatriation, and this poem in particular, as marking a decline in his "poetic powers" (Smith, 7). However, Auden understands mobility as being as central to his project and to the American landscape. The United States is congenial not only because it offers refuge from the war, but because it is "This raw untidy continent / Where the commuter can't forget / The pioneer" (237). In spite of the reference to pioneering, which we have already seen plays a mystifying role in Post, Auden is careful not to transform American mobility into a national myth. Thus he pairs his description of African Americans migrating north for better wages with an account of "Resourceful manufacturers" moving production facilities south to cut down on the cost of labor (ibid.). Auden's reflections on the economics of mobility lead him to very different conclusions on the scenic value of place. He is particularly concerned with the impact of Hollywood on the West:

> and kids,
> When their imagination bids,
> Hitch-hike a thousand miles to find
> The Hesperides that's on their mind,
> Some Texas where real cowboys seem
> Lost in a movie-cowboy's dream.
> More even than in Europe, here
> The choice of patterns is made clear
> Which the machine imposes, what
> Is possible and what is not,
> To what conditions we must bow
> In building the Just City now.
>
> However we decide to act,
> Decision must accept the fact
> That the machine has now destroyed
> The local customs we enjoyed,
> Replaced the bonds of blood and nation
> By personal confederation.
>
> (238)

Auden takes the United States as his example because, he argues, the impacts of modernization are visible there "more even than in Europe." He carries his analysis a step farther than Post, and in a different direction from Asch and the WPA, attributing both patterns of production ("the machine") and consumption ("movie-cowboy") with the weakening of "local customs." One result is alienation—"Compelling all to the admission, / Aloneness is man's real condition"—but a potentially positive effect is the erosion of blood and soil nationalism (ibid.). The standardization Asch and Alsberg hope to transform into progressive politics, Auden sees as clearing a space for cosmopolitanism.[5] Thus he represents American salesmen and commuters as a worldly roster of latter-day Gawaines, Quixotes, Ishmaels, and Henry Jameses. This sort of cultural travel, an aesthetic version of the daily commute, is supposed to replace disappearing local and community feelings (200). Art becomes the new place, the non-geographical locus of an imaginary community, and the preferred medium of social intercourse or "personal confederation."

This brief genealogy of road books and representations of travel—from Post through Cowley, Asch, and the WPA to Auden—is intended to demonstrate what might be polemically called "the disappearance of place." Places do not disappear, of course, but the systems binding them together—symbolic and geographical—become increasingly prominent in the age of interstate and international travel. Disembedding has a visible geographical form—the interstate—which finds its narrative and poetic forms in the twentieth-century travelogue. Of course, the term "interstate" predates automobile travel. Frederick Jackson Turner uses the term in his famous essay "The Significance of the Frontier in Western History," for instance, to describe a "Mobility of population [that] is death to localism" (53). However, automobility and its visible impacts on the landscape lend a more general credibility to the concept of networked space always central to Turner's theories.[6] In the twentieth century it becomes increasingly clear to even the most casual tourists that regional customs and architecture are taking a back seat to wider national and ultimately international trends, whether these

[5] Auden also sees the mobile American society as more conducive to religion, but this point lies beyond the scope of my analysis.

[6] Turner consistently described the lines of commerce and communication binding the nation together as a network, although his metaphors reflect the organicist orientation typical of nineteenth-century nationalism: "It is like the steady growth of a complex nervous system for the originally simple, inert continent. If one would understand why we are to-day one nation, rather than a collection of isolated states, he must study this economic and social consolidation of the country" (41).

be understood in terms of the technological (Post), political (Cowley, Asch, WPA), or artistic (Auden) sublime.

Henry Miller rejects the circular itinerary of the return, progressive politics, and even the possibility of locating a particular American culture, whether grounded in mobility or other social factors. We might sum up his perspective as a total negation of the scenic—the homology of site and sight—and with it the subject position of the tourist and consumer. Miller is in short an unhappy tourist. His perspective is surprisingly similar to that of the writer Auden mentions near the end of his long list of traveling literati: Henry James, who "learn[s] to draw the careful line," but perhaps not in the way Auden meant (239). In *The American Scene*, a book recounting his travels through the United States in 1904-05, James opposes those internationalizing trends that are connected to modernization—especially industry and immigration—while remaining committed to his own literary cosmopolitanism. When praising a new enclosure around Harvard Yard, James is clear about the defects of the alternative: "The open door—as it figures here [in America] in respect to everything but trade—may make a magnificent place, but it makes poor places"(62). James's travelogue is a half-hearted attempt to "draw the line" or close the American door through observations that are undeniably anti-Semitic, xenophobic, and elitist in tone. James shows more sympathy for personified old buildings, through which he ventriloquizes his concerns, than for the immigrants who are typically denied voices. However, his elegy for disappearing "places," and especially old buildings, must also be understood as a (conservative) protest against standardization and all attempts to recuperate it in terms of a national whole. Bill Brown argues that James "fashion[s] a scene that, despite or because of its assumed agency, is anything but scenic, anything but scopic," placing this anti-touristic style in a long tradition of modern and post-modern critics of the standardized capitalist landscape, from Simmel to Baudrillard (Brown, 178).

Miller also belongs on this list. His landscape is anti-scenic and anti-patriotic, although he achieves this effect without recourse to talking buildings. Rather, he provides what is perhaps the first sustained account of the two themes that would preoccupy post-war literature: alienation and the reign of the image. Miller is an early example of what Don Slater calls the "post-tourist":

> unlike the romantic traveller in search of authenticity, or the mass package holiday maker who captures a 'real' experience in snapshots, the 'post-tourist' *knows* that she or he is a tourist and 'that tourism is a series of games with multiple texts and no single, authentic tourist experience.' (198)

He travels because he does not belong anywhere, and he resists scenic moments of transcendence that would establish him as a "loyal" American subject in relation to a meaningful place. When his experiences are momentarily transcendental or sublime, the overall tone of his travelogue evacuates them of their significance. Miller's domestic travels document the disappearance of place, and with it the belief that interstate and international travel can be recuperated by a national whole.

II. The Unhappy Tourist

The outbreak of World War II forced Miller, like Auden, to return to the United States in 1939. He was reluctant to leave Europe and desperately short of funds. His best-known works like *The Tropic of Cancer* were at that time unavailable in his country of birth, but censorship gave him a reputation he hoped to capitalize on with a book that deliberately steered clear of prurient themes (Ferguson, 273-274). With this in mind, Miller accepted a commission from Doubleday to write a domestic travelogue from the perspective of the returning expatriate (Dearborn, 220). The book emerging from these travels, *The Air-Conditioned Nightmare*, was subsequently rejected by that press and has never received sustained critical attention. The painful contradictions everywhere evident in the narrative suggest that thc project—not to mention the cash advance from a major American publisher—placed Miller in an uncomfortable role. Buying a car, for instance, meant giving into the mass opinion and mass technology Miller so famously resents: "The only way to see America is by automobile—that's what everybody says" (14). Miller suspects that driving makes him more American than he wants to be—part of the "new world" that had become a world of things (17). The car is a "lethal comfort" and "the very symbol of falsity and illusion" (33). Newer cars are the air-conditioned nightmares of the title (242).

Miller's project was to compose an anti-automobile road narrative. This is also of course the book's central paradox. He is certainly not the first motorist to harbor a secret resentment against automobility. However he no longer believes that the authentic can be distinguished from the standard or that local traditions exist, at least not in the United States. His pessimistic account inverts the sublime encounter typical of much American landscape writing, moving him to terror rather than ecstasy, or to a terror that is itself ecstatic. Summarizing his impression of St. Louis he says, "my disgust grew so great that I passed over into the opposite—into a state of ecstasy" (69). Describing the country in general: "Topographically the country is magnificent—*and* terrifying. Why terrifying? Because nowhere else in the world is the divorce between man and nature so complete" (19). The geographical homogenization that "divorces" man from nature also estranges Miller from his muse: "I had to travel about ten thousand miles before receiving the inspiration to write a single line" (19-20).

The regional homogenization Miller understands as a symptom of consumer culture is relentlessly practical and therefore anti-aesthetic:

> Nothing comes to fruition here except utilitarian projects. You can ride for thousands of miles and be utterly unaware of the existence of the world of art. You will learn all about beer, condensed milk, rubber goods, canned food, inflated mattresses, etc., but you will never see or hear anything concerning the masterpieces of art. (157)

American commercial culture appropriates even avant-garde art, transforming surrealist provocations, for instance, into advertising gimmicks for Nylon stockings (257). It is this ability to capitalize on art for the purposes of advertising that renders the United States, in Miller's opinion, one of the least inviting lands for true artists (24, 49). In his diagnosis of the problem he significantly pairs Henry Ford with Walt Disney, the automobile industry with the image industry: "He's [Disney] the master of the nightmare. He's the Gustave Doré of the world of Henry Ford & Co., Inc." (40). Disney's unreal creatures reveal certain uncomfortable truths about modern life:

> Of course they're not real men and women: they're dream creatures. They tell us what we look like beneath the covering of flesh. A fascinating world, what? Really, when you think about it, even more fascinating than Dali's cream puffs. (41)

The implication is that we are all "dream creatures," constructed in the same standardized way that Ford constructs automobiles and Disney films:

> The typical American everywhere [. . .] looks as though he were turned out by a university with the aid of a chain store cloak and suit house. One looks like the other, just as the automobiles, the radios and the telephones do. (45)

The form of the road book opens up certain predictable avenues of escape from the world of standardized production methods (Ford) and cultural products (Disney). Miller, like Post, turns to nostalgia for relief; France, in its absence, still manages to inspire his poetic fancy. Describing the small town of Auxerre, Miller says,

> It was more French [than was Paris], more authentic. It created another kind of nostalgia, the nostalgia which later I was to discover in certain French books or through conversation with a whore in bed while quietly smoking a cigarette. (67)

Miller's nostalgia is not scenic. Unlike Post he does not see himself as the master or consumer of a site presented in visual terms. His brand of nostalgia leads him to the conclusion that the German invaders cannot succeed in conquering France, and then through the associative logic that is characteristic of his narrative, to the reflection that he and his former wife, in their role as American tourists, were also "invaders":

> With our dirty American dollars we were buying the things we wanted. But with every purchase we were given something gratuitous, something we had not bargained for, and it ate into us and transformed us, until finally we were completely subjugated. (68)

The gratuitous feeling that cannot be bought or sold is what makes France—and its "invaders"—French.[7] Authenticity is a shadow

[7] Here Miller is adapting an economic formula from the standard text of exilic literature, Hemingway's *The Sun Also Rises*, which establishes Spain as a place of

economy operating outside and through economic exchange; in fact, it seems at first glance to be the opposite of a consumer economy since it "eats" or consumes its invaders rather than the reverse. However, a closer look reveals this shadow economy to be merely the inverted or mirror image of American consumer capitalism. The world of Ford, Disney, and interchangeable citizens is, after all, not only a landscape of consumption but also a *consuming* landscape: one that regurgitates art and, ultimately, personality in its own standardized image.

The limitations of the shadow economy of authenticity become clear when Miller attempts to apply his model of escape to the American scene. Like many other tourists, he locates authenticity in pockets of "otherness" or ethnic difference, both living and monumental. Miller seeks to escape the commercialism of American life, for instance, by strongly identifying with Native Americans, even wishing at one point for an Indian sidekick. He also claims that the only place nature still exists in its undefiled form in the United States is on the reservations ("The little Cherokee reservation is a virtual Paradise," [36-37]). Any ethnicity, so long as it diverges from what Miller considers the American standard, can serve as a marker of authenticity. This becomes clear when Miller compares Native American cliff dwellings to a ninth-century Chinese Buddha he encounters in Louisiana. The following paragraph is taken from the penultimate page of his travelogue:

> It is doubtful if this continent will ever bequeath the world the deathless splendor of the holy cities of India. Only in the cliff dwellings of the Southwest, perhaps, does the work of man here in America arouse emotions remotely analogous to those which the ruins of other great peoples inspire in the traveler. On Avery Island, in Louisiana, I ran across a massive statue of the Buddha, brought over from China, which was protected by a glass cage. It was startling to look upon in its bizarre setting. It dominated the landscape which was in itself a work of art in a way that is difficult to describe. (287)

---

ritual violence and spirituality irreducible to the "clear financial basis" of France (Hemingway, 233). For Miller it is not France but the United States that is guilty of crass commercialism. That the accusation varies, even amongst expatriates, suggests the imprecision of commercialism as an invective.

The significance of this monument marking the end of Miller's journey is doubly inscribed; not only does the statue stand out from the landscape as something unique and worth visiting, it also, by virtue of its distinctiveness, helps transform nature into what Miller terms "a work of art," and transform his narrative—placed as it is at the end—into a discovery of something worthwhile. This is the transcendental and aesthetic moment usually denied to him by the "terrifying" and "disgusting" American landscape responsible for divorcing him from his muse. The surplus cultural value he attributes to the French shadow economy is embodied here in "the shadow of the Buddha [. . .] as if defining with unerring exactitude the utmost limits of hope, desire, courage and belief" (288). This sublime moment is established through a string of ethnic associations—American Indian, Eastern Indian, Chinese—whose common denominator is their presumed externality to the American commercial system. What might, for lack of a better term, be called the "ethnic sublime" corresponds to the technological, political, and poetic varieties we have explored in other travel narratives. Because the Buddha transforms nature into art "in a way that is difficult to describe," the artfulness of the scene begins to seem natural rather than artificial, somehow apart from the "staged authenticity" of consumer culture. Miller finally discovers a locus of authenticity in America, and it is Chinese, or Native American, or French, etc.

However, there is more to the Buddha than meets the eye. What Miller does not tell us is that Avery Island is the site of the Tabasco pepper sauce factory, that the Buddha was acquired by the company's owner E.A. McIlhenny in 1936, only shortly before Miller's visit, and that he was the one who had it placed in the artfully natural setting of the bird sanctuary, which was built on the grounds of the factory. Today the statue and sanctuary are major regional tourist attractions that can be combined with a Tabasco factory visit.[8] Even in Miller's day the Buddha betrayed a connection (however mystified) to the consumer capitalism from which he hoped it offered an escape.[9]

---

[8] "Tabasco Pepperfest Info Booth." Tabasco. http://www.tabasco.com/info_booth/faq/avery_visit.cfm.

[9] The Tabasco Buddha could productively be analyzed as a variation on the kind of roadside colossus explored at length by Marling: "Colossi are advertisements that point to commodities for sale" (67).

Here Miller's narrative, presumably against his will, demonstrates the risks of grounding a critique of consumerism in the touristic search for authenticity. The flight from standardized landscape and culture, represented by Ford and Disney, leads to the epiphany of pepper sauce.

The Air-Conditioned Nightmare falls into an obvious contradiction: critical of consumer culture, and of the standard travel narratives seeming to buy into it, Miller's book ultimately commits the same literary and touristic sins. What makes Miller's narrative different, and also indicative of emergent literary trends, is the way it figures its unresolvable contradiction in imaginary monuments and endless journeys. His chapter on "Arkansas and the Great Pyramid" describes a kind of counter-monument to the Buddha (although the effect was probably unintentional). The pyramid is the pipe dream of one William Hope Harvey, who, in the words of his own promotional pamphlet, wanted to erect a great pyramid "to attract the attention of the people of the world to the fact that civilizations have come and gone attended with untold suffering to hundreds of millions of people, and that this one is now in danger—on the verge of going" (Miller, 137). The pyramid exists in Miller's narrative but not his itinerary; it is a ruin never built. Its particular form of unreality, existing in a symbolic field but not in space, is opened as a possibility by the unique spatial-textual nexus of the travelogue, which assumes homology between narrative and itinerary on the one hand, and depends on the "trope of use" on the other (Bold, 17-18). The difference between travel narratives and their fictional counterparts is not merely their "origin" in reality—an ontological relation notoriously difficult to define—but their utility as guides for subsequent journeys. The pyramid, however, cannot be visited, although other sites in Miller's travelogue could be. It is an *image* because it does not exist, but an image of a different order from the Buddha, which can be visited but appears to be outside consumer culture, other than what it actually is. Here one sort of image, made possible by the structure of the travelogue, offers a critical alternative to the kind of image or illusion internal to the structure of tourism. On the one hand this uncanny moment is symptomatic of the contradictions Miller's travelogue cannot escape, on the other it points to an emergent cultural paradigm: the postmodern preoccupation with images.

Miller's narrative keeps returning to non-places which, due to their position in the travelogue, must be distinguished from fictional settings. Analogous to the absent monument is a destination endlessly deferred. In the chapter "My Dream of Mobile," which rehearses, as litany, many of the canonical introductory clauses to travel narratives—"*When Marco Polo went to the East* [. . .]" "*When the Spaniards sailed West,*" etc. (181 ff.)—Miller never actually reaches his destination because Mobile, as the name suggests, is always moving: "It is mobile, fluid, fixed, but not glued" (192). Miller provides the key to this narrative strategy of deferral when he argues that the imagination and imaginary communities are more important than geographical locations:

> I know that there are maps of the earth which designate a country called America. That's also relatively unimportant. *Do you dream?* Do you leave your little *locus perdidibus* [sic] and mingle with other inhabitants of the earth? (189)

In these rhetorical questions Miller comes extremely close to Auden's literary cosmopolitanism, except that his narrative denies the possibility of realizing art in the United States. The destination never reached is an alternate version of the imaginary monument: an ideal that can exist only in its absence. However, it also figures the tourist as an eternal exile, someone who travels because he does not fit in. This is also symptomatic of Miller's critical stance vis-à-vis the consumer culture he cannot escape, and indicative of another post-war cultural paradigm: existential alienation.

III. Let's Get Lost

Post-war critics such as Sidney Finkelstein saw Miller as "a direct link between the two generations of 'disillusion,' that which followed the First World War and that which followed the Second" (121). One chapter in *Air-Conditioned Nightmare* describes how Miller, whom Wilson described as composing the elegy to the expatriates, comes into contact with other travelers who can only be described as "Beat." Miller attributes his decision to purchase a car to two acquaintances who could have served as character models for *On the Road* (151). The protagonists are dressed in proto-Beat fashions ("a blue denim suit, big boots, horn-rimmed specs, long hair and a goatee," [156]); they are self-conscious drop-outs (158); they experiment with the spontaneous autobiographical writing that would later serve as the paradigm of Beat prose (159); and they are even engaged in the search for a lost father as are Kerouac's heroes Sal Paradise and Dean Moriarty (*Air-Conditioned Nightmare*, 162; Kerouac, 307). The main difference between Miller's acquaintances and Kerouac's characters is that the latter only purchase cars as a means of last resort, preferring instead to hitchhike, ride the bus, ride-share, or, like Dean, simply steal.

Finkelstein argued that Miller's and the Beat's rebellion was meaningless because anti-social and ultimately criminal. Less conservative critics tended to focus not on their distance but their proximity to mass or consumer culture. Isaac Rosenfeld's negative review in *Partisan Review* in 1946 accuses Miller of "falling so readily into the American stride, the tricky, self-advertising gait, that he becomes merely a conscious citizen, disgusted by his society, but by no means dissociated from it" (382). This line of critique was perhaps most forcefully articulated by Irving Howe in "Mass Society and Post-Modern Fiction" (1959), an essay dealing not with Miller but the Beats and the "Angry Young Men" in Great Britain. Howe, more dismissive of the Beats, decries rebellion against society as the flip-side of conformity:

> These writers, I would contend, illustrate the painful, though not inevitable, predicament of rebellion in a mass society: they are the other side of the American hollow. In their contempt for mind, they are at one with the middle class suburbia they think they scorn. (435)

His polemic culminates in a description of the alienated rebels as "veritable mimics of the American tourist" (ibid.). Howe's argument about "the predicament of rebellion in a mass society" brings us face to face with one of the core elements of consumer culture, although perhaps in a way opposite to what he intended. Howe thought the Beats were imitating the middle class, but current criticism focuses on how consumer culture cashes-in on the kind of rebellion evident in Beat literature by appropriating protest as stylistic innovation. As Don Slater has pointed out, "youth culture" can be understood as a form of rebellion that directly feeds into the planned obsolescence central to any consumer-driven economy; consumer goods "became the vehicle for expressing a seemingly total dissatisfaction or provocation to the society that produced them" (165). We need look no further than the discomfort of somebody like Kerouac who becomes trendy in his rebellion against prevailing trends, or at the popularity of "bop," "Beat," and later "hippy" looks, which might have begun as forms of protest but were proliferated as fashions.

To escape consumer culture the Beats follow the same path as Miller and other tourists: romancing the "other" as somehow immune from market forces. Kerouac's narrator Sal Paradise, for instance, achieves his hard-won moments of transcendence mainly when the road brings him into contact with African American jazz musicians, Latino migrant workers, or the Mexican Indians he characterizes as "the source of mankind and the fathers of it" (280). Transcendence is virtually assured for Kerouac when "otherness" is doubly embodied as race and gender; encounters with dark-skinned women are either over-sexualized ("remembering I was in Mexico after all and not in a pornographic hasheesh daydream in heaven" [289]) or deliberately asexualized ("They were like the eyes of the Virgin Mother when she was a child" [297]), both representational strategies firmly locating the locus of authentic difference in the body.

This explicit eroticism is another link to Miller, who repressed sex in his travelogue in order to capitalize on the infamy of his more controversial books (Finkelstein, 121). Rosenfeld criticized Miller for deviating from his trusted provocation:

> It may be the onset of age, or in some way the effect of America, but Miller back-at-home is not the same old Henry. The image he created for himself, among the despairing ecstasies of *Tropic of Cancer*, of a man, flashlight in hand, peering down a vagina, is hard to connect with the motorist peering through a windshield. (382)

However, subsequent road books by other authors were not so demure. Kerouac's moments of transcendence are often defined erotically. The sexual element of the journey is evoked, for instance, when Dean, Marylou, and Sal remove their clothes and nakedly "drive west into the sun" (161). Nabokov's *Lolita* (1955), to take the most famous example, depends on an extended analogy between the girl's body and the

American landscape: "The *tour* of your thigh, you know, should not exceed seventeen and a half inches [. . .] We are now setting out on a long happy *journey*" (Nabokov, 209, italics added). In John Updike's *Rabbit, Run* (1960), the protagonist Rabbit Angstrom is in permanent flight from the uniformity in his landscape and life: "The land refuses to change. The more he drives the more the region resembles the country around Mt. Judge [his hometown]" (37). The novel deliberately plays up the parallels between his quest for difference and his sexual encounters. As a minister says to Rabbit in the first book of the series, "of course, all vagrants think they're on a quest. At least at first"; he later adds, "it's the strange thing about you mystics, how often your little ecstasies wear a skirt" (120, 121). Sexuality, of course, is not necessarily or even usually a place of protest. David Lewis has convincingly shown that it has been involved in the marketing of cars "from the dawn of the auto age" (123). Nevertheless, it was intimately bound up with youth protest movements, which by the late 1950s were already drawing an analogy between sexual and political repression on the authority of Herbert Marcuse's *Eros and Civilization* (1955).[10] The body becomes the non-place of post-war travel narratives, a locus of experience rather than a destination or location.

Morris Dickstein's essay on post-war fiction in the *Cambridge History of American Literature* argues that most novels of the 1950s can be understood as road books, couching their social protest explicitly in the lyrical or picaresque modes, or operating as counter-road books by presenting their critique through the irony of *not* setting out (175). *Catcher in the Rye*, *On the Road*, and *Lolita*, and *Rabbit, Run* are complemented—not contradicted—by suburban novels like *Revolutionary Road* (1961) and *The Man in the Grey Flannel Suit*

---

[10] Marcuse's own summary of his project in his "Political Preface" to the 1966 edition of the book gives a concise formulation of the relation between political and sexual repression: "It was the thesis of *Eros and Civilization*, more fully developed in my *One-Dimensional Man*, that man could avoid the fate of a Welfare-Through-Warfare State only by achieving a new starting point where he could reconstruct the productive apparatus without that "innerworldly asceticism" which provided the mental basis for domination and exploitation [. . .] 'Polymorphous sexuality' was the term which I used to indicate that the new direction of progress would depend completely on the opportunity to activate repressed or arrested *organic*, biological needs: to make the human body an instrument of pleasure rather than labor" (xiv-xv).

(1955). In the 1950s the road becomes the universal expression of American culture, and also the primary "vehicle" of its critique.

The ambiguous symbolism of the road in the 1950s is not only the result of the constitutive contradictions of consumer culture; it also stems from contradictions within the automotive industry itself. James Flink argues that "in the late 1950s [. . .] many Americans began to have critical second thoughts about the automobile industry and its product" ("Three Stages of Automobile Consciousness," 468). Flink's primary piece of evidence is the increasing number of industry-critical books to come out in the late-1950s, beginning with John Keats' *The Insolent Chariots* (1958), Jane Jacobs' *The Death and Life of Great American Cities* (1961), and culminating in Ralph Nader's *Unsafe at Any Speed* (1965). He also points out, however, that this critical period follows fast on the heels of the record-breaking auto sales after World War II (467). The car, in other words, is both symptom and solution, the source of major social problems but also the only evident means of escape, not only for the economy but for the individuals who would hold themselves aloof from it: "automobility is one of the last bastions of individualism in our increasingly bureaucratized, collectivized and conformist society" ("Three Stages," 468). In his 1988 monograph *The Automobile Age*, Flink points to other factors contributing to the ambiguous status of the automobile in the 1950s. The nearly universal ownership of cars even among urban poor (359) coupled with technological stagnation (278) made planned obsolescence and stylistic change the determinate factors in motivating new sales (234-235). Hence the elaborate tail fins, chrome detailing, fancy accessories, etc. typical of post-war automotive production. Innovation becomes a matter of stylistic change, or image, at the moment that driving becomes nearly universal, both in fact and fiction.

IV. Postmodernism: Imagining the Interstate

The image and alienation are two of the most salient features of road literature. Howe's analysis of "post-modern" fiction provides a useful starting point for theorizing the relation between them. It is important to recognize that what Howe calls "post-modern" differs from dominant accounts of postmodern literary form, as well as from current descriptions of postmodernity as a stage of economic and cultural development. To take one example, Fredric Jameson defines postmodernism as a proliferation of subjectivities and styles symptomatic of a fast-paced consumer society that depends on the rapid recycling of fashions. Thus the dominant postmodern narrative forms are characterized by pastiche and schizophrenia: the proliferation of styles corresponding to the multiplicity of subjectivities necessary to consume them (Jameson, 111). Howe, on the other hand, is concerned with an earlier moment in consumer culture when individualism is still a trend. These are two distinct moments in the development of consumer subjectivity: the heroic-rebellious and the performative-ironic, the era of alienation and the era of image. However, the two moments should be understood along a continuum. Slater, summarizing postmodern critiques of "the society of spectacle," states the relation between alienation and image in the following terms:

> Because everything can be commodified and objectified—including all forms of opposition (the very idea of "revolution" can be packaged as a subcultural style, an advertising slogan, an urban guerrilla clothing fashion)—everything can be absorbed into the spectacle. Alienation, then, has spread from the work-place to absorb all of life, above all the free spaces of leisure, consumption and culture in everyday life. (127)

Spectacle or image, in this analysis, means both style and illusion. Consumers become increasingly dependent on images to define their identities (style) at the moment it becomes clear that images do not refer to a reality outside of consumer capitalism (illusion) (Slater, 30). The impossibility of "locating" alienation, say in the workplace, is the flip side of the impossibility of locating authenticity in the landscape. Miller's post-tourism, seeking a destination he knows to be false or nonexistent, forges the link not only between exilic and Beat literature, but between existential alienation and the postmodern preoccupation with the image.

Daniel Boorstin offered one of the first analyses of the link between alienation and the image in his 1962 book *The Image*, which laments "the thicket of unreality which stands between us and the facts of life" (3). Unreality is connected in several ways to consumer culture, but Boorstin clearly identifies the automobile as "one of the chief insulating agencies" (111), along with the "super highways," which he sees as "the climax in homogenizing the motorist's landscape" (112). As I have already pointed out, this complaint about standardization dates back as early as 1915. However, Boorstin is concerned with a stage at which standardization not only replaces but begins to *reproduce* local difference: "One thing motels everywhere have in common is the effort of their managers to fabricate an inoffensive bit of 'local atmosphere'" (113). Boorstin's image takes the place of the real. In the age of interstate highways and organized tourism even nature seems fake:

> The pre-eminence of Yellowstone National Park as a tourist attraction is doubtless due to the fact that its natural phenomena—its geysers and 'paintpots' which erupt and boil on schedule—come closest to the artificiality of regular tourist performances. (111)

Boorstin's fake Yellowstone is closer to Miller's Tabasco-brand Buddha than Post's "authentic" West. It emerges in an interstate landscape deprived of the recuperative energy of the national sublime, and hence without any distinct sense of the nation as place or space. Thus Boorstin sees his contemporary Americans, denied contact with reality and the nation, as tourists in their own land (109), permanent itinerants whose destination is "no place in particular—in limbo, en route" (114).

Boorstin's preoccupation with alienation and the image was anticipated by a book that proved extremely important in city planning and urban development circles, *The Image of the City* (1960) by M.I.T. professor Kevin Lynch. The difference in tone between the two analyses could hardly be overstated. For Lynch the image is not a symptom of general detachment but an urban design objective, one that is supposed to restore a sense of security, depth, and intensity to inhabitants who might otherwise be "lost" in the chaotic urban landscape (4-5). Nevertheless, the fact that Lynch characterizes city inhabitants as potentially "lost" reveals the deeper congruities between his argument and Boorstin's. Like his contemporary, Lynch sees modern building practices, and particularly freeways, as potentially alienating structures; he points out that pathways for motorists constitute "edges" or barriers for pedestrians (64), and that highways quickly become complete and self-enclosed structures in cities like L.A. (60), often seeming to exist without relation to the city at all (65).

Lynch hopes to correct this confusion by "forming our new city world into an imageable landscape: visible, coherent, and clear" (91). The use of the term "image" is deliberate; Lynch advocates redesigning cities with aesthetic criteria in mind, and particularly with concepts taken from music, narrative, and drama. It is these kinesthetic arts, rather than static ones like sculpture, that provide the vocabulary for Lynch's argument; they articulate the sequencing he sees as central to the modern experience of the city, which is primarily an experience of motion (Lynch, 113). Lynch is not one of those critics, just beginning to raise objections at that time, seeking to banish freeways from the urban setting.[11] On the contrary, he argues that the technology rendering "The underlying topography, the pre-existing natural setting [. . .] perhaps not quite as important a factor in imageability as it once used to be" actually brings about new possibilities for ordering the landscape (110). Lynch and Boorstin represent two sides of the same debate: Lynch's dream of an image adequate to modern mobility is the optimistic version of Boorstin's nightmare that the image is replacing reality. What the latter disparages as permanent tourism, the former praises as "the potential drama and identification" of modern life.[12]

In Boorstin and Lynch, as in Miller, the image is no longer *expressive* of place or space. The collapse of this symbolic relation entails the end of the *scenic* understood as a correlation between subject, site, and nation. Site is no longer related to sight through synecdoche but through distortion or illusion, or at best through artful

---

[11] One noteworthy anti-freeway exponent at the time was Lewis Mumford: "By allowing mass transportation to deteriorate and by building expressways out of the city and parking garages within, in order to encourage the maximum use of the private car, our highway engineers and city planners have helped to destroy the living tissue of the city and to limit the possibilities of creating a larger urban organism on a regional scale" (510).

[12] Lynch, 49. It would be possible to trace the development of the linked concepts of alienation and the image from Boorstin's and Lynch's seminal analyses to more recent theories of spectacle and postmodern architecture and city planning. Robert Venturi, Denise Scott Brown, and Steven Izenour's famous argument—"If you take the signs away, there is no place. The desert town is intensified communication along the highway" (18)—stands as paradigmatic for the semiotic turn typical of more contemporary accounts of postmodernism. What early theorists represent as total alienation from an illusory landscape on the one hand, or the deployment of artistic forms to reimagine landscape on the other, contemporary critics and architects understand as the transformation of space into text.

city planning; the part no longer "naturally" represents the national whole. The sublime vision of the nation gives way to suspicion of the fake. The subjectivity corresponding to this landscape of illusion is alienated, displaced, no longer "at home" in any particular location, and reduced to the mobile body as a site of "authentic" experience. The interstate provides a spatial figure for this situation, and also a convenient time frame in which to date its emergence. On June 29, 1956 President Eisenhower signed the Federal-Aid Highway Act allowing for the creation of the interstate highway system. The first stretch went under construction in Missouri that year, and the final segment was completed in Los Angeles in 1993 (Weingroff, "Three States Claim First Interstate Highway"). While the idea of a limited-access, high-speed network of roads goes back to at least the 1920s, and while the American freeway system continues to grow today, the opening and closing dates of the Federal-Aid Highway Act provide useful markers for locating postmodernity. The interstate, in other words, is postmodern space.

Bibliography

Asch, Nathan. *The Road: In Search of America*. New York: W.W. Norton & Co., 1937.

Auden, W. H. *Collected Poems*. Edited by Edward Mendelson. New York: Vintage International, 1991.

Bold, Christine. *The WPA Guides: Mapping America*. Jackson: University Press of Mississippi, 1999.

Boorstin, Daniel. *The Image*. New York: Atheneum, 1962.

Brown, Bill. *A Sense of Things: The Object Matter of American Literature*. Chicago: University of Chicago Press, 2003.

Cowley, Malcolm. *Exile's Return*. New York: Viking Press, 1956.

Dearborn, Mary. *The Happiest Man Alive: A Biography of Henry Miller*. New York: Simon & Schuster, 1991.

Denning, Michael. *The Cultural Front: The Laboring of American Culture in the Twentieth Century*. London: Verso, 2000.

Dickstein, Morris. "Fiction and Society, 1940–1970." In *The Cambridge History of American Literature*, edited by Sacvan Bercovitch, 7:101-310. Cambridge, MA: Harvard University Press, 1999.

Ferguson, Robert. *Henry Miller: A Life*. London: Hutchinson, 1991.

Finkelstein, Sidney. "Alienation and Rebellion to Nowhere." In *Henry Miller: Three Decades of Criticism*, edited by Edward B. Mitchell, 121-128. New York: New York University Press, 1971.

Flink, James. *The Automobile Age*. Cambridge, Massachusetts: The MIT Press, 1988.

—. "Three Stages of Automobile Consciousness." *American Quarterly* 24, no. 4 (Oct. 1972): 451-473.
Giddens, Anthony. *Modernity and Self-Identity*. Cambridge: Polity Press, 1991.
Gross, Andrew. "The American Guide Series: Patriotism as Brand-Name Identification."*The Arizona Quarterly* 62, no. 1 (Spring 2006): 85-112.
—. "Cars, Postcards, and Patriotism: Tourism and National Politics in the United States, 1893-1929." *Pacific Coast Philology* 40 (2005): 77-97.
Hemingway, Ernest. *The Sun Also Rises*. New York: Scribner Classic, 1986.
Howe, Irving. "Mass Society and Post-Modern Fiction." *Partisan Review* 36 (Summer 1959): 420-436.
Jakle, John A. *The Tourist: Travel in Twentieth-Century North America*. Lincoln: University of Nebraska Press, 1985.
Jakle, John A. and Keith A. Sculle. *The Gas Station in America*. Baltimore: Johns Hopkins University Press, 1994.
—. *Fast Food*. Baltimore: Johns Hopkins University Press, 1999.
Jakle, John A, Keith A. Sculle, and Jefferson S. Rogers. *The Motel in America*. Baltimore: Johns Hopkins University Press, 1996.
James, Henry. *The American Scene*. Bloomington: Indiana University Press, 1968.
Jameson, Fredric. "Postmodernism and Consumer Society." In *The Anti-Aesthetic: Essays on Postmodern Culture*, edited by Hal Foster. Port Townsend, WA: Bay Press, 1983.
Kerouac, Jack. *On the Road*. New York: Penguin, 1976.
Lewis, David. "Sex and the Automobile: From Rumble Seats to Rockin' Vans." In *The Automobile and American Culture*, edited by David Lewis and Laurence Goldstein. Ann Arbor: The University of Michigan Press, 1983.
Lynch, Kevin. *The Image of the City*. Cambridge, MA: The MIT Press, 1960.
MacCannell, Dean. *The Tourist*. New York: Schocken, 1989.
Marcuse, Herbert. *Eros and Civilization*. Boston: Beacon Press, 1966.
Marling, Karal Ann. *The Colossus of Roads: Myth and Symbol along the American Highway*. Minneapolis: University of Minnesota Press, 1985.
Marx, Leo. *The Machine in the Garden*. New York: Oxford University Press, 1970.
McDonald, William F. *Federal Relief Administration and the Arts*. Columbus: Ohio State University Press, 1969.
Mertz, W. L. and Joyce Ritter. "Building the Interstate." U.S. Department of Transportation Federal Highway Administration. http://www.fhwa.dot.gov/infrastructure/build.htm.
Miller, Henry. *The Air-Conditioned Nightmare*. New York: New Directions, 1970.
—. (Reply to Edmund Wilson's Review of "Air Conditioned Nightmare.") *The Shores of Light*. Edmund Wilson. London: W.H. Allen & Co., Ltd., 1952.
Mitchell, W. J. T., ed. *Landscape and Power*. Chicago: Chicago University Press, 1994.
Mumford, Lewis. *The City in History*. New York: Harcourt, Brace & World, Inc., 1961.
Nabakov, Vladimir. *Lolita*. New York: Vintage, 1997.
Nye, David E. *Narratives and Spaces: Technology and the Construction of American Culture*. Exeter: University of Exeter Press, 1997.

Post, Emily. *By Motor to the Golden Gate*. New York: D. Appleton and Company, 1916.
Rosenfeld, Isaac. "Return to the U.S.A." *Partisan Review* 13, no. 3 (Summer 1946): 380-382.
Schaffer, Marguerite S. *See America First: Tourism and National Identity*. Washington, D.C.: Smithsonian Institution Press, 2001.
Schnapp, Jeffrey T. "Crash (Speed as Engine of Individuation)." *Modernism/modernity* 6, no. 1 (1999): 1-50.
Slater, Don. *Consumer Culture and Modernity*. Cambridge: Polity Press, 1997
Smith, Stan, ed. *The Cambridge Companion to W. H. Auden*. Cambridge: Cambridge University Press, 2004.
Turner, Frederick Jackson. "The Significance of the Frontier in Western History." In *Rereading Frederick Jackson Turner*, edited by John Mack Faragher. New York: Henry Holt and Co., 1994.
Updike, John. *Rabbit, Run*. New York: Fawcett Crest, 1991.
Van Den Abbeele, Georges. *Travel as Metaphor: From Montaigne to Rousseau*. Minneapolis: Minnesota University Press, 1991.
Venturi, Robert, Denise Scott Brown, and Steven Izenour. *Learning From Las Vegas*, rev. ed. Cambridge, MA: The MIT Press, 1977.
Weingroff, Richard F. "Three States Claim First Interstate Highway." U.S. Department of Transportation Federal Highway Administration. 8 May 2005 (a). http://www.fhwa.dot.gov/infrastructure/rw96h.htm.
—. "Zero Milestone, Washington, D.C." U.S. Department of Transportation Federal Highway Administration. 8 May 2005 (b). http://www.fhwa.dot.gov/infrastructure/zero.htm.
Wilson, Edmund. (Review of "The Air Conditioned Nightmare.") *New Yorker* 21 (29 Dec. 1945): 62-66.
—."Twilight of the Expatriates" (1938). *The Shores of Light*. London: W.H. Allen & Co., Ltd., 1952.

Writing Grounds: Ecocriticism, Dumping Sites, and the Place of Literature in a Posthuman Age

Klaus Benesch

This essay is about public spaces largely unknown to the public: dumping grounds and garbage sites, by-products of an ever-increasing global economy of waste; it is also about the potential of literature to create certain kinds of sensibilities in relation to these "unknown" places. The discussion begins with an overview of eco-criticism and of how literary texts, especially during the nineteenth and early twentieth centuries, were keen on translating place into symbolic, abstract spaces. It then moves to analyze three contemporary authors who, on the contrary, foreground the physical, corporeal aspects of concrete public places. As we become fully involved in computerized worlds that replace physical proximity by mere images of closeness, these writers address the widening gap between the real and the virtual by proposing a shift from *space*-consciousness to *place*-consciousness. In so doing, they also carve out a new and important place for the literary in a posthuman age.

In evolution's beginning was not the word but the place we learned to pin the word to.
Richard Powers, *Galatea 2.2*

...and where there was litter, there was life.
T. C. Boyle, *Drop City*

This essay is both about public spaces that ironically have become *terra incognita* to a larger public and about the potential of literature to create certain kinds of sensibilities in relation to these "unknown" places. The spaces I am referring to here are dumping grounds and garbage sites, the places where we bury the staggering mounds of refuse that come with the large-scale manufacture of material goods and, more specifically, an ever-increasing global economy of waste.

The issue of garbage, of what we as a culture value and what we *de*value, what we are keen on keeping or what we rather refuse to keep around, involves a number of serious questions concerning the world we live in and our own unrelenting urge to discard things. Though I will return to some of these questions later, I am interested here, more specifically, in the ability of literary texts to capture and create a strong "sense of place." In what ways and to what degree fiction can and should deal with the environment has been a contested issue within the burgeoning field of ecocriticism or, as some of its practitioners prefer to call it, environmental criticism. I will thus begin with a brief overview of how the concept of an environmental literature evolved and how literary texts, especially during the nineteenth and early twentieth century, were keen on translating place into symbolic, abstract spaces. In a second step, I then discuss three contemporary texts that, on the contrary, foreground the physical, corporeal aspects of concrete public places (including such posthuman "non-places" as dumping grounds). As we become increasingly involved in computerized worlds that replace physical proximity by mere images of closeness these (and other) texts may be taken to mark the beginning of a shift from *space*-consciousness to *place*-consciousness in recent American literature. In so doing, they also carve out a new and important place for the literary in a posthuman age.

Even though the wilderness or, more broadly, natural landscapes have always played a major role in American literature (just think of Cooper's Leatherstocking tales, *Moby-Dick*, *Huck Finn*, or Faulkner's fictional Yoknapatawpha County) it is only recently that environmental criticism as both a style of writing *and* a specific way of reading has gained more than marginal importance among professional critics and scholars.[1] If judged by the staggering number of publications in this new field and the widespread activities promoted and organized by the Association for the Study of Literature and Environment (ASLE) environmental criticism has by now become a thriving academic

---

[1] There were, of course, exceptions: William Carlos Williams, for example, clearly affirms the role of the environment in Poe's (and any other American poet's) work when he claims that "Poe could not have written a word without the violence of expulsive emotion combined with the driving force of a crudely repressive environment. [. . .] Typically American—accurately, even inevitably set in his time" (Williams, 111).

discipline: ecocritics proudly blaze their own canonical texts (with Rachel Carson's 1962 classic *Silent Spring* evolving as the modern founding text of the movement), they repeatedly lock horns in methodological disputes and debates, and—as with other critical movements before—at least one scholar, in this case Lawrence Buell, a professor of English at Harvard University, emerged as ecocriticism's major theorist and, increasingly, its critical conscience.[2]

Though appealing to a staggering number of scholars and students, ecocriticism is not quite yet a movement comparable in scope and impact to feminist criticism, postcolonialism, race studies, poststructuralism, or new historicism. Rather it is an aggregate of very different, heterogeneous attempts at reading "environmentally," that is, to read with an eye on the literary representation of setting and context rather than on character and plot. One way to define ecocriticism is to say that it describes a movement among the four Aristotelian components of literature—character, plot, theme, and setting—from the first three to the latter. Thus, it may be seen to follow in its own peculiar way the shifting of interest in literary theory from author (Romanticism, hermeneutics, interpretation) to text (structuralism/post-structuralism) to reader (reader response theory) and, eventually, to context (as in new historicism, postcolonialism, and cultural studies, etc.). Since literary texts, if to varying degrees of intensity, had always already something to say about the physical context from which particular actions evolve, to foreground the environment as sole foundation of an eponymous literary practice (that is, to write "environmentally") poses a number of methodological problems. Perhaps the most far-reaching of these problems is linked to the issue of representation. Some ecocritics—especially those connected to the so-called deep ecology movement[3]—tend to assume that there is a

[2] Buell appears to be at once the most theory-minded and most prolific writer on ecocritical themes. In his latest book, *The Future of Environmental Criticism,* he not only reviews the current state of environmental writing and criticism but provides a comprehensive ecocritical glossary of its major themes and topics. Other recent critical views on ecocriticism include Phillips; and Armbruster and Wallace.

[3] The philosophy of deep ecology originated in the early 1970s with the Norwegian philosopher Arne Naess who envisioned a radically inclusive biotic universe that inextricably binds humans to nature and natural processes. Among its basic premises are a belief in the intrinsic value of "nonhuman" nature and an emphasis on human

special relationship between the environmental text and its topic, the environment itself. The claim here is that ecologically conscious writing has to be truthful (or mimetic) with respect to representing non-human environments and it has to be discursive, that is, it should communicate an unmistakable message on behalf of the environment thus represented. While the first contention seems to conflict directly with poststructuralist critiques of representation (any form of re-presentation, poststructuralists kept reminding us, always also alters and manipulates that which is represented), the second clashes head-on with the modern notion of authorship which places the artist on the outside of political and social discourse. Art never speaks directly to social issues, and it thus cannot be expected to have an opinion on, say, toxic waste management or global warming.

Another important issue is the definition of the environment itself. What kind of environment should be covered by that term and to what degree would non-human environments (streets, cities, manufactures, dumping grounds, etc.) qualify as environmentally viable sites?[4] This is also why ecocritics traditionally emphasized a genre known as "nature-writing" and why environmentally conscious texts are poised to foreground "natural" landscapes rather than the proliferating technoscapes of urban environments. To add to the obvious shortcomings of such myopic definitions of ecology, there is also a problem with the very notion of "nature" itself.[5] To think of ecology and environment in terms of an opposition between the natural and the human (with "nature" as a constitutive element of what surrounds human life and what is thus—by necessity—outside of it) would mean to omit the precarious ideological history of the concept of nature itself.

It could be argued, for example, that to theorize a separate "natural" sphere beyond human intervention, which then takes on the role of total other to civilization, is always already a form of

---

non-interference with the natural world. For a recent critical assessment of its strengths and shortcomings, see Katz, Light, and Rothenberg (eds.) *Beneath the Surface: Critical Essays in the Philosophy of Deep Ecology* (2000).

[4] Bennett and Teague, the editors of a collection of essays, have made the as yet most radical attempt to wed nature writing, urban theory, and ecocriticism.

[5] As Buell points out, in the original Greek version *oikos* means household or home in the comprehensive sense of both residence and grounds (*Future of Environmental Criticism*, 140).

perceptional-ideological framing and thus implies an act of intrusion and interference with the assumed other. What we call "nature" has always been a cultural construct that helped to measure human progress and civilization rather than denote an independent realm outside or beyond the human. From this perspective, scientific discovery and technological progress do not mark a further alienation from the natural but an ever-greater involvement in it. To stress that human agency and manipulation have always shaped and re-shaped the "natural," ecocritics such as Buell and Dana Phillips often refer to the French philosopher of science Bruno Latour who argues that:

> nature seemed to be held in reserve, transcendent, inexhaustible, distant enough. But where are we to classify the ozone hole story, or global warming, or deforestation? Where are we to put these hybrids? Are they human? Human because they are our work? Are they natural? Natural because they are not our doing. Are they local or global? Both. (Latour, 50)

Following Latour, Buell claims that for millennia "nature has been subject to such modification. So to insist that a thing is natural in the sense of being primordial is arguably to mythologize or obfuscate" (Buell, *Future of Environmental Criticism*, 143). After two centuries of industrialization and unimpeded technological progress, it has become virtually impossible to distinguish between a "first" or primordial and "second" nature, an already manipulated "middle landscape" created by humans through irrigation, damming, deforestation, and so forth. It is with this difficulty in mind that Buell proposes to replace the term ecocriticism with the more open concept of "environment," as it comprehends both built and natural worlds (even though it still assumes the human as center and reference point of a *sur*-rounding sphere). Since not all of us are constantly aware of our own embeddedness in the environment as a condition of our personal and social being, Buell also introduces the term "environmental unconscious," which is to refer to both the potentiality for a fuller coming-to-consciousness and a limit to that potentiality. Buell, who is foremost a literary critic, ascribes to the imagination a residual capacity "to awake [us] to fuller apprehension of physical environment" and to achieve a breakthrough, despite whatever cultural blockages there may be, of our unconscious realization of place and environment as crucial foundations of human life (*Writing for an Endangered World*, 22).

True, the concept of an "environmental unconscious" can lead to new and quite astounding insights (as, for example, in Buell's own reading of Richard Wright's *Native Son* against the theories of the Chicago School in urban planning and ecological organization). By and large, however, it turns out a heuristically soft, rather problematic critical tool: because it allows us to group together texts that are either deficient of any environmental concerns if not even inimical to them (which then makes them negative manifestations of the environmental unconscious) or to represent nature in such a way that they eventually activate ecological considerations (so-called potential environmental breakthroughs), the term loses its descriptive and imaginative power. Its connotation of an intricately woven skein of interrelations between the psyche and the environment notwithstanding, the notion of an "environmental unconscious" only provides a very thin cover for quite divergent fictional and non-fictional texts which have no more in common than a rather general interest in the environment. Since environment may refer to the "natural" (as in the novels of Faulkner or the non-fictional writings of conservationist Aldo Leopold), to highly organized forms of "culture" (such as urban environments and their determinist effects on Dreiser, Sinclair, Wright and other modernist writers), or even to economic conditions at large (toxic discourse and the exploitation of oceans and whales), we are still in need of an encompassing analytical tool that allows us to chart more convincingly the shifting environmental awareness in modern and postmodern texts.

One possible solution to this problem is to stress the critical function of the literary text itself. Literature is always more than merely a mimetic reflection of the real. If literary texts reproduce the various historical forces at work in a given society, they also deconstruct these forces thereby exposing latent or emerging tendencies toward cultural change.[6] As forms of representation that arise from specific historical,

[6] The potential of literature to expose and deconstruct the dominant ideology of a given society has for long been a controversial issue in literary theory. While more orthodox Marxist critics often question the critical function of the aesthetic, both new-critical and, later, poststructuralist approaches to literature undermine the critical capacity of the literary text by emphasizing either its aesthetic function or the elusiveness of any form of linguistic statement. For a nuanced and valuable contribution to this debate,

cultural, and ideological patterns, literary texts also invite readings that run counter to these very patterns. Put another way, even though an individual text may choose to deemphasize its critical function vis-à-vis the physical environments in which the fictional characters navigate, it can still be read as making important—if, in this case, negative—propositions with respect to the connections between people and places.[7] Interestingly, this latter connection has often been weakened by a transformation of geophysical into symbolic or aesthetic spaces. Recall, for example, the need to create a national cultural space in antebellum American literature. If America's earlier culture wars with Europe have prompted the aesthetic transformation of the wilderness into a new form of "civilized" nature (Leo Marx's "middle landscape"), the ensuing abundance of symbolically loaded "natural" landscapes in early American texts was accompanied by an almost total eclipse of the actual geographical site and its conflictive status within modern capitalist society. As I argue in the remaining parts of this essay, it was only recently that American literature moved from an emphasis on *space*-consciousness to what we may call *place*-consciousness, from a constant effort to transcend the here and now to a renewed understanding of the social and aesthetic value of concrete places.

---

and the one that supports the notion of literature as an intrinsically critical corrective to reigning forms of ideology, see Macherey.

[7] William Carlos Williams seems to even reach beyond this rather loose link of literature and places, when he calls the imagination itself a form of "place": "Sometimes I speak of imagination as a force, an electricity or a medium, a place. It is immaterial which: for whether it is the conditions of a place or a dynamization its effect is the same: to free the world of fact from the impositions of 'art' [. . .] and to liberate the man to act in whatever direction his disposition leads. The word is not liberated, therefore able to communicate release from the fixities which destroy it until it is accurately tuned to the fact which giving it reality, by its own reality establishes its own freedom from the necessity of a word, thus freeing it and dynamizing it at the same time" (Williams, 150).

II

If the tension between nature and culture informs many of the core studies on nineteenth-century America, environmental issues often figure in these works as a mere setting or backdrop against which the more important drama of nation-building unfolds. Henry Nash Smith's *Virgin Land* (1950), Richard W.B. Lewis's *The American Adam* (1955), Perry Miller's *Errand Into the Wilderness* (1956), Leo Marx's *The Machine in the Garden* (1964), Richard Poirier's *A World Elsewhere* (1966), Cecelia Tichi's *New World, New Earth* (1979), and Richard Slotkin's *The Fatal Environment* (1985)—all of these foundational texts in early American Studies foreground ideological constructions of nature in America. In so doing, however, they also relegate the environment to a malleable, symbolic foil. That the bulk of critical studies on the formative years of American literature is marked by a glaring indifference to the environmental imaginings of writers such as Thoreau, Melville, Whitman, or Twain, to name just the more obvious cases, is actually quite astounding (see Buell's *Environmental Imagination*). Not only has the rhetorical appropriation of nature been exposed as crucial for the exploration and eventual settling of the American continent itself (for example, in Annette Kolodny's groundbreaking study *The Lay of the Land*), but what we now call classic American literature is particularly known for having adopted nature as its preeminent theme: in many "great American novels" of the nineteenth century, nature figures as a cultural resource that drives both the narrative plot and the progressive establishment of America as one nation. "American literary history," as Buell rightly claims,

> thus presents the spectacle of having identified representation of the natural environment as a major theme while marginalizing the literature devoted most specifically to it and reading the canonical books in ways that minimize their interest in representing the environment as such. (*Environmental Imagination*, 8)

The glaring misrepresentation of environmental concerns in the classic American writers of the nineteenth century may have been caused, however, by their own inveterately ideological stances toward the natural sphere. Just consider Emerson's essay "Nature," a pioneering American text to establish a firm link between nature and poetics. "The beauty of nature," Emerson writes, "reforms itself in the mind, and not for barren contemplation, but for new creation" (Emerson, 16). In art, nature works through the will and imagination of the poet to herald and fulfill her works, thus turning the artist into a mediator between the diverging forces of the natural and the cultural. While Emerson's view of the work of the poet as predicated on natural processes appears to be fully in tune with Coleridge's romantic organicism, his was a new and genuinely American contribution to the nature/culture debate. If the latter accompanied the establishment of modern capitalist societies on both sides of the Atlantic, Emerson's poetic theory was designed to deflate rather than enhance the tensions between the nonhuman and the man-made worlds. To Emerson there is no opposition between nature and culture as such, only a widespread misconception of the former by the latter. It therefore becomes the task of the poet to reunite "things to nature and the Whole – re-attaching even artificial things and violation of nature [such as the factory-village and the railway], to nature, by a deeper insight" (11). Given that America's future depended on an ever-increasing intrusion of the cultural sphere into nature—a process succinctly epitomized by the image of the frontier as contact zone between the two forces—Emerson theorizes the American poet as a powerful mediator to bring forth the hidden beauty of the natural world and to negotiate the processes of modernization in a reconciliatory manner. Sharing not only God's laboratory (nature) but also his capacity to see through the thin veil of appearances and the mere surface of things, it is the poet (rather than "nature" itself) that looms large in Emerson's proto-modernist rendering of the nature/culture conflict.

The shift from the site of the natural to the person, from the landscape to the writer, who struggles to regain the dwindling authority of his work within an evolving technological society, possesses seminal significance for the development of the modern concept of authorship. Significantly, if also somewhat paradoxically, the modern empowerment of the author goes hand in hand with the increasing importance of nature as a site of aesthetic relief and cultural regeneration. An interesting case in point here is Kafka's unfinished first novel *Amerika* (1917, posthumously), originally titled *Der*

Verschollene ("The Man Who Disappeared"). In this exemplary modernist text Kafka, who had been a secretary and lawyer at an accident insurance company for industrial workers, juxtaposes the restlessness and fast-paced rhythm of the modern bureaucratic state with an imaginary natural counter-space opaquely called the "The Nature Theatre of Oklahoma." Riding on a crowded train through the American West, the novel's picaresque hero conjures up a pristine fairyland untainted by both time and civilization, against which "everything [. . .] faded into comparative insignificance before the grandeur of the scene outside."[8] Kafka's naïve vision of the American West as "virgin land" clearly shares an effort of many writers to reverse the changing meaning of time and space in modern society. Their juxtaposition of natural spaces of *being* (that is, outside of time) and the cultural spaces of *becoming* (those geared to historical progress), adds to a metaphysical tradition in Western philosophy of writing nature off as the other of culture and society that runs from Rousseau to Heidegger. Along similar lines, Whitman dismissed the constrained interior spaces of the house (culture) to embrace the "open air" (nature) as the only space conducive to personal and poetic communion.[9] Social philosopher William James, in a public lecture titled "What Pragmatism Means" (1910), compares the new pragmatist method of reasoning to "the open air and possibilities of nature, as against dogma, artificiality, and the pretense of finality in truth," the latter of which he associates with—mostly European—rationalism and empiricism (25).

As historians have repeatedly noted, industrialization and its attendant urbanization have lead to an increasing interest in natural spaces as sites of cultural regeneration and redemption. (Just think of the establishment of national preserves and parks at the turn-of-the-twentieth century or the modern appeal of *going-native* and *back-to-nature* movements, which all attest to the increasingly symbolic nature of "nature.") One might even argue that the current extension of geophysical space into cyberspace is merely a continuation of—rather than a break with—this modern tendency to constantly transform place

[8] Kafka, 297. For a more detailed reading of Kafka's "American" novel, see Benesch.
[9] "I swear I will never again mention love or death inside a house," Whitman boasts in "Song of Myself," "And I swear I will never translate my self at all, only to him or her who privately stays with me in the open air" (Whitman, 85).

into (symbolic) space. And yet, as we lose our sense of place in the virtual worlds of instant electronic communication and connectivity, we increasingly yearn for a tangible, physical equivalent to rest our worn-out senses, to lay down the burden of constant global presence.[10] Thus, the tremendous appeal of popular nature writing in the vein of Barbara Kingsolver or of eco-activist romances such as Ruth Ozeki's *All Over Creation* (2004) and Carl Hiaasen's *Hoot* (2004). And, as another example, there had been the renaissance of natural disaster movies such as Emmerich's *The Day After Tomorrow*, that turn on Freud's notion of the "Uncanny" or the "*Un-Heimlich*." These depict a once well-known place that has become strange, alienated from the people who live there, and is now perceived in a quintessentially paranoid manner as threatening.

If there is, on one side of the spectrum, a wave of nature writing with an arguably neo-Victorian, retro ring to it, we also find, on the other, a number of texts—both fictional and non-fictional—that turn to the environment in surprisingly new ways. They do so by foregrounding the geophysical foundation of today's virtual worlds—the physical places where computer access is provided or from where global players conduct their business transactions—or by foregrounding ominous "non-places" such as arid zones (deserts), dumping grounds, or car-packed urban roads.[11] To be sure, the issue of

[10] While physical spaces are constantly shrinking, virtual space, the electronic spaces of the World Wide Web and other global networks proliferate. If geophysical space has thus become a limited resource, cyberspace at once expanded and subverted our traditional sense of place. Although we can freely roam the as yet uncharted sites of the internet, as cybernauts we are everywhere and nowhere at the same time. As the German art historian Bernd Meurer perceptively noted, "the place which we perceive as telereality and the place where we do the perceiving are synchronous. Real proximity is replaced by the image of closeness. [. . .] Space and time disconnect" (15).

[11] In shifting attention from the symbolic to the physically concrete, that is, from space to place, contemporary American writing adds to the increasing importance of place-oriented paradigms in the humanities at the end of the twentieth century. As noted theorist of place Edward Casey points out, the revival of place theory, as it were, can be seen in the tendency to think of knowledge as "situated" (as, for instance, in Donna Haraway's cyborg theory) or of *bodies-as-places* (a concept that informs the work of French feminist critic Luce Irigaray). It also drives much of the renewed interest in phenomenological philosophy (in particular of the kind associated with the French phenomenologist Maurice Merleau-Ponty), as phenomenologists make a strong case for the fact that a body is always tied to a certain world and that *being* always already

waste and junk has been a topic of modernism all along, has formed even a kind of ambivalent counter discourse to the future-oriented progressivism of the modern era. As Miles Orvell, in an epilogue to his magisterial study of authenticity in modern American culture remarked, the fascination with discarded things of artists such as painter Charles Burchfield or photographer Ralph Steiner (who were asked to illustrate an article on junk-filled backyards that appeared in 1930 in the magazine *Fortune*) actually reveals a more general tension "between an ideology of progress that pushes the past from our views as fast as possible, and a contrary, deep attachment to things as they were, to old things, to junk" (288). Orvell also registers a shift in the artist's sensibility toward waste from a pre-war "critical" one to post-apocalyptic representations of the world-as-junk in more recent postmodern literature. While the former appears to be driven still by an urge to regain control over the constant flow of garbage, the latter embraces the dump as a source of artistic and linguistic rejuvenation: "a mountain of resources—discarded images and styles—out of which to fashion a whole new vocabulary" (Orvell, 295). If this account is right on target with respect to the mostly ironic, playful stances of the first wave of postmodern literature, more recently writers such as Paul Auster or Don DeLillo have explored the issue of garbage in quite new and—I believe—more reassuring ways. To these and other contemporary writers garbage neither represents a reminder of what has been nor merely a toolbox for artistic renewal but a way, arguably the only way, to reconnect both humans and ideas to concrete places in posthuman society.

III

I will now turn to three specifically literary incidents that underline the capacity of literature to create and strengthen an awareness of place.

---

implies a form of *being there* (Casey, "How to Get From Space to Place in a Fairly Short Stretch of Time").

My first example of this new *platial* sensibility in American writing is a short essay by novelist Jonathan Franzen, whose award-winning family novel *The Corrections* (published in 2003) earned him a reputation as one of America's foremost contemporary writers of realist fiction. The essay, titled "First City," appeared as part of a collection of short prose, published the same year as *The Corrections*, which turns on the idea of using language and literature to evoke, in Franzen's own terms, a new sense of place. "First City" is above all a dithyramb to New York, an American megalopolis that, for Franzen, is "the next best thing to Europe" (181). It's also a city with two faces, a city of progress and the center of global electronic capitalism and a very real, physical city, full of people, noises, smells, and the irresistible charm of a many-layered history written in stone. At one point, Franzen describes the atmosphere of so-called Silicon Alley, a district in Lower Manhattan, where software companies, cyber arcades, and internet shops abound. After a short visit to a cybercafé on Lafayette Street, Franzen is reminded of William J. Mitchell's cyber-classic *City of Bits* (1996):

> This is the new architectural promenade [. . .] a city unrooted to any definite spot on the surface of the earth, shaped by connectivity and bandwidth constraints rather than by accessibility and land values, largely asynchronous in its operation, and inhabited by disembodied and fragmented subjects who exist as collections of aliases and agents. Yet the Cyber Café – to say nothing of the thousands of clubs and galleries and bookstores and non-cyber cafes doing business within a mile of it – resembles nothing so much as an old-fashioned see-and-be-seen promenade. (Franzen, 180)

Franzen's New York thus is twofold: the virtual city of the cyber generation and a real, physical place, "a definite spot on the surface of the earth," as Franzen has it, "populated by young people who even as they disembody and fragment themselves cannot resist the urge to Be There" (180). "First City" is primarily a plea for the latter ("to be there") and for literature to enable both the writer and the reader to experience aesthetically the concrete places of human interaction, to immerse the self in the fictional space where outer and inner worlds merge:

> A city lives in the eye, ear, and nose of the solitary beholder. You turn to literature to find the interior point of intersection between subject and city, and as a living connection to New York's history a few lines of Herman Melville or Don DeLillo outweigh whole pages of an encyclopedia. (189)

For Franzen literary texts capture the synaesthetic ambience of New York in such an encompassing way that they allow the reader to resist the "Disneyfication" of the city and to develop one's own sense of place. What is more, we may use them as templates to engage a direct, physical relationship with the city's places. When reading on the floor of his apartment, Franzen tells us, I relax "by listening to my breathing, I can hear the slower respirations of the city itself, a sound like the rumble of a surf: subway trains crowded with people who are teaching themselves how to be here" (194). From yet a different angle Paul Auster's short novel *City of Glass* (1985) also investigates the intricate relationship of words and places. Auster's complex and challenging detective story has often been read as a straightforward postmodern meta-narrative about both the act of writing and the profession of the writer. Its self-referential implications notwithstanding, *City of Glass* is, however, primarily a city novel, a text about the city as storehouse of "shattered things" and about the task of the writer to confront the brokenness and disarray typical of metropolitan areas such as Manhattan. Moreover, in Auster's somber vision "our words no longer correspond to the world" (92). Stillman, one of the author's several doppelgängers in the novel, is obsessed with an almost Emersonian effort to reattach things to the world: "You see, the world is in fragments, sir. And it's my job to put it back together again" (91). He is equally involved in a linguistic project of much smaller scale. The latter revolves around Stillman's attempt to confine himself "to physical things, to the immediate and tangible" (92) and thus carries utmost importance with respect to the topic of platial awareness and the capacity of writing to resuscitate our dormant sense of place.

Stillman finds that the city, in particular New York, is "a junk heap" that suits his purpose of tracking down broken things admirably (94). His intention is to collect discarded items such as broken umbrellas and then invent a new language that allows us to capture the transformation of a once useful object into something that no longer performs its function. What is striking about Stillman's collection of junk is that it serves the purpose of ultimately realigning words and things. When whole, things appear to correspond to the world: we feel confident "that our words could express them" and we thus rarely question their place within our daily lives (93). Little by little, however, these things fall apart while our words remain the same and, as a consequence, our language becomes trite and dated. To remedy the pervasive lack of "grounding," of being in touch with both things and places, Auster proffers the quintessentially literary activity of "naming." In the novel Stillman, the junkman, sets out to "invent new words that will correspond to the things" (94). His is the writer's task to draw attention to the crucial importance of language as a link between the world outside and the human mind. If this task involves the juggling of words on a page, it also means to implicate the reader in a series of transformations—from word to page to place—that lay bare the vital connection between the "houses" of fiction and "real" places.[12]

Platial or, to use Buell's term, environmental consciousness also looms large in Don DeLillo's novel *Underworld*. Though DeLillo's *magnum opus* bustles with references to space, place, and environment, it is impossible to identify a single framing story or argument with regard to place politics in the novel. For lack of space I will focus here on one, particularly revealing episode that occurs midway into this meandering, multileveled text. The incident I have in mind involves a small group of ecological activists and environmentally conscious artists who set out to repaint military debris in the Nevada desert. The idea is to turn the discarded B-52 bombers that are scattered throughout the sublime landscape into a monumental piece of land art. As one of the artists points out, their project is not about war and peace but rather about the tangled relations of our postmodern culture of garbage and how we imagine and construct "nature":

> This is an art project, not a peace project. This is a landscape painting in which we use the landscape itself. The desert is central to this piece. It's the surround. It's the framing device. It's the four-part horizon. [. . .] It's so old and strong, I think it makes us feel,

[12] In his preface to *The Portrait of a Lady,* Henry James compares his fictions quite literally to a "house" inhabited by both author and readers of the novel (James, 40-58). James' notion of a "literary architecture" is discussed in detail in Frank, 167-217.

> makes us as a culture, any technological culture, we feel we mustn't be overwhelmed by it. [. . .] The desert bears the visible signs of all the detonations we set off. All the craters and warning signs and no-go areas and burial markers, the sites where debris is buried. (71)

That garbage has become a driving component of economic structure and rationale in the United States (as in most Western societies) constitutes a major theme of the novel: landfills and dumping grounds, DeLillo seems to suggest, are the "Underworld," the environmental "unconscious" of our postmodern imagination. The novel thus adds to a postmodern discourse on waste and garbage that runs from Pynchon's *The Crying of Lot 49* (1966) to A. R. Ammons, whose award-winning 1993 prose poem "Garbage" claims that "Garbage has to be the poem of our time [. . .] to get our attention, getting in the way" (18). In the above-quoted scene in DeLillo's *Underworld* the ever-expanding world of garbage, that scars our burgeoning economic landscapes from both above and beneath ground, is turned—by way of the imagination—into a metaphor of a bleak, alienated form of living. If it is true, as the novel suggests elsewhere, that technological progress "out-imagined the mind" (76), then it becomes the task of the literary, of fiction, to irrigate anew the arid "nether lands" of our postmodern minds and thus, perhaps, reverse the process of ecological and imaginative devastation. Standing on top of a gigantic landfill one of *Underworld's* many characters ponders about the sacred/secret meaning of this mysterious, artificial mountain:

> He'd seen a hundred landfills but none so vast as this. [. . .] Brian took a deep breath, he filled his lungs. This was the challenge he craved, the assault on his complacency and vague shame. To understand all this. To penetrate this secret. The mountain was here, unconcealed, but no one saw it or thought about it, no one knew it existed except the engineers and teamsters and local residents, a unique cultural deposit, fifty million tons by the time they top it off, carved and modeled, and no one talked about it but the men and women who tried to manage it [. . .] The landscapers who would build hanging gardens here, make a park one day out of every kind of used and lost and eroded object of desire. (185)

True, this man-made mountain of waste and refuse is a far cry from the ones Thoreau regularly climbed "to know what a world he inhabits"

(*Thoreau on Mountains*, 1). Not unlike the New England philosopher and naturalist, however, what the environmental fictions of DeLillo, Auster, Ammons, and numerous other contemporary writers seem to articulate is a lack of platial awareness, particularly with respect to places we believe to know best. If, as DeLillo reminds us, "the biggest secrets are the ones spread open before us" (185), it still takes the "burrowing" mind of the writer to unearth layers of cultural debris, to *re*-collect the trash and garbage off our littered streets and, much as Auster's waste-driven linguist, invent a "new" language and lend new meaning to both the things themselves and the environment they compound. More than ever, it is the task of the writer to go on this "mining" operation, to take us by the hand and make us, as Thoreau poignantly put it:

> work and wedge our feet downward through the mud and slush of opinion, and prejudice, and tradition, and delusion, and appearance, that alluvion which covers the globe, through Paris and London, through New York and Boston and Concord, through Church and State, through poetry and philosophy and religion, till we come to a hard bottom and rocks in place, which we can call *reality*, and say, This is, and no mistake. (*Walden*, 97-98)

If Thoreau is yet torn between a poetics of place (as in the latter quote) and the thing, i.e., *place* itself (as in his minute descriptions of New England topography), his postmodern fellow writers seem to propound a platial consciousness that is at once more literally geared to concrete physical spaces and more gropingly uncertain as to the tenuous relationship between literature and places.

Bibliography

Ammons, A. R. *Garbage*. New York: Norton, 1993.

Armbruster, Karla and Kathleen R. Wallace, eds. *Beyond Nature Writing: Expanding the Boundaries of Ecocriticism*. Charlottesville: University Press of Virginia, 2001.

Auster, Paul. *The New York Trilogy (City of Glass; Ghosts; The Locked Room)*. New York: Penguin, 1987.

Benesch, Klaus. "*Writing Machines*: Technology and the Failures of Representation in the Work of Franz Kafka." In *Reading Matters: Narrative in the New Ecology of Media,* edited by Joseph Tabbi and Michael Wutz, 76-95. Ithaca, NY: Cornell University Press, 1997.

Bennett, Michael and David W. Teague, eds. *The Nature of Cities: Ecocriticism and Urban Environments*. Tucson: The University of Arizona Press, 1999.

Bryson, Michael A. *Visions of the Land: Science, Literature, and the American Environment from the Era of Exploration to the Age of Ecology*. Charlottesville: University Press of Virginia, 2002.
Buell, Frederick. *From Apocalypse to Way of Life: Environmental Crisis in the American Century*. New York: Routledge, 2003.
Buell, Lawrence. *The Environmental Imagination: Thoreau, Nature Writing, and the Formation of American Culture*. Cambridge, MA: Harvard University Press, 1995.
—. *Writing for an Endangered World: Literature, Culture, and Environment in the U.S. and Beyond*. Cambridge, MA: Harvard University Press, 2001.
—. *The Future of Environmental Criticism*. Malden, MA: Blackwell, 2005.
Casey, Edward. "How to Get From Space to Place in a Fairly Short Stretch of Time: Phenomenological Prologemina." In *Senses of Place,* edited by Steven Field and Keith Basso, 13-52. Santa Fe, NM: School of American Research Press, 1996.
—. *The Fate of Place: A Philosophical History*. Berkeley: University of California Press, 1997.
Coupe, Laurence, ed. *The Green Studies Reader: From Romanticism to Ecocriticism*. New York: Routledge, 2000.
DeLillo, Don. *Underworld*. New York: Scribner, 1997.
Emerson, Ralph Waldo. "The Poet." In *The Collected Works of Ralph Waldo Emerson*, edited by Joseph Slater, 3:1-24. Cambridge, MA: Harvard University Press, 1983.
Fletcher, Angus. *A New Theory for American Poetry: Democracy, the Environment, and the Future of Imagination*. Cambridge, MA: Harvard University Press, 2004.
Frank, Ellen Eve. *Literary Architecture: Essays Toward a Tradition*. Berkeley: University of California Press, 1979.
Franzen, Jonathan. "First City." In *How To Be Alone. Essays*. New York: Farrar, Straus and Giroux, 2003.
Freud, Sigmund. "The Uncanny." In *The Standard Edition of the Complete Psychological Works of Sigmund Freud,* 17:217-256. London: The Hogarth Press and the Institute of Psychoanalysis, 1961.
Garrard, Greg. *Ecocriticism*. New York: Routledge, 2004.
Gatta, John. *Making Nature Sacred: Literature, Religion, and Environment*. New York: Oxford University Press, 2004.
James, Henry. *The Art of the Novel: Critical Prefaces*. New York: Charles Scribner's Sons, 1937.
James, William. "What Pragmatism Means." In *Pragmatism and Other Essays*. New York: Washington Square Press, 1972.
Kafka, Franz. *Amerika*. Translated by Willa and Edwin Muir. New York: Schocken Books, 1974.
Latour, Bruno. *We Have Never Been Modern*. Translated by Catherine Porter. Cambridge, MA: Harvard University Press, 1993.
Love, Glen A. *Practical Ecocriticism: Literature, Biology, and the Environment*. Charlottesville: University Press of Virginia, 2003.
Macherey, Pierre. *A Theory of Literary Production*. Translated by Geoffrey Wall. New York: Routledge, 1992.
Mazel, David, ed. *A Century of Early Ecocriticism*. Athens: University of Georgia Press, 2001.

McKusick, James C. *Green Writing: Romanticism and Ecology*. London: Macmillan, 2000.
Meurer, Bernd. "The Future of Space." In *Die Zukunft des Raums/The Future of Space*, edited by Bernd Meurer, 13-36. Frankfurt: Campus, 1994.
Meyrowitz, Joshua. *No Sense of Place: The Impact of Electronic Media on Social Behavior*. New York: Oxford University Press, 1985.
Mitchell, W.J. *City of Bits*. Cambridge, MA: The MIT Press, 1995.
Nietzsche, Friedrich. *The Gay Science*. Translated by Arnold Kaufmann. 1882. New York: Vintage Books, 1974.
Orvell, Miles. *The Real Thing: Imitation and Authenticity in American Culture, 1880–1940*. Chapel Hill: The University of North Carolina Press: 1989.
Phillips, Dana. *The Truth of Ecology: Nature, Culture, and Literature in America*. New York: Oxford University Press, 2003.
Rozelle, Lee. *Ecosublime: Environmental Awe and Terror from New World to Oddworld*. Tuscaloosa: University of Alabama Press, 2006.
Thoreau, Henry David. *Walden*. *The Writings of Henry David Thoreau*. Edited by J. Lyndon Shanley. Princeton: Princeton University Press, 1971.
—. *Thoreau On Mountains*. Edited by J. Parker Huber. New York: Houghton Mifflin Company, 1999.
Whitman, Walt. *Leaves of Grass*. Edited by Sculley Bradley and Harold W. Blodgett. New York: Norton, 1973.
Williams, William Carlos. *Imaginations*. Edited by Webster Schott. 1923. New York: New Directions, 1979.

Contributors

KLAUS BENESCH is Professor of English and American Studies at the University of Munich (Germany). He was a 2004 Mellon Fellow at the Harry Ransom Humanities Research Center of the University of Texas (Austin), and has taught at the University of Massachusetts (Amherst) and Weber State University (Utah). Previous publications on the topic of literature and public spaces include *The Power and Politics of the Aesthetic in American Culture* (editor/2007); *Space in America: Theory, History, Culture* (editor/2005) and *Romantic Cyborgs: Authorship and Technology in the American Renaissance* (2002).

TIMOTHY DAVIS is the senior historian for the U.S. National Park Service's Park Historic Structures and Cultural Landscapes Program. He is the co-editor of *America's National Park Roads and Parkways: Drawings from the Historic American Engineering Record.* His writings on parkways and other aspects of the American landscape have appeared in various journals and anthologies including *Everyday America*, *Inventing for the Environment*, *Landscape Journal*, *Perspectives in Vernacular Architecture*, and *Studies in the History of Gardens & Designed Landscapes*.

KAY EDGE teaches in the graduate architecture program at Virginia Tech. She holds a Bachelor of Arts in Philosophy from The University of the South, a Master of Architecture from Virginia Tech and a Master of Environmental Design from Yale University. She has worked professionally for Pelli Clarke Pelli Architects of New Haven, Connecticut.

ANDREW GROSS is assistant professor in American literature at the John F. Kennedy-Institut für Nordamerkastudien at the Freie Universität Berlin. In 2001 he received a PhD in English from the University of California at Davis. His most recent publications include "Cars, Postcards, and Patriotism" in *Pacific Coast Philology* (2005), "The American Guide Series: Patriotism as Brand-Name Identification" in *The Arizona Quarterly* (Spring 2006), and "Counter-tourism in Jean Améry and W.G. Sebald" in a collection of essays entitled *Representing the Unimaginable: Narratives of Disaster* (2007). He is currently working on two books on representations of the Holocaust, one (with coauthor Michael Hoffman) dealing with memorial architecture and commemorative practices, and the other (with coauthor Susanne Rohr) focusing on art and fiction of the 1990s.

PETER BACON HALES is Professor and Chair of the Art History Department of the University of Illinois at Chicago. His writing on American culture has concentrated on the built environment and its representation. *William Henry Jackson and the Transformation of the American Landscape* (1988) was nominated for a Pulitzer; *Atomic Spaces: Living on the Manhattan Project* (1997) won the Hoover and Kayden Prizes and was runner-up for the Parkman Prize in American History. His first book, *Silver Cities*, was recently reissued in a vastly expanded and revised form. *The Likes of US,* a study of Depression-era photography in America, coauthored with Stu Cohen, will be released by Godine in the Fall of 2008; a history of the public face of Chicago's architecture will be published by Acanthus Press in 2009. Two works on recent American cultural landscapes will be published by the University of Chicago Press in 2009 and 2010. For sixteen years, he has directed the American Studies Institute at UIC, a training center for international scholars and teachers.

NADINE KLOPFER is assistant professor of North American History and Canadian Studies coordinator at the John F. Kennedy Institute, Free University of Berlin. Previously, she taught North American Cultural History at the University of Munich, where she has recently completed her PhD in history with a thesis about "Space and Social Order in Montreal, 1880–1930." In 2004 she was a doctoral research fellow of the German Historical Institute, Washington D.C. She has received degrees in art history and history at the University of Tübingen (Germany) and Paris-Sorbonne. Her main field of interest is architectural and urban cultural history.

TORBEN HUUS LARSEN received his PhD from the Center for American Studies, the University of Southern Denmark in 2008. His dissertation "Revisiting the Middle Landscape: From Biltmore to Dollywood – the Pastoral in the Tennessee Valley" traces and analyzes the development of the modern middle landscape from an unintended to a designed space. He has published articles on American culture and environmental history, as well as on literature and film, and he is currently co-writing a book about the Western genre as a reflection of American society.

LAURA LAWSON is Associate Professor of Landscape Architecture at the University of Illinois, Urbana-Champaign. She received her PhD in Environmental Planning and MLA from the University of California, Berkeley. Her research includes historical and contemporary community open space, with particular focus on community gardens and the changing roles of parks in low-income communities. Her book, *City Bountiful: A Century of Community Gardening in America* (University of California Press, 2005) uses an historical analysis of urban gardening programs to explore values associated with gardening and the sustainability of user-initiated open space.

SARAH LURIA is an Associate Professor of English at the College of the Holy Cross in Worcester, Massachusetts. She is the author of *Capital Speculations: Writing and Building Washington, D.C.* (2005). This essay is part of a new book project, currently titled *Ruling Nature: Essays on the Literature of Land Surveying*.

JEFFREY L. MEIKLE is Professor of American Studies and Art History at the University of Texas at Austin. He is the author of *Design in the USA*; *American Plastic: A Cultural History* (which was awarded the Dexter Prize by the Society for the History of Technology); and *Twentieth Century Limited: Industrial Design in America, 1925–1939*; as well as numerous articles and catalogue essays. In 2003-04 he served as Bicentennial Fulbright Professor of American Studies at the University of Helsinki.

ANNA MINTA is Assistant Professor at the Department of Art History at Bern University, Switzerland. She has worked as postdoctoral researcher at the Collaborative Research Center "Institutionality and Historicity," Technical University Dresden, Germany, and as lecturer at the Department of Art History, Kiel University, Germany. Her PhD focuses on nation-building in Israel: *Israel bauen. Architektur,*

Staedtebau und Denkmalpolitik nach der Staatsgründung 1948 (Berlin: Reimer Verlag, 2004). She has published various articles on architecture, constructed identity, and assertions of power in the United States and conflicted territories like Israel/Palestine and East-/West-Germany. She is preparing a comprehensive book on representational architecture in Washington, D.C., comparing religious and federal buildings.

DAVID E. NYE is Professor and Chair of American Studies at SDU-Odense, Denmark. He has been a guest professor at the universities of Cambridge, Harvard, MIT, Leeds, Notre Dame, and Warwick. The most recent of his ten books is *Technology Matters: Questions to Live With* (MIT Press, 2006). In 2005 he was awarded the Leonardo da Vinci Medal, the lifetime achievement award of the Society for the History of Technology.

MILES ORVELL is Professor of English and American Studies at Temple University. He has written widely on modernist literature, visual culture, material culture and architecture, including *The Real Thing: Imitation and Authenticity in American Culture, 1880-1940* (1989), *After the Machine: Visual Arts and the Erasing of Cultural Boundaries* (1995); and *American Photography* (Oxford History of Art Series, 2003). He is the Editor in Chief of the online *Encyclopedia of American Studies* (Johns Hopkins University Press) and is presently working on a book called "Main Street in the American Mind."

ERIC SANDEEN received his PhD from the University of Iowa in American Studies (1977). Since 1982 he has directed the American Studies Program at the University of Wyoming. He is the author of *Picturing an Exhibition: The Family of Man and 1950s America.* Sandeen is fascinated by linear landscapes that are manifested throughout American terrain. His current obsession is the Lower Manhattan Expressway, an unbuilt Robert Moses project of the 1950s and 1960s. The prospect of this elevated motorway constructs a larger narrative of urban renewal and civic revival in the dynamic cityscape of post-World War II New York City.

RICKIE SANDERS is Professor of Geography/Urban Studies and former Director of Women's Studies at Temple University. During her tenure at Temple University she has served as both Graduate Chair and Chair of her department. Her publications include *Growing Up in America: An Atlas of Youth in America* (with Mark Mattson); Senior Consultant on *World Geography* (with James Petersen, Marci Smith Deal, and Daniel Arreola); and numerous publications in *Women's Studies Quarterly, Journal of Geography, Professional Geographer, Gender Place and Culture, Journal of Geography in Higher Education, Antipode, Urban Geography,* and *Legislative Atlas for the Commonwealth of Pennsylvania*.

KERSTIN SCHMIDT is Assistant Professor of American Studies at the University of Munich/Germany. She is the author of *The Theater of Transformation: Postmodernism in American Drama* (Postmodern Studies 37; New York: Rodopi, 2005) and has published on contemporary American drama, ethnic literatures in the United States and Canada, theories and cultures of diaspora as well as on media studies. She co-edited the essay collection *America and the Sea* (2004) and has edited and contributed to *Space in America: Theory History Culture* (2005) and *Re-Reading McLuhan* (2008). She is also a founding member of the women's studies journal *Freiburger FrauenStudien*. She is currently writing a book on *Negative Space: Concepts of Space in American Literature, Architecture, and Photography* (1850–1920).

JOHN F. SEARS is an independent scholar and the author of *Sacred Places: American Tourist Attractions in the Nineteenth Century*. From 1986 until 1999, he served as Executive Director of the Franklin and Eleanor Roosevelt Institute in Hyde Park, N.Y. where he initiated the New Deal Network website and supervised the restoration of FDR's Top Cottage. He edited and wrote the introduction to the Penguin Classics edition of Henry James's *American Scene* and has published essays on Robert Frost, William James, Nathaniel Hawthorne, Timothy Dwight, William F. "Buffalo Bill" Cody, Andrew Jackson Downing, Franklin D. Roosevelt, Eleanor Roosevelt, and Muir Woods.

BRYANT SIMON is professor of History and Director of American Studies at Temple University. He is the author of *A Fabric of Defeat: The Politics of South Carolina Millhands, 1910-1948* (1998) and *Boardwalk of Dreams: Atlantic City and the Fate of Urban America* (2005). His book, *Consuming Starbucks*, will appear in 2008.